World of Wildlife

ANIMALS OF THE JUNGLE

ORBIS · LONDON

From the original text by Dr Félix Rodríguez de la Fuente
Scientific staff: P. de Andres, J. Castroviejo, M. Delibes, C. Morillo, C. G. Vallecillo
English language version by John Gilbert
Consultant editor: Dr Maurice Burton
Creative director: Brian Innes

Contents

Illustration acknowledgments

E. Appel/Photo Researchers: 40, 53
R. J. Ashworth/Photo Researchers: 55
F. Ayer/Photo Researchers: 252
B. Barbey/Magnum: 70
Barnaby's Picture Library: 139
D. Bartlett/Bruce Coleman: 32
Bassot/Jacana: 15, 28, 178
F. Bel-G. Vienne/Jacana: 146
H. Beste/Ardea Photographics: 233
L. Bissell/N. Palmer: 194
J. Burton/Bruce Coleman: 16, 23, 37, 137, 172, 192, 199, 200, 205, 213, 224, 226
J. Burton/Photo Researchers: 202, 203
Miguel Ángel L. Castaños: 45, 103, 112
Ernesto Cerra: 43, 90, 91, 105, 136, 148, 149, 158, 159, 163, 235, 258
C. Channer/N. Palmer: 115
Bruce Coleman: 44, 64, 96, 113, 125, 143
L. R. Dawson/Bruce Coleman: 277
A. Ducrot/Jacana: 155
M. D. England/Ardea Photographics: 12, 238, 261
J. Englebert/Black Star: 3, 13
Antonio Escudero: 50, 73, 79, 179, 265
K. W. Fink/Ardea Photographics: 231
E. P. Gee/WWF: 120, 138, 237, 244, 249, 278
B. Glinn/Magnum: 14
B. Grandjean/WWF: 156, 235
Grossa/Jacana: 64, 69
J. A. Hancock/Bruce Coleman: 109, 143
J. A. Hancock/Photo Researchers: 253
E. Hanumantha: 70, 73, 104, 106, 107, 113, 119, 121, 128, 129, 187, 236, 238, 264, 270, 271, 273, 277, 279, 280, 282, 285, 289, 290, 295, 297, 300
E. Hanumantha/Photo Researchers: 9, 33, 76, 165, 282
R. Harris/F. W. Lane: 120
Harrison Forman: 113
E. Hosking: 143, 151, 157, 233, 250
D. Hughes/Bruce Coleman: 203, 204
P. Jackson/Bruce Coleman: 71, 107, 125, 262, 268, 287
P. Jackson/WWF: 169, 175, 177, 237, 243, 248, 257, 264, 299
R. Kinne/Bruce Coleman: 181
R. Kinne/Photo Researchers: 74, 134, 216, 224
C de Klemm/Jacana: 102, 176
P. Koch: 3, 251, 254, 255
José Lalanda: 5, 18, 58, 77, 80, 103, 122, 142, 162, 178, 180, 206, 213, 227, 228, 231, 235, 241, 265, 268, 269, 276

E. Lindgren/Ardea Photographics: 18, 20, 22, 159
A. Margiocco: 68, 75, 131, 182, 188, 247, 267, 293
R. D. Martin/Bruce Coleman: 30, 31
J. McCabe/F. W. Lane: 183
T. McHugh/Photo Researchers: 35
P. A. Milwaukee/Jacana: 35
E. Miyazawa/Black Star: 78, 81, 82, 83, 84, 85, 87, 88, 89, 92, 93, 95, 97
T. Molenaar: 100, 109, 117, 246
P. Morris/Ardea Photographics: 196, 212
R. Morse/Time Life: 50, 51
W. Müller/Len Sirman Press: 171
R. K. Murton/Bruce Coleman: 66
N. Myers/Bruce Coleman: 8
Okapia: 15, 58, 63, 135, 151, 154, 222
S. Ousoff/Jacana: 47, 272
P. Pfeffer: 7, 10, 19, 23, 24, 28, 29, 46, 47, 76, 123, 133, 142, 239
P. Pfeffer/Jacana: 122
J. Pickerell/Black Star: 174
G. Pizzey/Bruce Coleman: 240
I. Polunin: 195, 196, 201, 206, 207, 208, 209, 210, 212, 219, 221, 223
I. Polunin/NHPA: 220
E. Puigdengolas: 155, 176
C. Rabanit/Jacana: 160
E. A. Schelling: 19
W. Schraml/Jacana: 137, 297
E. Schumacher: 110, 111, 114
P. Scott/Photo Researchers: 9
J. Simon/Bruce Coleman: 144
J. Simon/Photo Researchers: 52
J. Six: 150, 181
Marcelo Socías: 44, 60, 166, 178, 197, 225, 228, 274
J. Solaro/Jacana: 49, 191
J. C. Stoll/Jacana: 52
J. X. Sundance/Jacana: 126
R. Thibout/Jacana: 42, 48, 55
P. Throckmorton/N. Palmer: 293, 295
University of Wisconsin: 97, 99
G. Vice/Jacana: 185
A. Visage/Jacana: 167, 168
P. Ward/Popper: 23, 26
J. V. Wormer/Bruce Coleman: 230, 231
Ag. Zardoya: 11
Zentrale Farbbild Agentur Gmbh: 39, 61, 62, 63, 65, 130, 140, 146, 152, 153, 236

The green jungle, bleak deserts and blue oceans of South-east Asia are the exotic settings of this seventh volume of *World of Wildlife*. The main emphasis is on the rain forest which covers large areas of the mainland and islands of the Malay Archipelago. Against a background of massive trees and a tangle of trailing plants, gorgeous orchids and other tropical flowers, a complex animal community struggles for survival. Enormous butterflies flit through the clearings, amazing flying reptiles, amphibians and mammals glide from branch to branch, apes and monkeys chatter in the treetops, magnificent birds strut through the undergrowth, and at night storms of bats obscure the stars.

In these pages the cameras of many of the world's leading natural history photographers reveal exciting and little-known aspects of the lives of the astonishing animals of rain forest, mangrove swamp, steppe and desert—acrobatic gibbons, radiant pheasants, deadly cobras, predatory tigers and man-eating crocodiles. Apart from the more familiar species we are introduced to many fascinating animals with unusual habits dictated by the need to survive. In the mangrove swamp lives the extraordinary mudskipper, a fish which can live both in and out of water; and here too are a monkey and a mongoose which feed almost exclusively on crabs. In the forest the sun bear, smallest member of its family, climbs trees and feeds, like another unfamiliar species, the sloth bear, on ants and termites.

South-east Asia, like other regions, has its share of threatened species—the great Indian bustard, the solitary orang-utan, the Javan rhinoceros and the blackbuck or Indian antelope, to mention just a few.

CHAPTER 82

The Oriental region: green jungle and azure seas

The natural disasters and devastating wars of recent decades have largely obscured the romantic image of the East traditionally conjured up in the Western mind by writers such as Rudyard Kipling. Yet our imagination can still be stirred by thoughts of ruined temples buried beneath a sea of vegetation, of jungles where troops of monkeys chatter in the treetops and black panthers prowl in the shadowy undergrowth, of bazaars and snake-charming fakirs, of holy men in search of higher knowledge and spiritual enlightenment–a world from which we in the West are separated not only by thousands of miles but even more by centuries of history and culture.

Alfred Russel Wallace, the English 19th-century naturalist who shares with the far more celebrated Charles Darwin the credit for the theory of evolution by natural selection, was principally concerned with the plant and animal life of Southeast Asia, drawn to the subject by eight years of travel in the region which he described in his book *The Malay Archipelago,* published in 1869. Wallace was the father of zoogeography (the geographical distribution of animals) and it was he who defined the bounds of the area in Asia which he called the Oriental region, buttressed to the north by the mighty mountain chain of the Himalayas and flanked in the south by the larger islands of the Indian and Pacific Oceans.

The Oriental region, with its rich and rare assembly of plant and animal life, is not as exclusive and self-contained as these mountain and ocean barriers might imply. Thus from southern China in the east as far as the Aralo-Caspian basin in the west, there are many areas of contact with the fauna of the Palearctic region, though the most important meeting points are to the north where a number of tropical species have colonised the

Facing page : The Oriental region has had an extremely eventful past, characterised by many volcanic eruptions and earthquakes. Even today parts of South-east Asia are occasionally ravaged by upheavals in which vast areas are devastated. Submarine volcanoes sometimes cause land to vanish below the surface or result in the emergence of a new island.

Eurasian steppes and mountains. The Arabian desert, however, appears to have played only an intermittent and hence less effective role in channelling exchanges between the species of the Oriental and Ethiopian regions.

Close study of the wildlife of the Oriental region shows, in any event, that there are links not only with the Palearctic, Ethiopian and Australian regions, but even (as in the case of the tapir) with the Neotropical region of South America. It is also noticeable that there are fewer species in the Oriental region. Some authors explain this by seeing it as the focal point for colonising other parts of the world. The alternative, seemingly contradictory, theory is that it has been the melting-pot where foreign groups have met and mingled. There is perhaps an element of truth in both hypotheses, suggesting that the region may have played a dual role in zoogeography.

Volcanoes and earthquakes

Although the area defined as the Oriental region appears to cover only a small part of the continent of Asia, it is in reality extensive in terms of land and sea combined. It has been formed by a succession of violent and dramatic physical upheavals, primarily volcanic in origin, and these processes are incomplete even today. South-east Asia is geologically unstable, its shape and structure in a constant state of flux, for volcanic eruptions and earth tremors can still cause islands to emerge and vanish overnight; others continue to grow, albeit very slowly.

The Oriental region began to assume its present recognisable shape approximately 63 million years ago in the Paleocene epoch. At that time India was itself an island and a chain of mountains and volcanoes covered a part of Burma. After a period of comparative calm the Miocene epoch (some 25 million years

Vegetational map of the Oriental region.

Semi-desert		Rain forest	
Savannah		Thorn forest	
Deciduous forest		Mangrove forest	
Monsoon forest		Swamp	

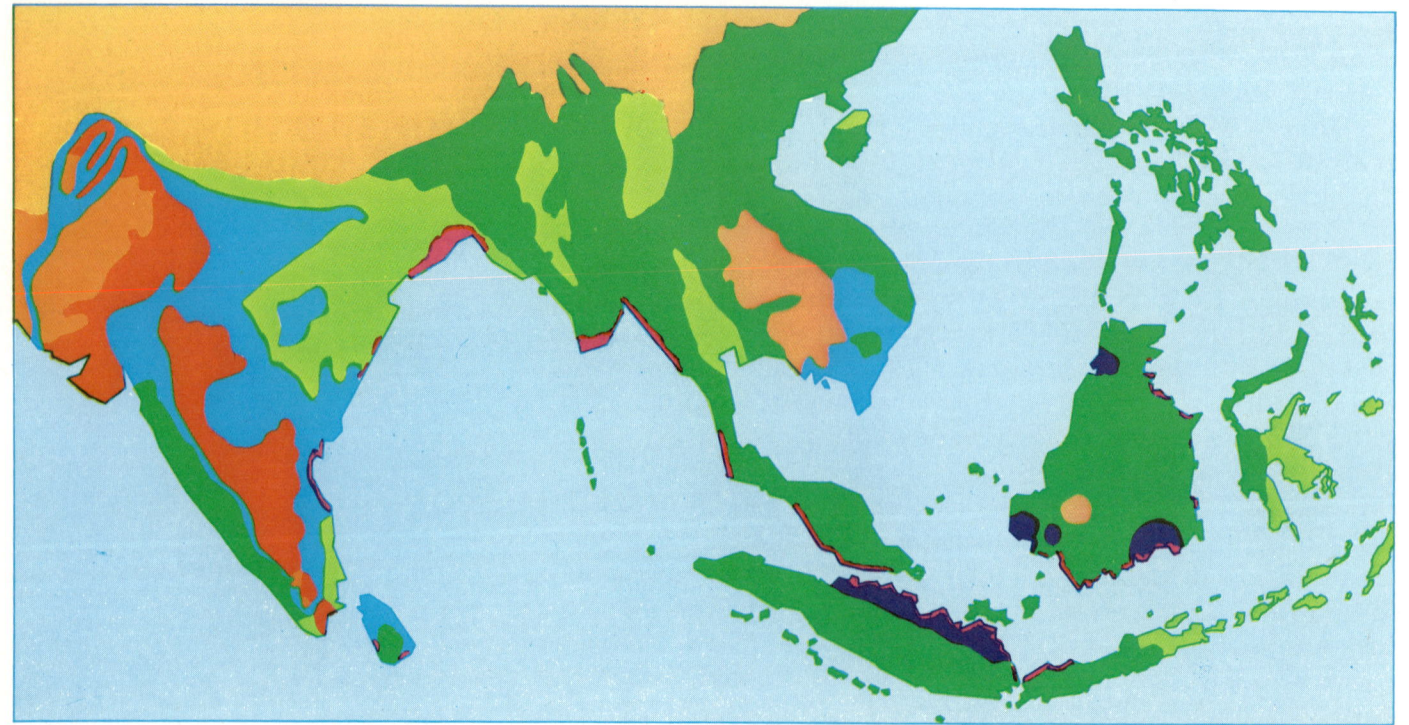

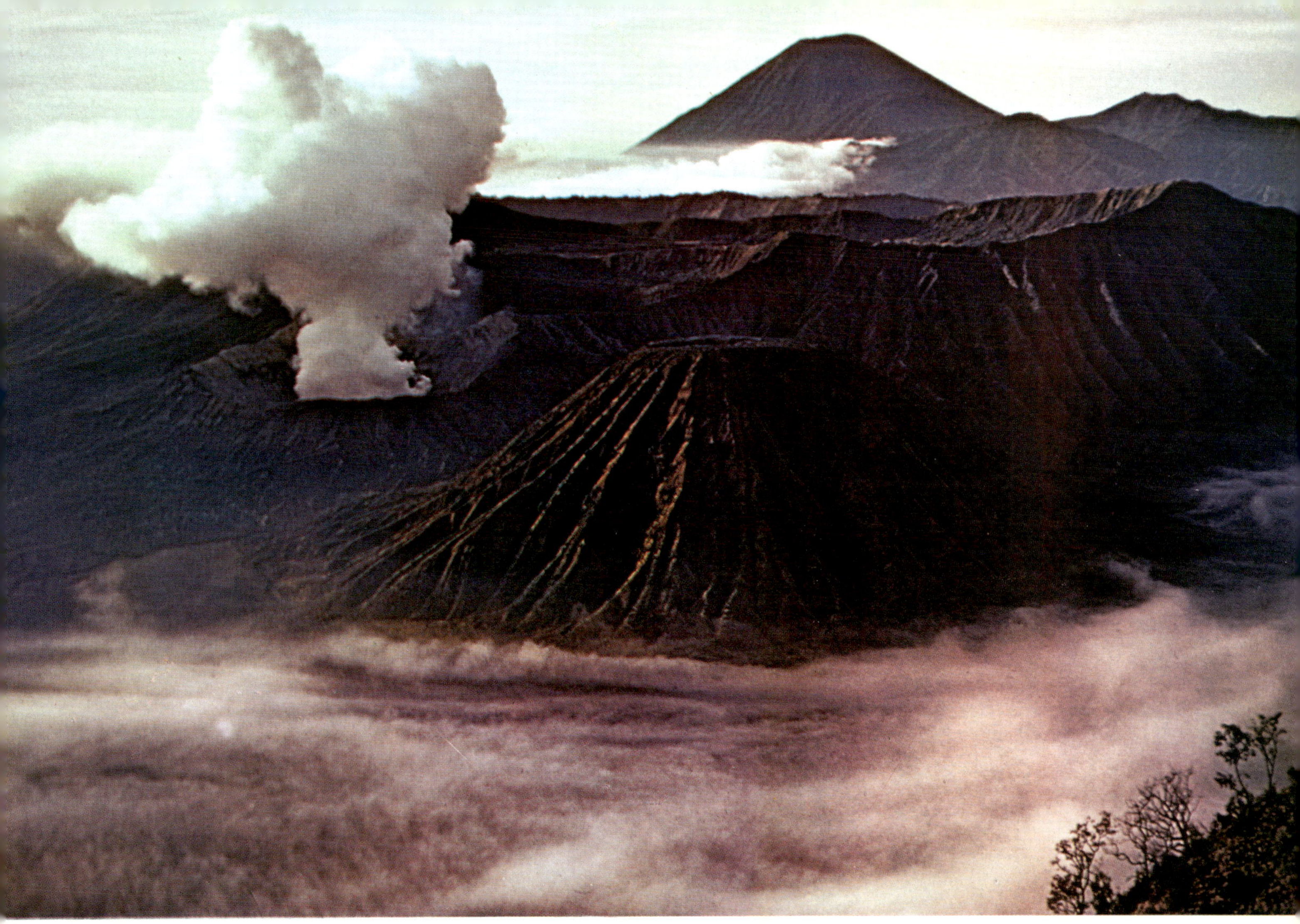

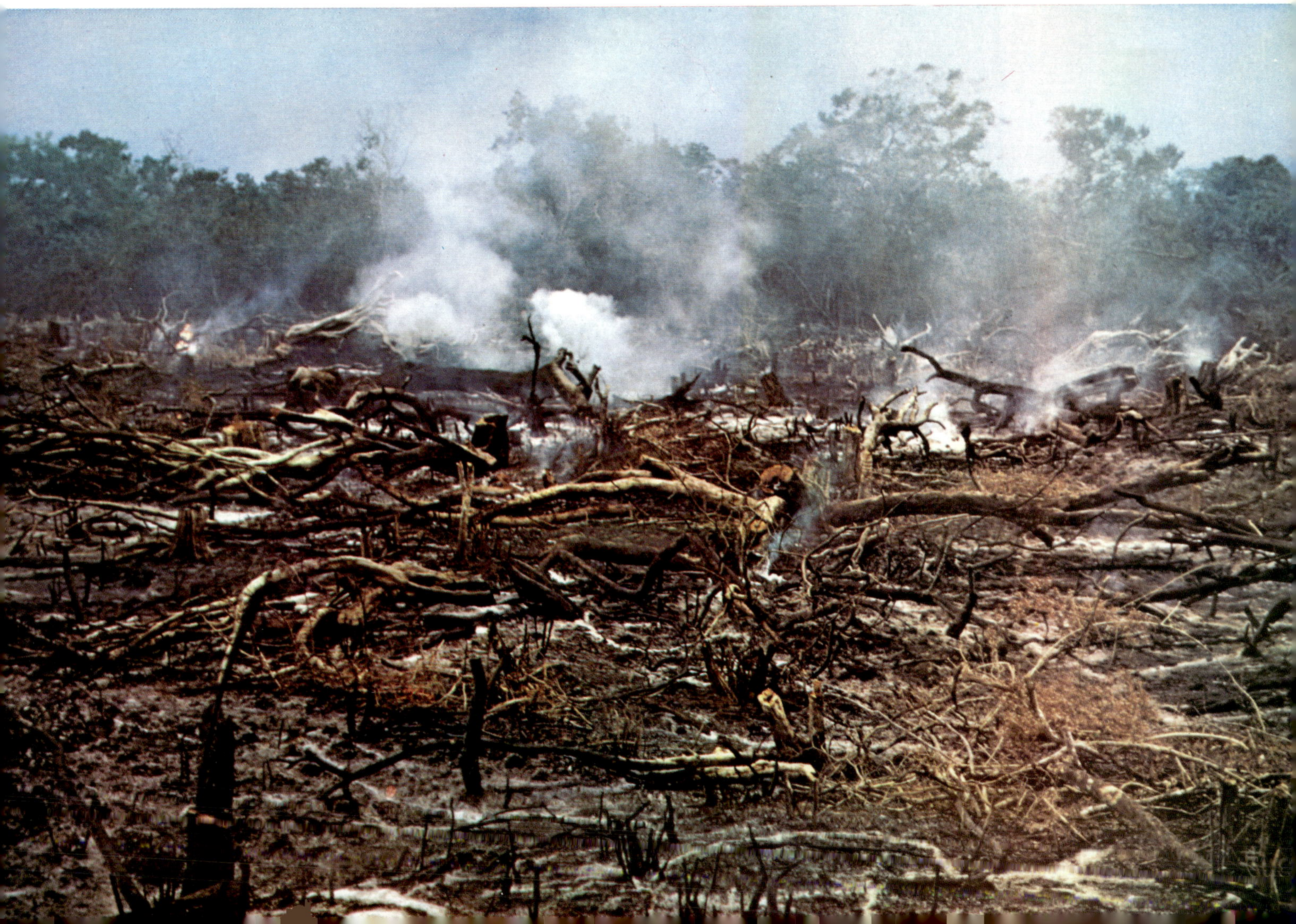

4

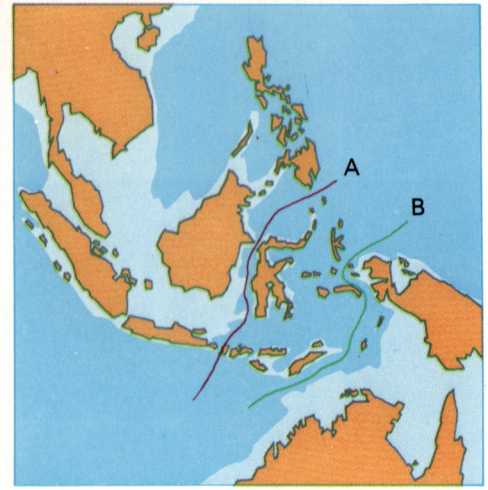

The similarities between the flora and fauna of some of the large islands of the Malay archipelago and species from the South-east Asian mainland are explained by the fact that the present-day structure of the continent and outlying islands has undergone many changes. During the Ice Ages of the Pleistocene frozen ice resulted in a lowering of the levels of the seas, exposing land links that have since been submerged. Investigation of the ocean bed shows that there are two distinct continental shelfs, one much deeper than the other. The naturalist Alfred Russel Wallace drew a line (A), known as the Wallace Line, through the island groups of the Malay archipelago, dividing the Oriental region from the Australian region. The rich and varied plant and animal life to be found to the west of this line is not present in the islands between lines A and B, because the latter have never been joined by land to neighbouring continents. Isolated on all sides by the sea, comparatively few species have been able to find their way to these shores.

ago) saw the first upheavals which created the Himalayas as they exist today, India beginning to throw out land links towards the continent. It was only after the Pliocene epoch, about 13 million years ago, that India was finally fused with the Asiatic mainland and that in the area which we know today as Assam the mighty peaks of the Himalayan massif merged with the mountain systems of South-east Asia to form a long, continuous chain, now partially submerged by the sea (the Malay archipelago). Subsequent upheavals occurred farther east and are indeed still taking place, producing an ever-changing pattern.

The so-called Wallace Line, which divides the Oriental and Australian regions, runs south of the Philippines, through the Macassar Strait (between Borneo and the Celebes) and then through the Lombok Strait (separating the islands of Bali and Lombok). Investigation of the bed of the ocean shows that each region rests on a different continental base. The Sunda platform is an extension of the Malay peninsula whereas the Sahul shelf joins New Guinea to Australia and Tasmania. Thus there were never any land links between the Celebes, the Moluccas and the Lesser Sunda Islands, surrounded as they are by deep water, and the Asiatic mainland.

Shifting seas

Such an eventful geological history could not fail to have a profound effect on living things; yet from the viewpoint of distribution the most significant changes have occurred only since the Pleistocene epoch, namely in the course of the last two million years.

Although this period includes the great Ice Ages, bringing about a general lowering of temperature and the formation at the poles of huge ice-caps, these climatic fluctuations had few direct repercussions on the equatorial and tropical regions. There was, however, an important indirect consequence of glacier formation. Immense quantities of frozen, locked sea water resulted in a general drop in the levels of the world's oceans. Bearing in mind that the Sunda platform is almost perfectly horizontal and that the depth of the sea above is on average a mere 180 feet (and in some places, as between Borneo and Java, no more than about 85 feet), it is easy to comprehend how from time to time, in the course of successive glaciations, the larger islands of the Malay archipelago were linked with one another and with the mainland.

The changing levels of land and sea provide an essential clue to the otherwise puzzling question of the distribution of flora and fauna in the Oriental region and justify a comprehensive survey of the plant and animal life both of islands and mainland, since they form a homogeneous whole. It is interesting to note that these changes, which have taken place only since the emergence of higher mammals, have imposed a continuous stream of adaptations affecting a large number of animal species, among which man himself must be included.

Scientists have long believed that volcanic action played a significant part in the formation of continents and the steady

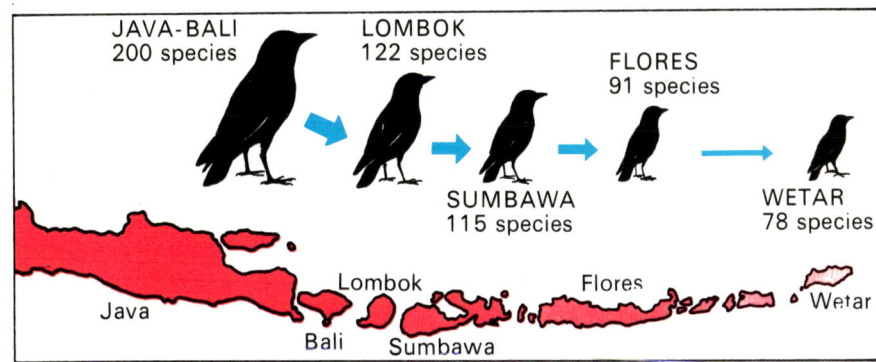

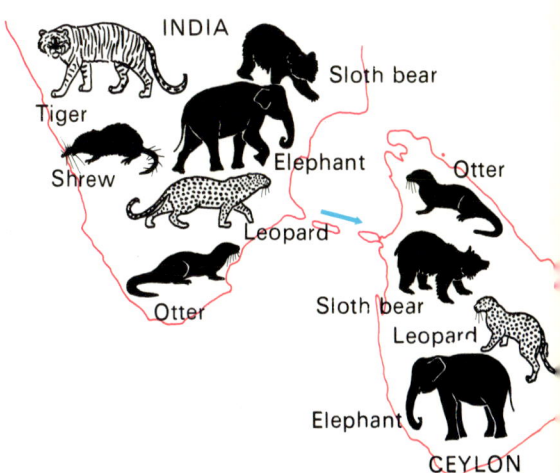

development of living creatures. Recent discoveries have confirmed this theory, showing the phenomenon to be especially marked in the Malay archipelago where the distribution and evolution of flora and fauna appear to have been determined not only by changes in sea level but also by volcanic action. Surveys have revealed the traces of ancient eruptions in countries where volcanoes were not believed to exist and many details of modern wildlife would remain obscure if scientists failed to take into account these bygone events.

Thus in the interior of several islands there are zones, once covered by rain forest but subsequently destroyed by fire, where the luxuriant primeval vegetation has never regrown and where, both in quality and quantity, plant life is poor. This devastation was undoubtedly caused by volcanic action.

In his book on tropical Asia S. Dillon Ripley points out that local folklore frequently alludes to natural phenomena which have affected the region throughout the ages. Traditional legends describe mountains of fire carrying death and destruction in their wake, and cataracts of water whose appearance heralded fertility and life. The mystic and the supernatural still loom large in the lives of millions of simple peasants. Although directly attributable to the stable, predictable climatic pattern of the Oriental region, the annual phenomenon of the rice crop is nevertheless hailed as extraordinary, even miraculous.

Land links created by the falling sea levels during the Ice Ages (and subsequently submerged) enabled some animal species to colonise new habitats, as happened, for example, in the case of India and Ceylon. But since Ceylon's climate and vegetational cover must have changed character between the successive periods when such land journeys were possible, the island would have attracted only those species suited to the habitat at a particular time. This is why there was not a general exodus and why certain species found today in India do not exist in Ceylon. In the islands of Indonesia (*top left*), the chain known today as the Lesser Sunda Islands was never linked by land to the Asian continent and these islands were thus colonised only by animals capable of crossing the ocean, namely birds. Distance, however, imposed checks to progress, and the farther east one travels the fewer are the bird species originating from Asia.

Rain the punctual provider

Broadly speaking, one might say that the climate of these tropical regions is governed by the winds which bring life-giving rain. Over vast areas, such as the plains of India and parts of Burma, Thailand and Indonesia, where the dry season may last six months, all forms of life – plant, animal and even human – seem to hinge upon the arrival of the rains. At this time of year local newspapers devote front-page headlines to weather forecasts, such information being of far more importance to peasants than any remote international crisis.

Rainfall in these parts is the direct result of the monsoon – a seasonal wind which blows with unfailing regularity. Meteorologists know what causes the wind but have not been able to explain satisfactorily why it occurs so punctually. In fact there are two monsoons, one from the south-west during the summer, the other from the north-east during the winter. In summer the sun travels across southern Asia, with areas of low pressure

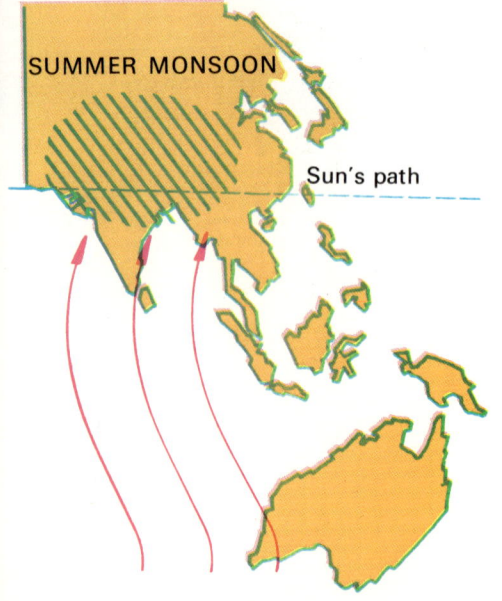

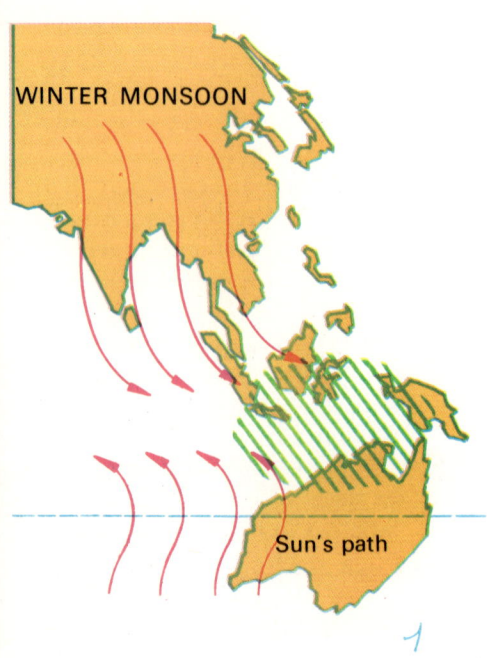

The greater part of the Oriental region is influenced by the seasonal phenomenon of the monsoons. In summer, when the sun follows a path across southern and south-eastern Asia, low-pressure areas drive the ocean winds northwards to discharge their humidity in the form of torrential rain (*hatched area*) over much of the continental interior. In winter the sun is below the equator, with low pressure north of Australia. The southerly monsoon winds bring drought to the continent and heavy rain to the Malay peninsula and Indonesian islands.

moving towards the tropic of Cancer. As a result the winds blow in from the Indian Ocean in a northerly direction, drenching the dry regions of the interior with torrential rains. It is this period of summer rainfall which is properly known as the monsoon season. In autumn, on the other hand, the sun sinks below the equator and crosses Australasia, with low pressure areas building up in the Pacific and north of Australia. As a result Asia, from north to south, is affected by the winter monsoon, with heavy rainfall in the Himalayas and a pleasant, dry autumnal climate in India, to be followed by a searing winter drought with high temperatures that crack open the ground, burn the grass and cause widespread suffering to animals and humans alike. The Malay peninsula and archipelago now receive torrential rainfall.

Nevertheless the incidence of annual rainfall varies enormously from one region to another. In desert areas hardly any rain occurs for the whole year round, yet in the parts covered by the tropical rain forest it is heavy and continuous, with scarcely a three-day dry interval. Thus in Assam the average annual rainfall is exceptionally heavy–equivalent in places to some 320 gallons per square foot every year.

Such rain is of an intensity seldom experienced in temperate climes and it invariably brings in its wake severe flooding, landslides and avalanches of mud. Raging water tears down the hills and mountainsides with incredible power and speed, laying bare the underlying rock and uprooting everything in its path, including all vegetation. Although such occurrences are very often disastrous on a local, individual level, causing widespread havoc, the rains are nonetheless beneficial and life-giving in the context of the ecology of the Oriental region as a whole, exposing immense areas of virgin terrain which are promptly invaded and colonised by new species of plants and animals. By such means (abetted of course by man's own activities of deforestation) an artificial, yet stable, layer of vegetation begins to take root, characterised principally by giant ferns, which effectively prevents the re-establishment of the primeval rain forest, perhaps for centuries to come.

The Asiatic rain forest

The vegetational cover of the Oriental region is extremely diversified, according to the climatic conditions of the region and depending principally on the amounts of rainfall. But the immediately striking aspect of this mosaic of plant forms is the widespread distribution of forests which have developed as a result of an ideal combination of temperature and humidity.

Climatically, the region may be subdivided into equatorial, characterised by continuous and abundant rainfall throughout the year, and monsoon, dominated by seasonal winds that precipitate alternating periods of rain and drought.

In areas where the equatorial climate prevails rain forests encroach almost everywhere, except in coastal districts where they merge with mangrove swamp forests. But elsewhere, in areas influenced by the monsoons, where annual rainfall ranges from 30 to 80 inches, two other types of forest are found–the

evergreen monsoon forest and the deciduous forest. If rainfall is below 30 inches—usually in inland districts—vegetation consists principally of various forms of acacia, such as *Acacia catechu* (producing the extract known as cutch, used in dyeing and tanning) and *Acacia leucophoea*. The drier the region the smaller the trees. This type of terrain looks very similar to the flat expanses of thorn and scrub common to the Ethiopian region. If the climate is even drier, with annual rainfall below the 20-inch mark, producing semi-desert conditions, typical plants

The rain forest (*left*) is the most characteristic form of vegetational cover in the very hot, perpetually wet areas of the Oriental region, with a rich variety of flowers. Coastal districts with a similar climate are more typically covered by belts of mangrove swamp forest (*right*).

As in Africa the trees of the Asiatic primary rain forest are so densely packed that very little sunlight penetrates the foliage. The undergrowth is therefore sparse and most of the animals are to be found nearer the treetops.

Facing page : The last surviving rhinoceroses and elephants of Asia are nowadays found in the Oriental region, including the wetlands of Assam.

include the shrubs of the genus *Euphorbia ;* and these coastal regions are dotted too with mangrove swamps.

In the mountain areas the scene changes. The upper level of the vegetational cover varies according to the latitude and the direction in which the slopes face. Because of the cool temperatures at high altitude two types of woodland are to be found, one comprising large, broad-leaved trees, mainly green oak, the other conifers, especially pine (*Pinus excelsa* and *Pinus longifolia*), spruce (*Picea morinda*) and fir (*Abies webbiana*).

South of the Himalayas

Our survey of the vast and varied Oriental region must take us from the northernmost ramparts of the Himalayas south through the Indian plains, south-east to the Malay peninsula and across to the outlying islands of the Malay archipelago. What we are doing, because it is simpler and more convenient for our purposes, is to divide this huge area into several distinct geographical regions (recognising that each contains a variety of physical features and natural biomes) rather than deal individually, for example, with the different types of forest which in any event are seldom to be found in a continuous belt or confined to a single part of the Oriental region as a whole.

The journey southward from the Himalayas brings us first to an immense expanse of low-lying terrain watered by two great rivers, the Indus and the Ganges. Here, in the Indus plain to the west and the Gangetic plain to the east, we already find almost all the most characteristic biomes of the Oriental region. A large part of the Indus plain is dry and arid, in fact semi-desert, although some thousands of years ago it was covered by dense forests where, judging by fossil discoveries, the rhinoceros, water buffalo and elephant roamed. Here too were the sites of the vanished cities of Mohenjo Daro and Harappa. But today it is an empty expanse of sand dunes and scrub, giving way, farther east, to an area of thorn forest and steppe.

The most characteristic ungulates of the Indus plain region are the nilgai (*Boselaphus tragocamelus*), the blackbuck or Indian antelope (*Antilope cervicapra*), the mountain gazelle (*Gazella gazella*) and the wild boar. Smaller mammals include hedgehogs, shrews, bats and rodents. Centuries ago the elegant cheetah lorded it over this region but today it has vanished and is, in fact, extinct throughout Asia. The Ghana reserve, south of Delhi, affords protection to the largest community of aquatic birds in India, and the Gir Forest reserve to the last of Asia's lions.

Travelling east across the Gangetic plain, the landscape changes completely. Deciduous forest gradually gives way to monsoon forest and finally to the rain forest proper. The wetlands of Assam and neighbouring territories are in fact almost entirely covered by rain forest – notable, above all, for the variety and abundance of its exotic flowers, especially orchids. In the state of Sikkim alone, for example, there are more than one thousand orchid species!

Typical animals include the elephant, Indian rhinoceros, deer, buffaloes, strange water mammals and a host of birds.

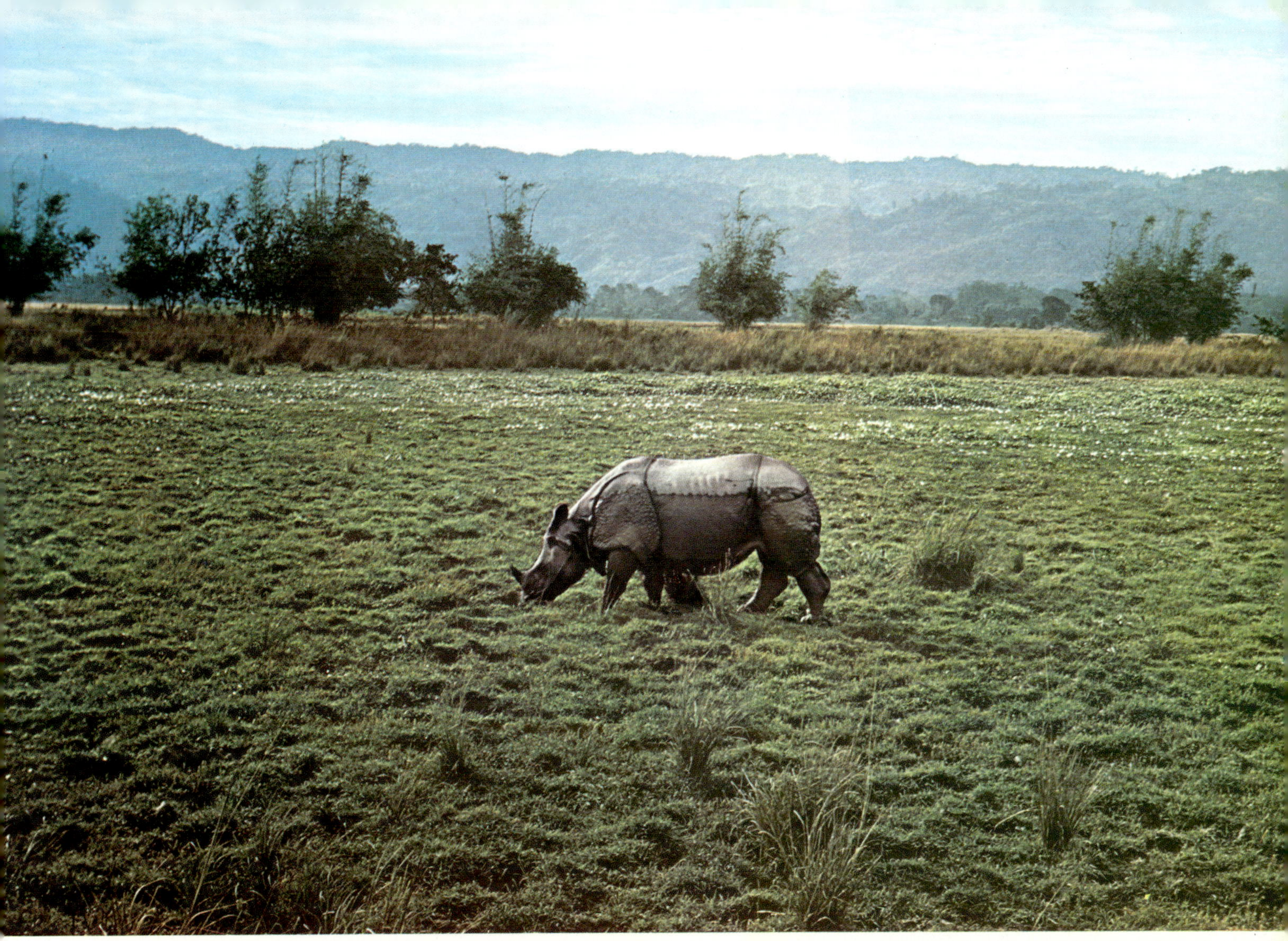

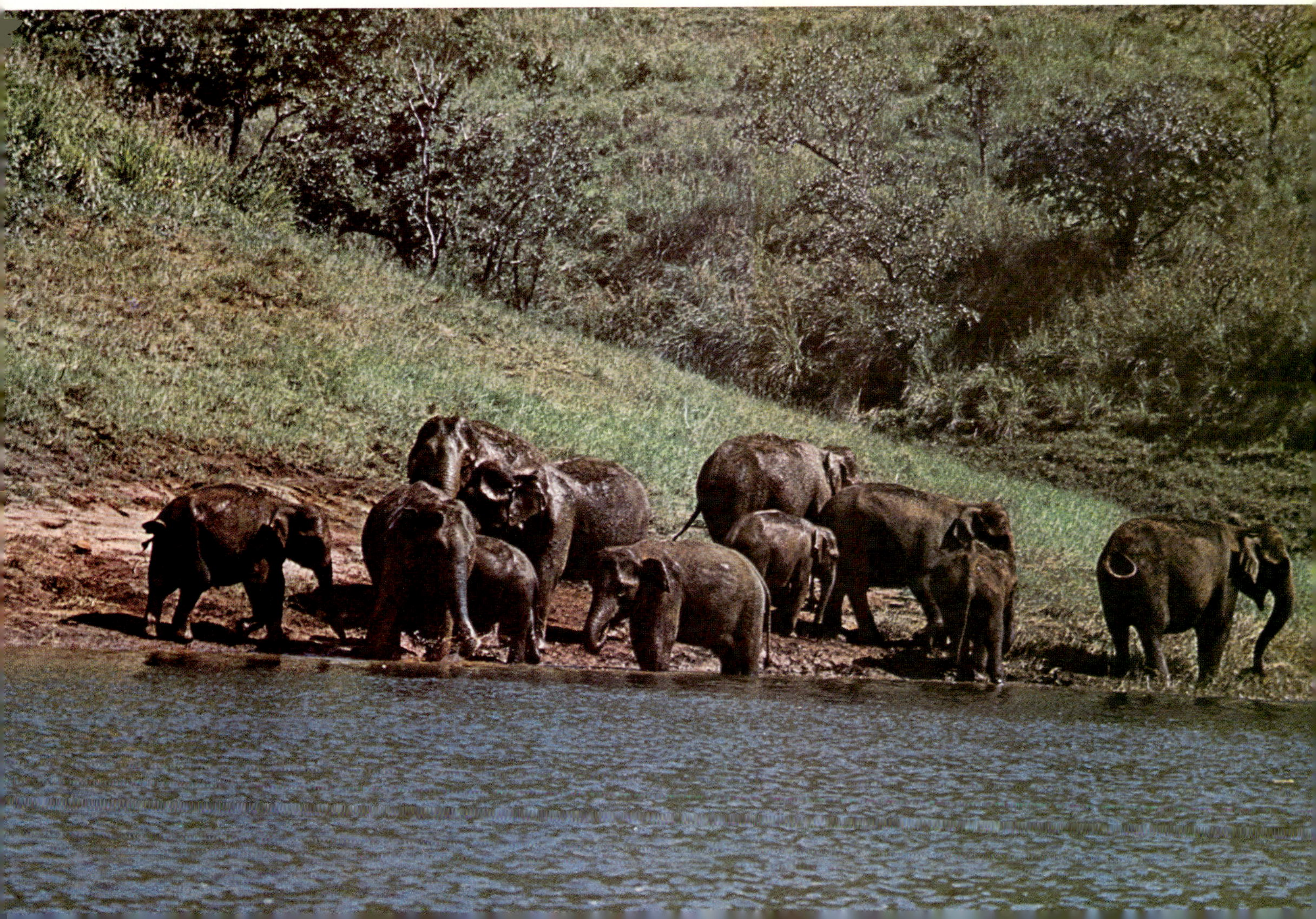

The Deccan triangle

The Indian peninsula lying south of the Indus and Ganges is separated from the Indo-Gangetic plain by the mountain ranges of Vindhya and Satpura. Apart from the Malabar coastal strip, peninsular India comprises the Deccan tableland, a triangular plateau dominated by mountain peaks. Climatically the Deccan is divided into three zones. The most westerly part, flanked by the Western Ghats, is very wet and covered by tropical rain forest; towards the centre the terrain is drier and more hilly with typical monsoon forests, and in the east, where the Eastern Ghats form a natural barrier halting the progress of the seasonal rains, vegetation consists in the main of thorny plants and trees.

Without any doubt the most characteristic and celebrated Indian mammal is the tiger. In the Sundarbans district, where the Ganges meets the Brahmaputra, the handsomely striped feline has the sinister reputation of being a man-eater. The tangled vegetation of the mangrove swamps conceals a number of huge amphibians such as the salt-water or estuarine crocodile (*Crocodylus porosus*) and the gavial (*Gavialis gangeticus*), the reptilian water monitor (*Varanus salvator*), and large mammals such as the water buffalo (*Bubalus bubalis*). Another typical animal of the region is the sloth bear (*Melursus ursinus*) which, although ruthlessly hunted in the past, is still found in the most densely forested areas of the Deccan. Among members of the Canidae are wolves, jackals, wild dogs and hyenas; and of course there is a tremendous variety of monkeys.

India has the unhappy reputation of being the country where more people are killed by animals than anywhere else in the world. According to official statistics, some 25,000 people die accidentally in this way every year, four-fifths of them as a consequence of snake-bite, the remainder as a result of being attacked by carnivores such as tigers and leopards.

The submerged world

Were it not for the savage wars which have recently ravaged the region, causing so much suffering and devastation, much of South-east Asia might have remained what it undoubtedly was in bygone times – a tropical paradise.

Flanked on one coast by the China Sea, on the other by the Bay of Bengal, and buttressed to the north by the mountains of Szechwan, Indo-China is comparatively young in terms of geological time. It is broken from north-west to south-east by a succession of low, narrow mountain ranges and irrigated by a number of rivers that rise in central Asia, Tibet and Yunnan. The flow of these rivers is very variable for they are periodically swollen by melting snows as well as by monsoon rains. The most important of them is the Mekong, whose source is in Tibet and which courses through the peninsula, finally forming a huge delta in South Vietnam. Before reaching the sea it divides into three branches, one of which, the Tonle Lap, forms a natural flood reservoir as it flows into a huge lake (Tonle Lap is in fact the local name for Great Lake). This is the only river

In the drier parts of the Oriental region the jungle is replaced by lighter, open forest with several species of acacia.

in the world which overflows its banks twice a year. In the dry season (November to May) it flows from the lake towards the Mekong delta and for the rest of the year in the opposite direction.

This region, therefore, is frequently under water. Yet flooding, as in the case of the Nile basin, is by no means catastrophic. Quite the reverse, in fact, for when the waters recede they leave behind a thick layer of fertile silt. At the peak of the floods hills are transformed into islands, forests are submerged and

Lord of the Asiatic jungle and one of the handsomest and most powerful felines in the world is the tiger.

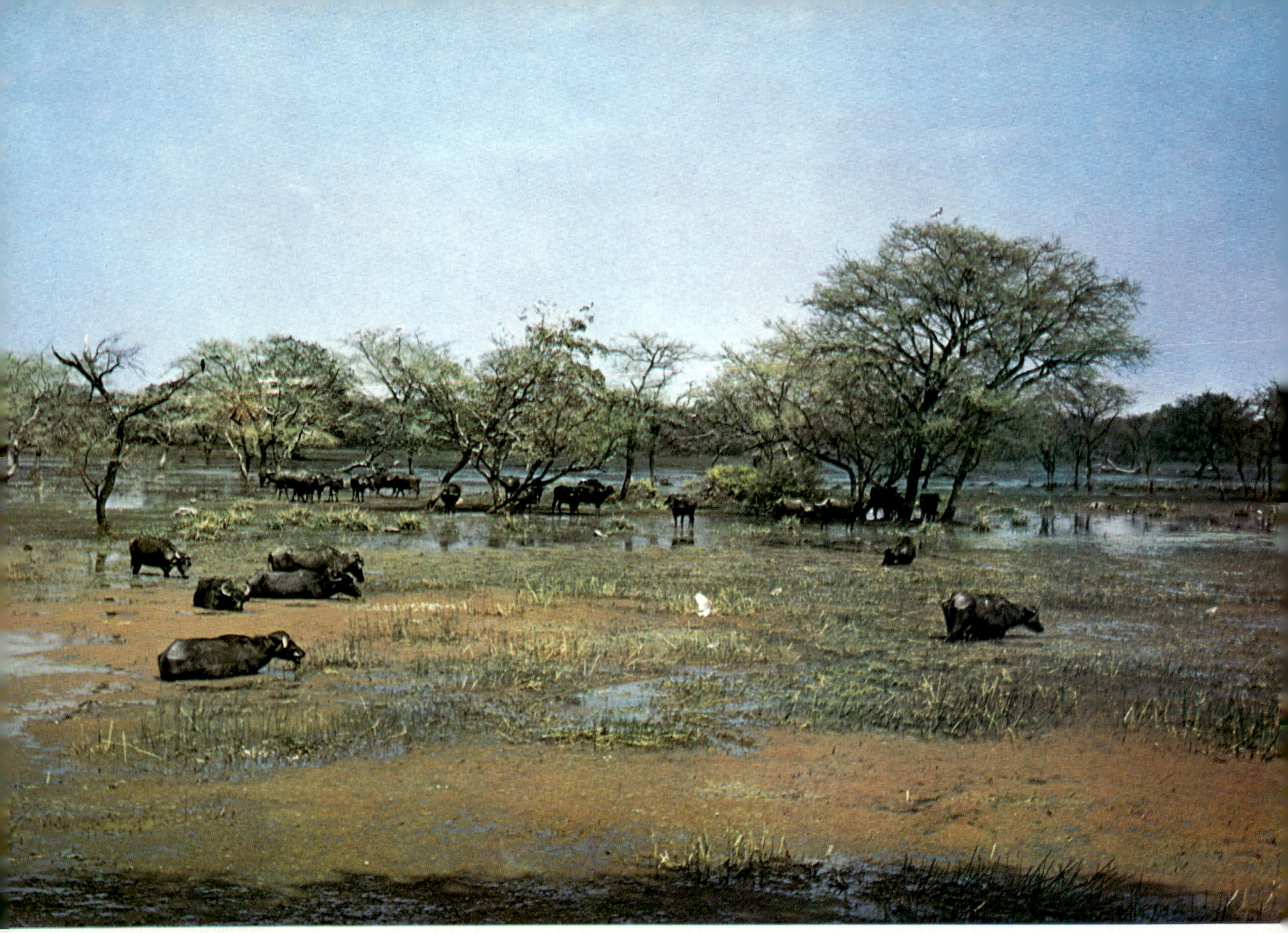

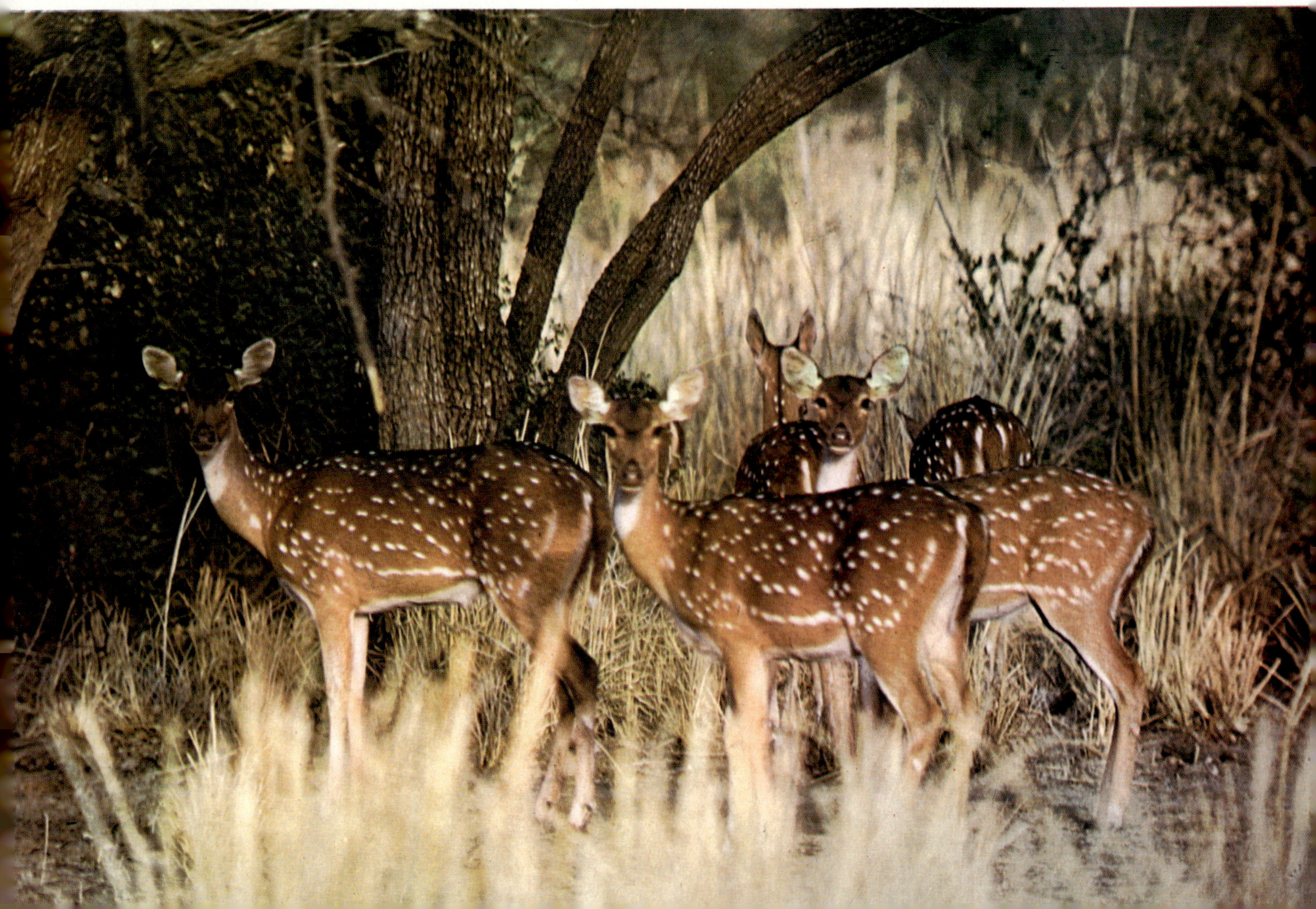

villages converted into extraordinary lake communities, each resembling a Venice in miniature, where the only means of transportation is a hollowed canoe known as the pirogue. The villagers turn their hands to fishing and reap an incredibly rich harvest from the flood waters.

When the water level drops the ebb-tide leaves behind a multitude of ponds and puddles swarming with animal life. So shallow are these pools that Pierre Pfeffer reports seeing the dorsal fins of large fishes sticking up above the surface and sandbanks littered with stranded species of all kinds, constituting a feast for predators as well as for the local peasants. It is hardly surprising that flocks of birds should also congregate around these parts at such seasons – notably pelicans, cormorants, skimmers, darters, marabou storks and many species of long-legged herons.

Buried temples

In the South-east Asian highlands there are belts of lighter, more open forest which have developed on poor, sandy soil. The plant species making up the dense underbrush may, during the rainy season, grow to the height of a man.

Such surroundings are ideal for a number of herbivores of all sizes, ranging from hares and dwarf antelopes to elephants and buffaloes.

One curious animal of these parts is the kouprey (*Bibos sauveli*), discovered in 1937 by the veterinary surgeon Dr Sauvel, which has been the subject of some controversy among taxonomists. It bears a close resemblance, for example, to the gaur (*Bibos gaurus*), the banteng (*Bibos banteng*), the zebu (*Bos indicus*) and the buffalo. Some authors claim that the animal is simply a hybrid form of the first two of the aforementioned species, others that it is a surviving descendant of a species of cattle domesticated by the Khmers, a now-vanished civilisation of Indo-China. In the view of Pfeffer it is, alongside the gaur, one of the species of the genus *Bos,* possibly the direct ancestor of the zebu, typical domesticated cattle of the Oriental region.

So many grass-eating animals inevitably attract hordes of carnivores. including tigers, leopards and wild dogs or dholes, which hunt in packs in the manner of wolves. These lighter forest regions also harbour an enormous variety of birds, ranging from diminutive sunbirds to enormous ibises.

In the mountains, where temperatures are considerably lower, there are a few pine forests but in general the tree cover is more typically that of the evergreen rain forest. In the midst of these jungles are relics of the splendid Khmer civilisation, the most famous of which is the lost city of Angkor which, then known as Yasodharapura, was the capital of the Khmer kings from the ninth to the fourteenth century. The magnificently carved temples of Angkor Wat have been partially rescued from the tangle of vegetation under which they were subsequently buried. For centuries these imposing ruins have been the haunts of bats and monkeys, especially the agile gibbons, noted for their acrobatics in the treetops.

The staple crop of Asia, ideally suited to the hot, wet conditions generated by the seasonal monsoons, is rice.

Facing page : Among the typical herbivores of the Oriental region are water buffaloes (*above*) and axis or Indian spotted deer (*below*) which are becoming increasingly rare in the wild.

The jungles of Asia conceal the ruins of many magnificent temples, almost buried under a mass of luxuriant vegetation.

Islands of the Malay archipelago

Our brief survey of the lands of the Oriental region concludes with glimpses of the countries that comprise the Malay peninsula and the islands lying between the mainland of Asia and the Wallace Line.

Many of the plant and animal species of these islands are obviously related to those of the Malay peninsula and thus offer both the geologist and the naturalist a rich, interesting field of research. The climate, as in all equatorial regions, is for the most part uniform, characterised by hot sun and rain throughout the year. The waters around the coasts are smooth and placid. But in some places, especially to the north of the Philippines and around the island of Timor, this pattern may be interrupted periodically by storms and hurricanes. The calm is also shattered from time to time by volcanic eruptions.

Tropical rain forests cover much of the interior of these islands. Among the rarer mammal species in the depths of the jungle are the Javan and Sumatran rhinoceroses, both on the verge of extinction, and that strangest of primates, the orang-utan (literally 'man of the woods'), nowadays confined to Sumatra and Borneo, though at one time an inhabitant of the Asiatic Mainland as well. The mangrove swamp forests lining the coasts also abound with animal life, including brightly coloured crabs, archer fishes which kill their prey by spitting out jets of water, and mudskippers which use their pectoral fins as forelegs for crawling or hopping, and which are capable of remaining out of water. Here too are otters, dugongs and many kinds of monkeys.

The landscape of the Malay peninsula is dominated on the one hand by rain forest and on the other by savannah, naturally or artificially formed. Life in the jungle is not made easier by the prevalence of blood-sucking leeches, some of which live on the ground, others in trees. In one painful experiment Pierre Pfeffer counted 175 leeches on his body in three hours.

The forests of the Malay peninsula are a treasure trove for the entomologist, with hundreds of fascinating species, including one enormous bird-wing butterfly. The many birds of the jungle are difficult to study in these surroundings and the thick canopy of foliage also conceals a multitude of monkeys. Even amphibians have adapted to the environment. The most astonishing is a tree-frog which uses its webbed feet as a form of parachute and is thus able to glide from branch to branch. Among other unusual residents of these rain forests are a giant squirrel, which may grow to three feet in length, and the shy tapir, whose only surviving relatives live far away in the jungles of the Amazon basin in South America.

The open savannahs are the home of many other interesting animals. Green peacocks, for example, strut through the clumps of Malayan rhododendrons, and typical mammals include wild boars, banteng and dark-coloured deer known as sambars.

Having given the reader some idea of the fascinating variety of biomes to be found in the Oriental region, it is now time to examine the wildlife in more detail.

Among the many primates to be found in the Asiatic treetops are the orang-utan (*left*), now confined to Borneo and Sumatra, and the curious crab-eating macaque (*right*), an inhabitant of mangrove swamp forests.

CHAPTER 83

Life in the jungle treetops

The larger part of the Oriental region is blessed with a hot, rainy climate favouring the growth of that dense, luxuriant form of forest which is commonly known as jungle – a world of giant trees, climbing plants, strange insects, dangerous reptiles, troops of monkeys and striped and spotted predators which lurk, almost invisible, in the shadowy undergrowth.

Of the world's three great forest masses – in the Amazon basin, Africa and Asia – it is probably the last which is most familiar to the European, if only through the descriptions contained in childhood books, particularly the wonderful *Jungle Books* of Rudyard Kipling. The tropical rain forest of Asia is the home of Shere Kan the tiger, Bagheera the panther, Akela the wolf, Baloo the bear, Kaa the python and Hati the wild elephant. But disappointment is in store for anyone who travels out to India, the Malay peninsula or the islands of the Malay archipelago, expecting Kipling's vivid evocation of eighty years ago to be faithfully reproduced before his eyes. The primary forest described by Kipling and other romantic story-tellers is nowadays difficult to find. The population explosion in these areas has forced man to cut back the frontiers of the jungle. To provide space for his domestic animals and his crops he has destroyed and burned almost at will. Yet the results, unfortunately, have not been what he expected. Torrential rainfall has swept away soil, leaving ground rich in iron and aluminium that has hardened to solid rock – the building material for ancient temples now concealed beneath a sea of encroaching vegetation. Having so heedlessly and pointlessly ruined the terrain, man then abandoned it and the primeval forest gave way to sterile savannah covered with coarse grass, such as the cogon which today covers almost 40 per cent of the land area of the Philippines and nearly

Facing page : Although the original primary forest has been devastated by axe and fire in many regions it has remained virtually intact in sparsely populated mountain areas.

Facing page : Despite the lack of sunlight many unusual flowers bloom in the Asiatic jungle. The most prolific are the numerous orchid species (*above, left to right*), and the strangest the carnivorous plants (*below*), which trap insects.

Torrential rain falling regularly on hills and mountainsides causes landslides, rooting up trees and, over a prolonged period, denuding the original forest cover. This is replaced by grassy steppe, attracting a multitude of herbivores and attendant carnivores. Although orang-utans and gibbons are confined to the jungle, monkeys such as langurs and macaques have extended their range to include these more open regions.

30 per cent of former Indo-China and Indonesia. But some of these grasses have developed naturally, especially above the forest level where rainfall, because of mountains barring the progress of the monsoon, is minimal.

Yet although in many of the best-favoured lowland areas only impoverished fragments remain of the primary forest, there are still vast untouched expanses of this forest, complete with original flora and fauna, in the mountains and in districts where the population pressure is of no significance.

For anyone who is familiar with the Congolese equatorial forest, the Asiatic rain forest appears, at first glance, very similar. All the world's jungles have certain resemblances but each has distinctive, unique features. Thus the Asiatic jungle, in contrast to the African, is not uniform. Amounts and distribution of rainfall, nature of the soil, altitude and location give rise to a number of different forest forms. To put it in the simplest terms, the true tropical rain forest occurs in Assam and the mountains of Borneo, where temperatures are high and rainfall heavy all year round. In areas where rain is seasonal and less frequent, so that six wet months are succeeded by six months of drought, there is monsoon forest; and in regions where the dry season is further prolonged there are broad stretches of deciduous forest.

The flowers of the Oriental region are so prolific that a lifetime would be insufficient to study them thoroughly. Animals are a rather simpler matter, and it is not all that important for the zoologist studying vertebrates, for example, to distinguish precisely between the different forms of forest. What he must know, however, is which is a primary and which a secondary biome, the distinction being between a forest that has remained intact since it originated and one that has been modified by human activities—these being factors which have had a profound influence on the numbers and distribution of animals.

Thus in the primary forest ground vegetation tends to be sparse, with the dense foliage of the treetops impeding the penetration of sunlight. Although the seeds of large trees germinate in heavy shade, neither grass nor underbrush can develop to any appreciable extent. The near-absence of grass explains why there are so few seed-eating birds as compared with large numbers of fruit-consuming and insectivorous species. Ungulates are equally scarce except for the odd solitary individual or small family group. Among the primates, both gibbons and orang-utans are to be found in the primary forest whereas the langurs and macaques are commoner in secondary forest areas. Here the sun's rays can reach through to ground level and stimulate the growth of underbrush, at the expense of the original tree species whose seeds only germinate in darkness. This type of forest has more seed-eating birds and ungulates.

It is in fact this secondary forest which most closely approximates to the familiar picture of the jungle as described in so many books and portrayed in adventure films. These are the surroundings where one literally has to hack a way through the impenetrable tangle of vegetation underfoot—in contrast to the primary forest where there is little save mud and dead leaves.

The changing face of the South-east Asian jungle has lately
been brought about by man; but long before he ever appeared
on earth natural processes were at work (as they still are) which
in some regions were responsible for the formation of forest
clearings. These play an important role in the ecology of the
tropical rain forest.

Rain water flowing down the wooded mountainside in cascades
easily penetrates the soil by virtue of the roots of trees that keep
the earth spongy, and works its way down to a layer of imper-
meable rock where it accumulates. The level steadily rises and
the sheer weight of water is soon carrying away trees by the roots
and bringing tons of soil and debris crashing down the slopes.
An aeroplane pilot over Borneo once counted 117 separate landslides of this nature in the course of half an hour which uncovered
more than two acres of bare rock.

The empty gaps left by these avalanches are quickly invaded
by plants which need plenty of light for growth. It is unlikely
that in these regions the original pattern of primeval forest will
be re-established for centuries, if indeed ever, as further slips
occur; and as new plants appear the forest which once harboured
only a relatively small number of specialised animals now
supports a much richer and more varied fauna.

One of the most astonishing flowers to be found in the jungle,
though seen by few people, is the rafflesia, named after Sir
Stamford Raffles. It is a parasitic plant, growing on tree roots,
having neither roots, stems nor leaves of its own; there is no
sign of it prior to the appearance of the small bud which eventually
develops into a magnificent red bloom, mottled with white, up

Jungle flowers are noted for their variety of strange shapes, large dimensions and unusual colours, all of which are evident in this specimen.

Facing page: The Asiatic jungle is a happy hunting ground for the entomologist. Many species of butterflies bear striking patterns (*above left*) and predators use background vegetation to conceal themselves from their victims, as is the case with the mantis (*above right*) lurking on an orchid. Among jungle butterflies the largest specimens belong to the genus *Ornithoptera*. Shown in the lower picture is the male of *Ornithoptera priamus*.

to 3 feet in diameter. The few botanists who have examined this rare flower are agreed on one point—that its perfume does not measure up to its beauty. The scent is in fact rather unpleasant.

The fig trees of the genus *Ficus* are always associated with hot countries (there are in fact some 600 known species) but not usually with the jungle. Yet some fig trees are native to Asia where they attract, as elsewhere, multitudes of fruit-eating birds and mammals. The typical jungle species have a somewhat sinister reputation and are commonly known as 'stranglers', the reason being that the sinuous roots of these epiphytic (but not parasitic) plants cling to the trunks and branches of other trees and may even cause damage to walls and buildings. In their early stages the hollow roots work their way down to the ground, in the process crushing the life from the host tree. These roots converge to form what looks like a trunk and snake out in all directions underground to form tree thickets.

Hide-and-seek in the jungle

The hot, moist conditions of the Asiatic jungle encourage the proliferation of myriad forms of tiny life. Although most of them have their counterparts in other lands nowhere else in the world are they to be found in such vast numbers or in such a diversity of forms and colours.

Leeches, as previously noted, are everywhere in evidence, and these tiny blood-sucking worms are a constant scourge to the entomologist, for example, who in all other respects finds the jungle an incomparably rewarding place for the detailed study of rare and beautiful insects. Many of these go unseen by the layman for they have adapted the most ingenious patterns of cryptic coloration; and even the larger species blend so cunningly with leaves, bark, branches and flower petals that they can deceive all but the most trained, experienced eye. Camouflage, whether by means of shape or colour, spares many insects the attentions of predators and facilitates their own hunting efforts. Thus the corolla of a magnificent orchid may be a death-trap for an insect which fails to detect the presence of a lurking mantis. Some harmless insects protect themselves by taking on the guise of similar but dangerous species, while others exude a disagreeable odour which keeps insectivorous birds at bay.

Particularly fascinating for the specialist are the nests of the spinning ant (*Oecophylla smaragdina*). The breaking of a leaf concealing one of these nests will provoke the exodus of a stream of worker ants who station themselves along the broken edge to repair the damage by mouth. These are followed by a second group of workers, each carrying a larva between its mandibles. The larvae secrete a silk substance which welds together the jagged edges of the leaf.

The moths and butterflies of the Asiatic jungle are singularly beautiful. A narrow, tree-lined track may suddenly open out into a clearing to reveal, on the ground, a group of Rajah Brooke's birdwings (*Ornithoptera brookiana*), butterflies whose black wings with metallic green reflections may be up to 8 inches from tip to tip.

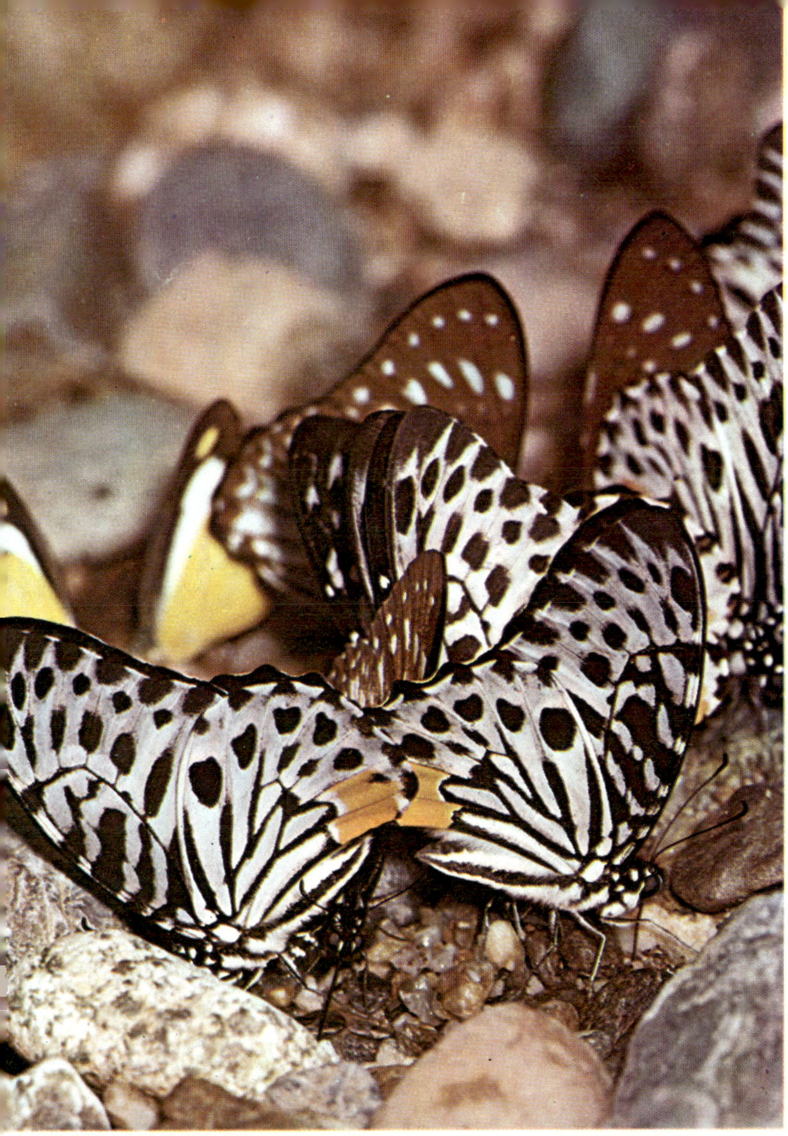
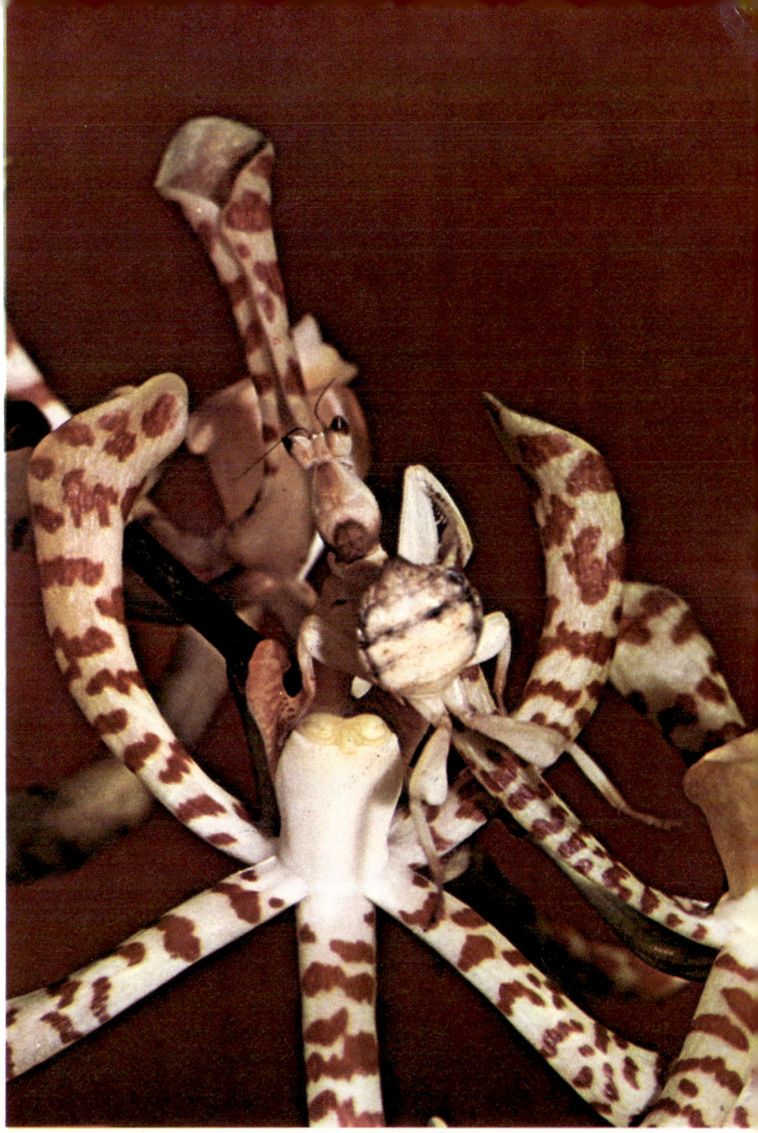
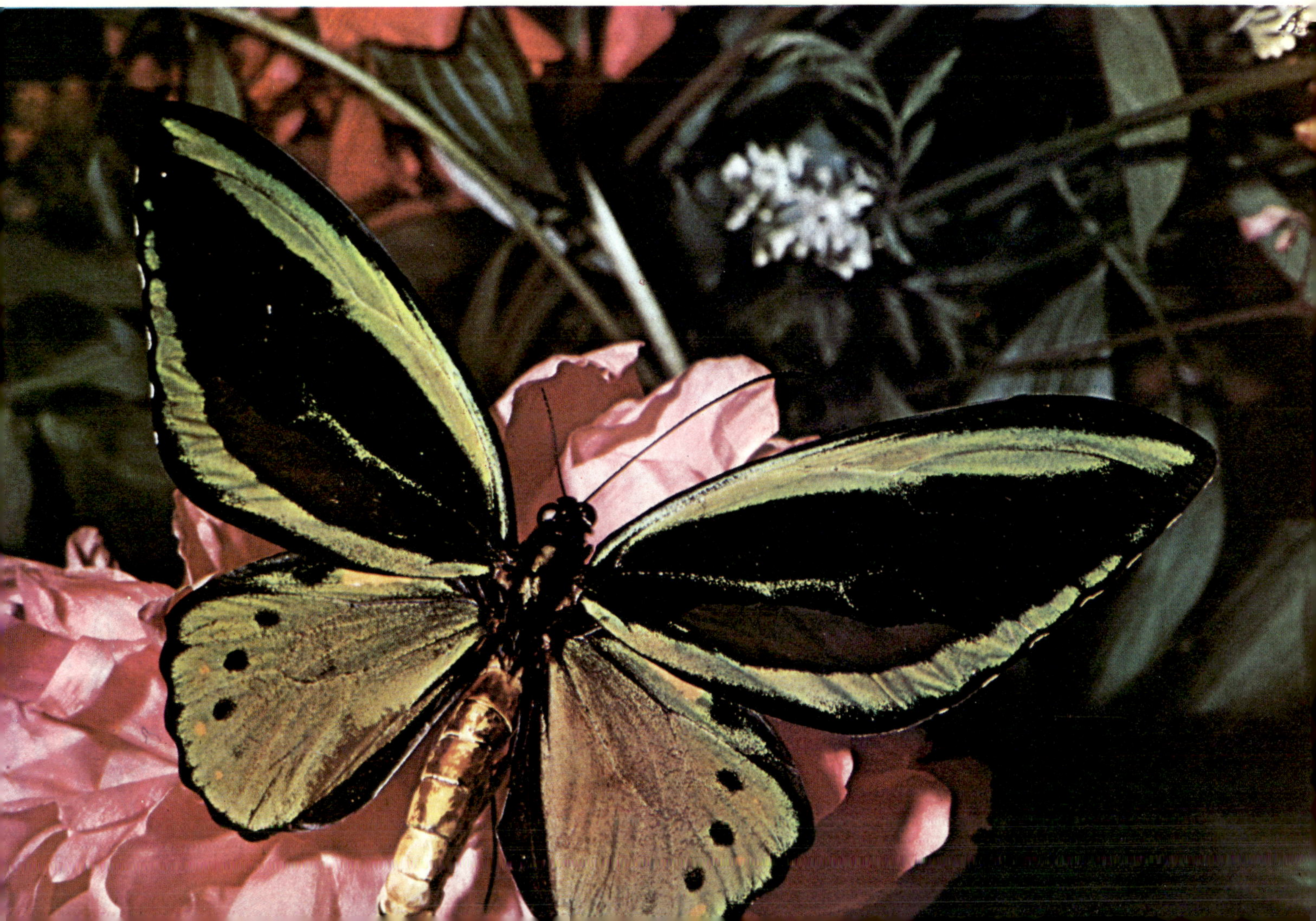

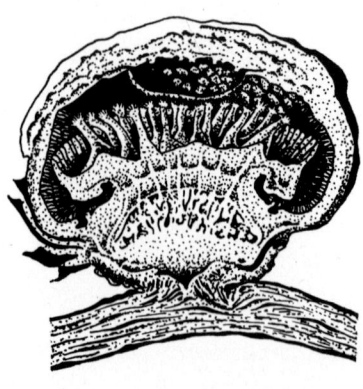

The largest flower in the world is the rafflesia, a parasitic plant with no proper roots, stem or leaves. It grows very slowly and although beautiful emits an unpleasant scent.

Facing page : Gibbons are the acrobats of the jungle, clinging to branches with their prehensile limbs. Accidents are nevertheless more common than might be expected, as examination of skeletons prove.

One point that used to puzzle entomologists was why male butterflies of the genus *Ornithoptera* tended to be more numerous than females, although the latter were sometimes larger. It was later discovered that this was an illusion arising from the females' habit of settling high up in a tree where they were difficult to see. These beautiful species are greatly prized by collectors, a single specimen fetching 2,500 dollars at an auction in Paris in 1967. Unhappily this kind of market demand inevitably leads to extensive hunting and the gradual disappearance of some particularly rare varieties.

For entomologists specialising in beetles, there is equally no lack of interest in the jungle. Among a wide variety of magnificent specimens are iridescent rose chafers, scarabs and, most striking of all, a multitude of long-horned wood boring beetles (Cerambicidae). There are some 20,000 known species of these beetles which are characterised by their long, backward-curving antennae. Alfred Russel Wallace, during his eight years of travels in South-east Asia, showed particular interest in these beetles and made a special point of looking for them wherever he went. He collected almost 2,000 species in Borneo alone and all but about a hundred of them were found in a restricted site, not much more than one square mile in area, close to the Sadong River in the western part of the island.

For sheer diversity and number – in this as in other regions – no zoological group can compare with the arthropods or jointed invertebrates, which include insects, arachnids and crustaceans; but the Oriental region has many surprising forms of vertebrates as well and without a doubt the most curious are the various 'flying' mammals which propel themselves through the air in different ways. Certain species of frogs and reptiles show similar gliding aptitudes, though some of these simply make use of their extensible skin membranes to break their fall rather than utilise them parachute-fashion as a means of moving about from place to place. Among the mammals are a number of flying squirrels, closely related to the squirrels of the Palearctic region, and the flying lemur, only representative of the order Dermoptera and confined to the Indo-Malayan zone. Primates of the region include orang-utans, gibbons, macaques, langurs, tree shrews and tarsiers.

The most characteristic birds of South-east Asia are members of the Phasianidae (also represented of course in the Palearctic and Ethiopian regions). But here in the jungles of the Oriental region are pheasants of incomparable beauty, including the magnificent peacock which struts so proudly across our own parks and lawns. Incidentally, the wild ancestors of our familiar domestic fowl came from Asia and in fact still flourish there.

For reasons already explained, there are comparatively few terrestrial mammals in the jungle, but characteristic ungulates are wild boars, water buffaloes, gaurs, bantengs and three species of rhinoceros, all becoming rare. Here too are the celebrated pangolins or scaly anteaters, tapirs and a variety of predators – wolves, wild dogs, tigers, clouded leopards, civets and small felines – which hunt their prey either on the ground or high up in the trees.

The flying gecko (*right*) can descend from tree trunks to the ground without using its limbs, thanks to folds of skin on either side of the body which enable it to glide through the air. Equally well camouflaged but even better equipped for gliding is the flying lizard (*left*). The extensible membranes are supported by enlarged ribs so that they open and close like an umbrella, serving as 'wings' that carry the reptile considerable distances.

Flying lizard
(*Draco volans*)

Wallace's flying frog
(*Rhacophorus nigropalmatus*)

Life in the treetops

The jungle floor, when not completely drenched by rain, is almost always wet and muddy. Grass and low plants cannot grow here because little sunlight filters through the dense, intertwined foliage of the tall forest trees. The consequent scarcity of insects and herbivores at ground level means a similar lack of predators; but the smaller hunters and their victims are to be found in large numbers higher up, especially near the treetops, where they act out a drama of life and death unseen from below.

Anyone venturing into the tropical rain forest will immediately be struck by this apparent absence both of mammals and birds; and even the animal inhabitants of the jungle are confronted with the perpetual difficulty of making themselves seen and heard in this world of shadow and darkness. Most of them locate others of their species by characteristic sounds which vary in sonority according to the circumstances or the time of day. The 'song' of cicadas, the croaking of frogs, the trumpeting of hornbills, the cooing of doves and the scream of gibbons combine, especially at sunrise and sunset, to produce an unbelievable, often terrifying, cacophony.

As for the larger herbivores and ground-dwelling carnivores (such as the tiger and wild dog) these are seldom found in the heart of the jungle but frequently congregate in the forest clearings which for the most part have been artificially formed by devastating fires.

The jungle species which live at varying levels above the ground, feeding as they do on leaves, wood, fruit, insects, nestlings and small mammals, have in the course of evolution acquired special mechanisms or anatomical peculiarities enabling them to master their environment. They include birds, of course, but also bats and a host of insects such as mosquitoes, flies, bees, butterflies and moths, flying beetles and ants. Yet there are other species which are equally at home in the treetops despite the fact that they lack wings. Some have prehensile limbs, others a long, similarly prehensile tail, by means of which they can maintain their balance up in the branches. Among such are the binturong (*Arctictis binturong*), related to civets and mongooses, the orang-utan and the gibbon, and a primitive mammal midway between an insectivore and a primate, the tree shrew. As in the Holarctic region, other tree-dwelling mammals include squirrels, martens and civets, all of which move rapidly along trunks and branches, using their tail for balancing.

Of all these animals the most spectacular and highly specialised are those that have skin folds or membranes which can either be used for 'parachuting' or gliding. Indeed the Indo-Malayan peninsula might be regarded as the realm of flying animals.

Flying squirrel
(*Petaurista petaurista*)

Animal parachutists and gliders

In 1855, during his visit to the Far East, Alfred Russel Wallace described for the first time a strange frog which a peasant had shown him, swearing that it had flown down from a tree. Wallace could neither prove nor disprove the claim, noting, however, that the digits of the frog's fore and hind feet were joined by a membrane of skin and that there were pads on the joints of both the arms and legs. The mystery of the 'flying' frog was not resolved until 1964 when Davis and Inger managed to photograph one individual actually in flight in the Borneo jungle.

The Wallace's flying frog (*Rhacophorus nigropalmatus*) does not, in fact, fly in the accepted sense but makes use of its interdigital membranes as a form of parachute for dropping from branch to branch. Never, however, does it alight on the ground, for it lives permanently in the trees, building nests of leaves so tightly packed that rain water collected therein does not seep through, enabling the female to lay her eggs in the customary amphibian fashion.

Certain reptiles have also acquired various adaptive features suitable to the environment and none is better equipped to lead an arboreal existence than an agamid known as the flying lizard (*Draco volans*). Crouching on a branch or clinging to the trunk, this reptile is almost invisible, so perfect is its protective colouring, except during the mating season. Yet when it launches itself through the air it takes on the appearance of a vividly coloured butterfly. This transformation is the result of spreading out, on either side of the body, orange-and-black 'wing' membranes, supported by five or six long ribs which open and close like the struts of an umbrella. In this manner the lizard achieves a form of gliding flight carrying it 30–40 feet. When at rest the ribs are folded along the length of the body.

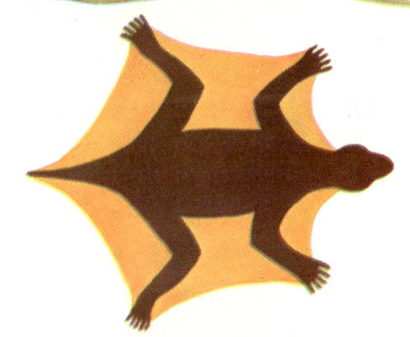

Flying lemur
(*Cynocephalus volans*)

Flying gecko
(*Ptychozoon homalocephalum*)

Another member of the lizard family with similar aptitudes is the flying gecko (*Ptychozoon homalocephalum*), although it cannot travel as great a distance through the air as the flying lizard or the various flying squirrels. This reptile has simple folds of skin on either side of the body which, in conjunction with the tail and interdigital membranes, can be employed as a parachute when dropping from one level to another, along an oblique path inclined at about 45°.

The brilliantly coloured paradise flying snake (*Chrysopelea paradisi*) is an accomplished climber and swimmer, which, although it cannot exactly be said to fly or glide, has an ingenious method of moving about from branch to branch. Extending its ribs to form a flattened plane, it retracts its ventral region into a kind of hollow dome, creating an improvised 'parachute' which makes it possible for the snake to fall from a moderate height without any risk of being hurt.

There are a number of flying squirrels living in the tropical rain forest, very similar to the species found in the European taiga. Most of them are concentrated in the Indo-Malayan peninsula. They are of varying size, some little larger than a mouse, others (belonging to the genus *Petaurista*) more akin to a small fox. All possess an extensible fold of skin on each side of the body, linked with both the fore and hind feet, leaving the tail free to serve as a balance.

Naturalists distinguish five species of giant flying squirrel, which also live in parts of Holarctic Eurasia, including, for example, the forest regions of China and Japan. Very little is known about these animals apart from the fact that they normally live some 70 feet or so up in the trees, never descending to the ground. Food consists of a mixture of leaves, fruit, young shoots and possibly insects, eggs and nestlings. According to Ernest P. Walker some squirrels belonging to the genus *Petaurista* are capable of gliding up to 500 yards. The females of these species give birth to one or two young.

The flying lemur

Certainly the most remarkable glider of the tropical Asian jungle is the flying lemur (genus *Cynocephalus*), only surviving representative of the order Dermoptera.

There are two species of flying lemur—*Cynocephalus volans*, found in the Philippines archipelago on the islands of Mindanao, Basilan, Samar, Leyte and Bohol; and *Cynocephalus variegatus*, an inhabitant of the South-east Asian mainland, including the Malay peninsula, of Borneo, Sumatra, Java and the surrounding islands. Slightly larger than rabbits, these animals seldom weigh more than about 3 lb. Clinging to the branches with their hands and feet, or sometimes only with their powerful claws, they spend their time reputedly feeding on leaves, flowers, fruit, buds and shoots. Nevertheless, after studying the stomach contents of eight individuals in the wild, Walker found only green vegetation; and in captivity they show a marked distaste for fruit. To drink they lap up water that accumulates in the hollows of leaves and the corollas of tropical flowers, the latter storing an

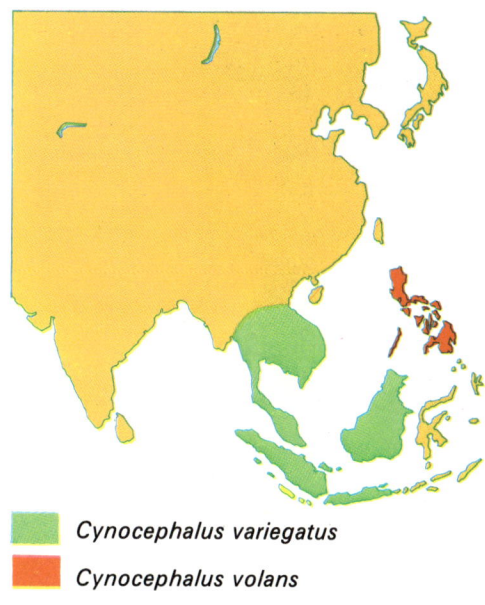

■ *Cynocephalus variegatus*

■ *Cynocephalus volans*

Geographical distribution of the two species of flying lemur, *Cynocephalus variegatus* and *Cynocephalus volans*.

FLYING LEMUR
(*Cynocephalus volans*)

Class: Mammalia
Order: Dermoptera
Family: Cynocephalidae
Total length: 20–27 inches (50–69 cm)
Length of tail: 8½–11 inches (22–27 cm)
Weight: 2¼–3¾ lb (1–1·75 kg)
Diet: vegetation, especially leaves
Gestation: 60 days
Number of young: usually one

Adults
Strange appearance due to fold of skin running from the neck to the forefeet, thence to the hind feet and so to the tail. Colour yellowish or brown with lighter marks down the back. Digits also linked by membrane; long claws. Short muzzle, huge eyes.

Young
Measure about 10 inches at birth, clutched to maternal belly.

Facing page: The flying lemur is a tree-dwelling mammal of the Oriental region which climbs trunks by gripping the bark with its claws and soars from tree to tree by spreading the extensible fold of skin (patagium) on either side of the body. The young lemur clutches its mother's belly during such gliding flights.

enormous quantity of liquid.

When the flying lemur moves from place to place or is obliged to do so when threatened by an enemy, it launches itself into space, stretching out limbs and tail and opening the broad, hairy membrane which extends from the sides of the neck to the hands, then links the hands to the feet and finally the feet to the tip of the tail. When this so-called patagium is fully outstretched, the animal is able to glide considerable distances without losing appreciable height. It has been estimated that over a distance of 150 yards the lemur drops only some 30–40 feet.

The flying lemur is a nocturnal animal and every night it settles on precisely the same portion of branch and hurls itself into the air, again always in the same general direction towards the trees which are likely to provide a meal. When it senses a hidden threat it begins clambering upwards (it is an expert, though not particularly rapid, climber), and when it reaches the treetop once again soars off to a safer vantage point. Should it accidentally fall, however, it is in serious trouble, being absolutely defenceless on the ground where it can only move about slowly and awkwardly.

It appears that the animals mate around the beginning of the year and that gestation lasts 60 days. As a general rule the female gives birth to one offspring (twins are rare) either in the crook of a tree or in a kind of nest made of branches, this serving equally as a place of refuge. The baby, which measures some 10 inches at birth, is held close to her chest and is not allowed to let go even when feeding. This additional burden does not seem to cause the mother any inconvenience when undertaking a gliding flight from tree to tree.

In dangerous situations the lemurs emit loud growling sounds which appear to serve as alarm signals. Otherwise they remain silent and for that reason are hard to locate. This used to give the impression that they were not very numerous but in fact local populations are comparatively large, despite the fact that the animals are intensively hunted both for fur and flesh.

Bats and flying foxes

Better adapted than any other mammals to life in the jungle treetops are the bats (order Chiroptera). There are innumerable species to be found in the tropical rain forests of Asia, varying greatly in size and feeding habits. The smaller bats are primitive, insect-eating species but the larger ones require a considerable quantity of more nourishing food consisting almost exclusively of vegetable matter, especially fruit.

The smallest bat of the monsoon forest is the long-winged bat, genus *Miniopterus*, which measures a mere 4 inches, tail included. Its widespread distribution also encompasses Europe and Africa. It spends its day clinging to rocks or tree branches, venturing out only at dusk to hunt insects 30–60 feet above the ground. The mating season is apparently towards the end of summer. In the tropics gestation lasts 120 days but in Europe this period is considerably extended because the bats go into partial hibernation. The female gives birth to a single offspring

Bats are characteristic inhabitants of the tropical rain forests of Asia and there are about one hundred and fifty species of *Megachiroptera*—large, fruit-eating bats sometimes referred to as flying foxes. These animals spend the day suspended from branches and venture out at dusk for food.

Facing page : Flying foxes assemble in large numbers in the jungle trees and may travel dozens of miles to the plantations where ripe guavas, mangoes and other types of fruit are to be found. In many areas they reach plague proportions.

and the animals live for up to nine years.

Some medium-sized bats supplement their insect diet with rodents, nestlings, amphibians, fishes and occasionally other bats. Among these are the false vampire bats of the genus *Megaderma* which measure about 6 inches long. Two species are known in the forests of the Oriental region and a third in Africa.

The most typical of the Asiatic forest bats, and by far the largest and heaviest, are the fruit-eating species belonging to the genus *Pteropus*, commonly known as flying foxes. The most imposing individuals have a 5-foot wingspan. These bats spend the day suspended, head-down, from the highest branches, wings wrapped tightly around the body, sleeping in this position for hours. Now and then they fan themselves with their wings but only when the sun goes down do they begin stirring purposefully, flitting about in groups and eventually swooping down in immense numbers—perhaps 500 or more—on plantations of fruit trees. If necessary the bats will travel up to 40 miles in quest of food (one naturalist has reported a group which flew an enormous distance of 200 miles), nosing out the ripe fruit and then landing among the branches in a noisy, squabbling throng. Some individuals are content to devour their fruit on the spot; others rip it off with their sharp teeth and carry it away for private consumption. Having satisfied their appetite, the bats gradually quieten down and as the first streaks of dawn appear they fly back to the trees which they occupied the previous evening.

Flying foxes often cause tremendous damage to village plantations and local farmers understandably regard them and treat them as pests. When fruit is locally scarce they evidently fly off to other districts in search of food but their return is predictable. The next fruit harvest will find them suspended from the same trees, ready for new invasions.

The curious tree shrews

The little mammals known as shrews are all classified as insectivores. So too, until quite recently, were the tree shrews of the genus *Tupaia*, residents of the rain forests of South-east Asia, Indonesia and the Philippines. These primitive tree-dwelling mammals were thought to be little different from other shrews, having a long snout, non-prehensile limbs and clawed toes, although the long bushy tail is admittedly more reminiscent of a squirrel. Yet some naturalists, including W. E. Le Gros Clark, expert on comparative anatomy and anthropology, pointed out that in some respects they seemed to have close links with primates and that by rights they should be classified in that order, together with lemurs, lorises, galagos and tarsiers. Among the reasons advanced by Le Gros Clark were the structure of the skull, the advanced degree of differentiation displayed in the visual mechanisms of the eye and the brain, the reduced dimensions of the olfactory organs, the dental structure (thirty-eight teeth, including two upper incisors and three lower incisors in each half of the jaw), the mobility of the big toes and the thumbs, the low position of the testes in a prominent scrotal sac, and the haemochorial type of placenta.

The tree shrew, formerly classified as an insectivore, is now acknowledged by most authors to be the most primitive of primates, belonging to the suborder Prosimii.

It was clear that these strange creatures played an important role in the history of evolution. The primates were known to have been descended originally from a branch of tree-dwelling insectivores which had acquired the ability to grip to trunks and boughs with hands and feet and which had developed keen powers of sight and smell. The tree shrews of the Oriental region obviously resembled the traditional insectivores and in outward appearance they are undoubtedly closer to these than to primates. But the arguments of Le Gros Clark and others were sufficiently persuasive to change the accepted classification and they were subsequently regarded as prosimians (suborder Prosimii) which have undergone little evolutionary change for millions of years.

One reason why tree shrews were formerly classified as insectivores was that they were assumed to feed exclusively on insects. It has now been revealed that in addition they eat fruit, seeds, leaves, worms, eggs, nestlings and even small rodents. Almost all are tree-dwellers although certain species spend prolonged periods on the ground. High in the trees, however, they scuttle along the branches, digging sharp claws into the bark.

Doubts about their correct classification were all the more difficult to resolve because of the problems of studying their behaviour in the wild. In captivity, however, males display such fierce aggressiveness towards one another that it has been assumed that intraspecific fighting is also a feature of life in the jungle and that territories are staked out and obstinately defended against rivals. Reproductive behaviour was likewise

The tree shrews of the Asian jungle possess anatomical features which resemble those of both insectivores and primates – which accounts for the difficulties of classification. All members of the family in fact consume insects but some supplement the diet with vegetable matter. Some are exclusively arboreal but others often come down to ground level.

Geographical distribution of tree shrews.

TREE SHREW
(Tupaia glis)

Class: Mammalia
Order: Primates
Family: Tupaiidae
Total length: 12–18 inches (30–45 cm)
Length of tail: 6–9½ inches (15–24 cm)
Diet: insects, fruit, leaves, etc
Gestation: 45–50 days
Number of young: 2, sometimes 3

Adults
Long bushy tail; elongated muzzle; dark eyes; rounded ears. The fairly long feet have five sharp, curved claws. Upper parts greyish with random ochre, chestnut, green or black marks. Underparts whitish or beige.

Young
Born blind, weighing about half an ounce. Leave nest after four weeks.

Facing page: The slow loris is a tree-dweller which moves around in a measured, deliberate manner. As its huge eyes suggest, it is nocturnal by habit, using its prehensile hands to capture insects and eggs.

shrouded in mystery until 1966 when Robert Martin published his findings on the sexual habits of the common tree shrew (*Tupaia glis*). According to this report the male constructs a nest in the thickets (unique behaviour for a mammal) some days before the female is due to give birth. This nest is made up of hundreds of leaves. After a gestation of 45–50 days two, sometimes three, naked, blind babies are born. The mother scrupulously cleans them for a couple of hours, suckles them and then leaves them in the nest.

In the case of some species, however, that have been bred and reared in captivity, it seems to be the female rather than the male who actually builds the leafy nest.

Within only a few hours of birth the young tree shrew is apparently capable of imbibing milk equivalent to one-third of its own weight so that it soon takes on a completely round shape. After that it does not need any further sustenance for a couple of days. Then it will once more drink a sufficient quantity, very rapidly, to permit another two days' fasting, and so on until it is one month old. During this time the mother only visits the nest at intervals to suckle the young, a process taking between four and ten minutes. After the first month, when it weighs little more than 3 ounces, the youngster will be abandoned by the parents and left to shift for itself; but by the age of four months it is sexually mature.

Scientists are not agreed as to the number of genera and species, nor their distribution. Most authors, however, accept the existence of thirty-two species under six generic headings—*Ptilocercus, Tupaia, Tana, Anathana, Dendrogale* and *Urogale*.

Loris: slow and slender

The topical forests of Africa and Asia are the home of lemur-like mammals (true lemurs live only in Madagascar) belonging to the family Lorisidae. Like the tree shrews they are primates, part of the suborder Prosimii, but they have all reached a higher level of evolution.

The slow loris (*Nycticebus coucang*) and the slender loris (*Loris tardigranus*) closely resemble the potto of Africa, and their slow, deliberate gait has likewise earned them the reputation of laziness. The slow loris, with a range covering all Southeast Asia, looks rather like a shaggy bear cub with its thick woolly coat but the rounded head and huge dark eyes, placed far forward, give this nocturnal animal an even more engaging air of playful innocence. The eyes, of course, are vitally important for distinguishing shapes and contours accurately. The long limbs are prehensile and all four are normally used for gripping branches since there is no tail. Yet although their length might seem to be suited to the most acrobatic pursuits, the slow loris is exactly what its name suggests. Every step is carefully measured and progress quite unhurried as the hands reach out for insects, eggs or nestlings that lie in the animal's path.

The shaggy coat of the slow loris would appear to play an important role in keeping the body temperature constant, for the metabolism is extremely low.

The slender loris, an inhabitant of southern India and Ceylon, has been ruthlessly hunted for the sake of its eyes, thought to be a cure for human eye diseases and to contain magical properties valuable for bringing love affairs to a successful conclusion.

TARSIER
(Tarsius spectrum)

Class: Mammalia
Order: Primates
Family: Tarsiidae
Total length: 10½–14 inches (27–35 cm)
Length of tail: 6½–10 inches (17–25 cm)
Weight: 3½–5½ ounces (100–150 g)
Diet: insects, eggs, nestlings; no vegetation
Gestation: about 6 months
Number of young: one

Adults
Small body and proportionately large, rounded head. Enormous round, orange eyes; large ears. Hind legs well developed, adapted for jumping, long tarsi. Digits terminate in fleshy pads. Long tail has hairy tip.

Young
Weigh under an ounce at birth but body is covered with hair and eyes are open.

The slender loris has a more limited range in southern India and Ceylon. Like its relative it is nocturnal by nature, with enormous eyes, but it spends more time moving slowly around rather than just clinging to the branches. Food consists in the main of insects, lizards, frogs, eggs and nestlings and the speed with which it sometimes strikes out for prey to some extent belies its apathetic air. One advantage of its slow gait is that it can approach victims without being detected.

Very little is known about the biology of the loris. Gestation lasts about six months and there is usually only one baby, although twins are not exceptional. In India the animal has long been hunted for its eyes which, according to superstition, help to cure human eye ailments and are an important ingredient in the preparation of love potions!

The tarsier

Another primitive, nocturnal representative of the suborder Prosimii is the tarsier. It too has huge eyes but its long tail is a clue to the fact that of all prosimians this is the most closely related to true monkeys. There are in fact other important links with primates, and because it obviously differs in many ways from the members of the Lorisidae it has been accorded a separate classification in the family Tarsiidae. Although some authors distinguish three species of the genus *Tarsius*, most accept only one (*Tarsius spectrum*).

Fossil remains indicate that the tarsiers are of extremely ancient stock and that they once inhabited much of Eurasia and North America, even though the few rare survivors are nowadays confined to some of the islands of the Malay archipelago, either in secondary jungle regions or on the outskirts of villages where they can obtain plenty of food. This consists principally of worms, insects, lizards, frogs, young birds and possibly eggs and fruit. They appear to be the only primates that do not consume any kind of vegetation.

The tarsier has a strange way of moving, not unlike that of a frog, due to the well developed hind legs (the animal's name derives from the length of the tarsi). These serve as springboards for leaping from branch to branch. A tarsier can cover up to 16 feet with a single jump and when landing clings to the tree surface with hands and feet which have five digits, each terminating in a fleshy, sucker-like pad.

Half the area of the tarsier's skull is occupied by the huge eye sockets and this has resulted in a regression of the muzzle so that the face is very flat. The animal has a poor sense of smell (the tip of the nose is not moist like that of most species with keen powers of scent) but its ears are very large and hearing is good. Another interesting feature of the animal is the remarkable flexibility of the neck, allowing the head to revolve through almost 180°.

The tarsier lives alone or with a mate within a territory that is fiercely guarded, its frontiers being marked by secretions of the anal glands and by urine. The female has a polyoestrus sexual cycle and mating may occur at any time of the year. The

male pursues his companion with bird-like cries and both animals spend much time grooming each other. Gestation lasts six months—an extremely long time for such a small animal. A fully grown adult weighs only $3\frac{1}{2}$–$5\frac{1}{2}$ ounces and the weight of a newborn baby is just under one ounce. Within a month it has learned to jump but up to that point has travelled about clutched to its mother's belly and sometimes on her back.

It is not known how long the animal lives in the wild but one female in the Philadelphia Zoo lived twelve years.

The tarsier is a nocturnal inhabitant of the rain forests of some of the islands of the Malay archipelago. Distinctive features are the size of the eyes in relation to the skull, the long tail, the well developed hind legs and the sucker-like pads on the feet which help to maintain a grip on trees.

ORDER: Dermoptera

The only modern representatives of the order Dermoptera are the flying lemurs, otherwise known as colugos or caguans. The order comprises the one family Cynocephalidae and the single genus *Cynocephalus*. There are two species, *Cynocephalus volans*, living in the Philippine Islands, and *Cynocephalus variegatus*, found in the countries of South-east Asia, the Malay peninsula and the islands of Indonesia.

The flying lemurs are curious animals and comparatively far removed from all other forms of mammal. Linnaeus originally included them among the Lemuridae. Cuvier subsequently classified them in the order Chiroptera (bats); and eventually Illiger, in 1811, allotted them a separate order.

Some confusion has arisen over the fact that the term *Cynocephalus* ('dog-headed') is commonly applied to baboons, the colugos at one time being distinguished by the generic name of *Galeopithecus*.

From the evolutionary point of view it is virtually certain that, like the representatives of the Chiroptera and the suborder Prosimii, these interesting mammals had their derivation in a relatively unspecialised branch of the Insectivora.

At the commencement of the Paleocene epoch, approximately 60 million years ago, these various orders were already well differentiated so that there were animals belonging both to the Chiroptera and Dermoptera, in addition to the earliest primates which were probably represented by the tree shrews.

Flying lemurs are inhabitants of the forests of the Oriental region. They are roughly the size of cats and their characteristic feature is a membrane known as the patagium which extends on either side of the body from the neck downwards, linking the fore and hind limbs as well as the individual fingers and toes. The patagium, completely covered by hair, and larger than the corresponding membrane of the flying squirrels and flying phalangers, is subdivided into three sections—the propatagium, between the neck and the arms; the plagiopatagium, between the arms and the feet; and the uropatagium, between the hind legs and the tail. This extensible fold of skin, which is quite different from the wing membrane of a bat, enables flying lemurs to launch themselves into space and to make long gliding flights from one tree to another.

The fingers and toes are furnished with curving claws that make it possible for these animals to climb trees and cling to branches. They are, however, exceptionally clumsy and defenceless on the ground. When threatened, they climb to a higher part of the tree and soar to safety.

Flying lemurs feed exclusively on leaves and fruit. The dental structure is unusual. There are as many milk teeth as adult teeth, both looking alike; the upper incisors and canines have double roots and the large, flattened lower incisors have enamelled folds, giving a comb-like surface pattern. The lower canines too are untypical, more closely resembling premolars. For these reasons and also because of the appearance and position of the incisors, canines and first upper premolars, it is not easy to establish the dental formula, but it is usually given as follows, totalling 34 teeth:

$$\text{I}: \frac{2}{3} \quad \text{C}: \frac{1}{1} \quad \text{PM} + \text{M}: \frac{5}{5}$$

The hemispheres of the brain are poorly developed, without many convolutions. The stomach is of the simple type, though more specialised than that of insectivores. The testes are contained in a scrotal sac and the caecum is large and complex.

The uterus of the female is bicornute and the placenta discoid. The mother gives birth to one baby which is wrapped inside the patagium and which clutches her chest when she glides.

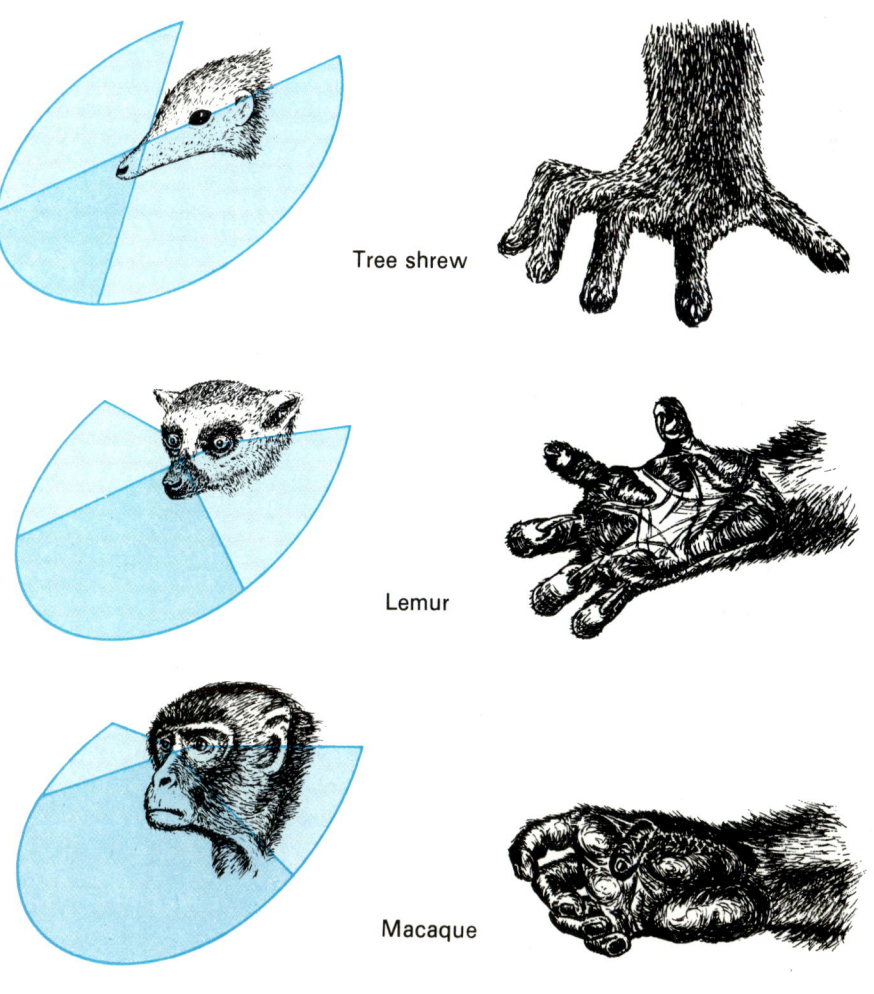

Tree shrew

Lemur

Macaque

The tree shrew, lemur and macaque represent three phases of primate evolution, as indicated by the changing position of the eyes from the sides to the front of the head and the increasing mobility of the thumb in relation to the other fingers.

The tree shrew has now been shown to be the most primitive of primates, which has evolved little in millions of years.

FAMILY: Tupaiidae

The tree shrews of the family Tupaiidae were formerly classified by many authors as members of the order Insectivora. Recent investigations, however, including those by George Gaylord Simpson and W. E. Le Gros Clark, have tended to alter this opinion, assigning the tree shrews to the primate suborder of Prosimii; other authors such as Robert Martin, have gone even further and given them a separate order – Tupaioidea. The low degree of specialisation shown by these animals has suggested that they are of very primitive stock, similar to the tree-dwelling ancestors of monkeys.

The shape of the body, the elongated muzzle, the non-prehensile limbs and certain aspects of their teeth structure are among the features which appear to link the animals with elephant shrews, the latter being true insectivores. But more detailed study of the anatomy, initiated by Doran in 1789, drew attention to the structure of the ear and the presence of ossicles, revealing striking analogies with the prosimians. Thus there are a number of features relating to sight, hearing, skull, brain, musculature, limbs and reproductive organs which bear a close resemblance to similar anatomical characteristics in lemurs.

Such arguments have persuaded the majority of modern authors that the tree shrews are in fact the most primitive – though least evolved – of all living primates, and very possibly the original stock from which the higher primates sprang, subsequently branching out in different directions.

Tree shrew
(*Tupaia glis*)

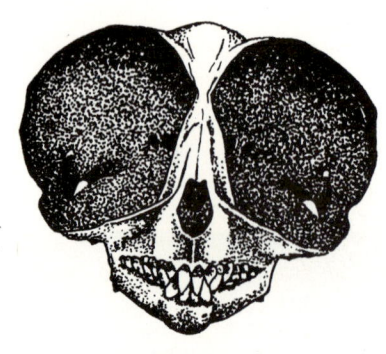

The huge eyes of the tarsiers have modified the shape of its skull. Its convex form is an important characteristic which differentiates the animal from the lemur on one side and the true monkey on the other.

Facing page : The Tarsiidae were once widely distributed through Eurasia and North America as well as the Oriental region, to which they are nowadays confined. Of very primitive stock, as fossils have revealed, the tarsier is more highly evolved than the tree shrew.

Simpson went so far as to suggest that if all the higher primates were somehow to disappear, leaving vacant all the ecological niches thay currently occupy, the surviving tree shrews alone would be capable, in time, of engendering a diversity of primate species which would closely approximate to those known today.

The tree shrews are for the most part arboreal mammals although some species are to be found living at ground level. Food consists either of insects or fruit. Their present-day habitat is restricted to the forests of South-east Asia.

Scientists have distinguished two separate subfamilies—Ptilocercinae and Tupaiinae. The former comprises only one species, *Ptilocercus lowi*, known as the pen-tailed tree shrew, displaying more primitive features than any other member of the family. The Tupaiinae are made up of five genera—*Tupaia, Tana, Anathana, Dendrogale* and *Urogale*—and a total of thirty-one species, the majority of which belong to the genus *Tupaia*.

FAMILY: Tarsiidae

In 1846, thirteen years before Charles Darwin published his *Origin of Species*, the naturalist Carl Burmeister wrote, in the preface to a monograph on the *Tarsius*, that he was dealing with one of the transitional phases of the animal kingdom, an early type of monkey which was possibly the first of a long evolutionary line of primates culminating in man. Today the tarsiers of the genus *Tarsius* are the sole survivors of the family Tarsiidae, confirmed as one of the most primitive of the Prosimii and certainly related to the true monkeys.

Fossil remains discovered both in Eurasia and North America strongly suggest that the present-day Tarsiidae, confined to some of the islands of the Oriental region, are the last surviving remnants of a vast group which had a much wider distribution about fifty million years ago.

The tarsier is a rat-sized mammal with a rounded head and enormous, frontally positioned eyes. The eye sockets, as in the case of monkeys, are separated from the temporal fossa by a bony partition. Other features recalling those of the higher primates are the short muzzle, flattened face, mobile lips and poorly developed olfactory organs. Despite the fact that the ears themselves are large and mobile, the external auditory duct has a similar structure to that of monkeys. The brain, however, is of the primitive type, with smooth hemispheres. The uterus of the female is bicornute.

Tarsiers move about in the trees in a curious way, rather in the manner of frogs, largely as the result of well developed hind legs which enable them to make 6-foot jumps. The feet are furnished with big toes and opposable thumbs; furthermore the second and third nails of the digits are curved, insectivore-fashion, whereas the others are flat, more akin to those of monkeys and man. Large, sucker-like digital pads help the animals to retain their foothold when landing.

Tarsiers are nocturnal, tree-dwelling mammals which hunt insects (though not, as some authors have claimed, exclusively), the diet being supplemented by nestlings and small rodents, and possibly eggs as well. The dental formula is as follows:

$$I: \frac{2}{1} \quad C: \frac{1}{1} \quad PM: \frac{3}{3} \quad M: \frac{3}{3}$$

Borneo, Sumatra, Celebes and the southern Philippines are the principal habitats of the tarsier. Most authors recognise only *Tarsius spectrum*; others distinguish also *Tarsius syrichta* and *Tarsius bancanus*.

CHAPTER 84

The tree-dwelling apes of the Asiatic jungle

The tropical rain forests of South-east Asia differ from the deciduous woods and forests of temperate zones by virtue of the fact that vegetational growth, instead of being more or less evenly distributed from the ground upwards, is essentially restricted to the upper strata. The forest floor is little more than a sterile morass of mud, almost empty of wildlife, whereas high above in the trees a specialised community of strange and fascinating animals thrives in a wonderland of green leaves, succulent fruits and vividly coloured flowers.

We have seen how some of these animals have adapted to an arboreal life in the most ingenious manner, nature having provided them with the means of leaping, parachuting or gliding from tree to tree. But there are other inhabitants of the jungle treetops which have succeeded in mastering their environment simply by using their prehensile arms and legs. Some of these are self effacing, almost inconspicuous—tarsiers and lorises, for example—while others are more spectacular, such as the gibbons, swinging acrobatically from branch to branch. They are one of the two species of man-like apes to be found in the forests of the Oriental region.

There is no place in the Asiatic jungle for predominantly terrestrial apes such as the gorilla and chimpanzee of Africa. Both the small, long-limbed gibbon and the much heavier and slower orang-utan are tree-dwellers. It is not surprising that the former should have retained the arboreal habits of most Old and New World monkeys of similar size; and if at first glance it seems puzzling that the far more powerful orang-utan should have failed to evolve in the same way as the large anthropoid apes of Africa, which at some stage took the vital transitional step from tree to ground, this is explained by the simple fact that absence

Facing page : The gibbon, smallest of the man-like apes, is a characteristic inhabitant of the jungles of the Oriental region. The species is well distributed through South-east Asia.

The gibbon spends the greater part of its time high in the trees of the jungle where it feeds on fruit, leaves and flowers as well as insects, nestlings and eggs.

of light, excessive damp and lack of food would have made survival at ground level virtually impossible.

The contrast between the two species, however, could hardly be greater. The gibbon is noisy, agile and sociable, its biology and habits fairly well, if not thoroughly, recorded, its population flourishing. The orang-utan, on the other hand, is silent, solitary and retiring. Little is known of its behaviour and its future is precarious in the extreme.

The gibbon: smallest of the apes

The anthropoid or man-like apes are subdivided into two families. The Pongidae comprise chimpanzees, gorillas and orang-utans; the Hylobatidae are represented only by the gibbons. The last are by far the smallest of the four types and are distinguished from the others principally by their method of locomotion, known as brachiation, whereby the exceptionally long arms are employed for grasping branches and lianas and swinging the body forward. Although gibbons sometimes descend to the ground—and in fact resemble man in their upright, two-legged manner of walking—their true habitat is high in the trees of the Asiatic jungle, that mysterious, dark world so vividly described by Alfred Russel Wallace. The combination of shadow and silence, even at midday, conveys an awesome impression of infinite age and vastness, a fitting background for creatures which are probably not far removed from man's own distant ancestors.

An animal which lives a hundred feet or more above the ground and which can neither fly nor glide runs a constant risk of grave injury or death if it makes the slightest false step. Leaping from tree to tree and clinging to flimsy, swaying branches demand split-second timing and exceptional agility and such attributes are essential factors of survival for the gibbon. Tailless, like other anthropoid apes, the gibbon hangs down by its arms and brings them forward in turn so rapidly that the animal really appears to be flying through the air. The forelimbs are much more highly developed than the hind legs and since the total weight of the slender body seldom exceeds 15–18 lb it is immediately obvious that the gibbon has the build of a natural athlete.

Although, in comparison with other apes, all gibbons are small and lively, there is some diversity among them in size and colour. No great difference exists between male and female but there is considerable individual variation, partially based on geographical distribution. At one time it was generally assumed that only one gibbon species existed – *Hylobates lar* – and this so-called lar gibbon has a fairly wide range, including north-western Thailand, the Malay peninsula, Java, Sumatra and Borneo. It may be either black or buff, with white facial surrounds and white hands and feet. But scientists now distinguish several other species as well. *Hylobates hoolock* (commonly known as the hoolock) is black with white brows only and inhabits forests from Burma to Assam and Yunnan. Vietnam, northern Laos and the island of Hainan are the habitats of *Hylobates concolor*, the black gibbon, distinguishable from the hoolock by white side-whiskers. *Hylobates agilis*, the dark-handed gibbon, grey with black hands, is found in Sumatra; and the wau wau, *Hylobates moloch*, grey with a continuous white band over the eyes, lives in Java and Borneo. The largest gibbons, weighing up to 30 lb, are the siamangs, classified separately in the genus *Symphalangus*, inhabitants of Malaya and Sumatra. The larger of the two recognised species is *Symphalangus syndactylus*, the other being the dwarf siamang, *Symphalangus klossi*.

Acrobats of the jungle

The speed and agility with which gibbons swing through the forests have aroused the astonishment and admiration of many a naturalist, including William Charles Martin who in 1840 could only liken it to the action of flying. C. R. Carpenter, about a century later, was equally intrigued but more scientific in his approach. He has described how, in the event of a branch or creeper breaking under its weight, a gibbon will make a kind of twisting turn in the air to grab hold of a branch lower down, interrupting its progress only for a fraction of a second. Impressive too is the way in which the animal sometimes uses only one arm to keep its balance, darting out the other to trap a bird or an insect in flight, again testifying to an uncommon degree of skill and flexibility. In the normal way progress from branch to branch is leisurely and relaxed but if the gibbon senses danger – perhaps from a prowling leopard – this is speeded up and prodigious leaps of 30–40 feet soon carry it towards the tree-tops and safety.

Another notable characteristic is that whereas a troop of gibbons may shatter the jungle calm at dawn or dusk with an ear-

Hoolock
(*Hylobates hoolock*)

Black gibbon
(*Hylobates concolor*)

Lar gibbon
(*Hylobates lar*)

Wau wau
(*Hylobates moloch*)

44

SUPER-PREDATORS

PREDATORS

PHYTOPHAGES

PLANTS

splitting chorus, the animals flit through the trees like silent shadows, never advertising their presence to an enemy.

The strength and flexibility of arms and shoulders enable the gibbon to soar through the air and cling to the slenderest branches; but agility alone would be of little avail without amazingly swift reflexes, predominantly visual, which make it possible for distances to be gauged with great precision. As is the case with other primates, the gibbon's eyes are placed in a frontal position and this is the secret of its excellent stereoscopic vision. Not that this is always infallible, however, for there is much evidence to show that even a gibbon may miscalculate and lose its balance. Falling from a tree, if it does not have fatal consequences, may prove a serious handicap. In Thailand A. H. Schultz examined 233 adult gibbons and remarked that a number of them showed signs of fractured limbs or were permanently deformed, apparently as a result of such accidents.

In order to hang from a branch for any length of time the limbs must obviously be fully prehensile. The hands of a gibbon are very narrow in comparison with those of other apes and the fingers are unusually long, except for the first which is much shorter and opposable to the others. When gripping a branch or liana the gibbon does not use this thumb, tucking it into the palm; only when climbing does the thumb come into play, helping to avoid slipping. Also of vital importance are the joints of wrist, elbow and shoulder, the last enabling the gibbon to make 180°-angle arm movements up and down, forwards and backwards and from side to side.

In some ways gibbons resemble baboons for both species have ischial callosities on their hindquarters and a relatively simple brain structure. But in other features they resemble gorillas, chimpanzees and orang-utans, notably in the length of their limbs, the absence of a tail, the method of locomotion and the comparatively small brain.

The gibbon (but not the siamang) is an inexpert swimmer, yet able to cope quite adequately on the ground. Significantly, it is the only primate, apart from man, which is capable of walking any distance erect on two legs, raising the arms over the head to maintain balance. An even more astonishing sight is a gibbon using this two-legged method of locomotion to scamper along a branch high above the ground or to cross a wire or rope slung tautly between two poles. The only individuals adopting a four-legged method of walking on the ground are gravid females.

The troop and its territory

Almost all the information we possess today about gibbon behaviour in the wild is due to the invaluable research work of Carpenter who for four months in 1937 made a detailed study of the lar gibbon in the rain forests of north-western Thailand. Although published more than thirty years ago, Carpenter's findings are still valid and the following sections are largely based on them.

Lar gibbons form family groups, usually consisting of one male, one female and offspring of varying ages. The smallest clan

Comparison between the hand of a gibbon (*left*) and that of an orang-utan (*right*) shows that the former is much narrower and that the long fingers are better suited for grasping branches. The thumbs of both species are relatively small but mobile.

Facing page : Ecological pyramid of Asiatic jungle vertebrates. Typical animals include 1. Wolf. 2. Tiger. 3. Leopard. 4. Leopard cat. 5. Common palm civet or toddy cat. 6. Malayan or sun bear. 7. Himalayan black bear. 8. Binturong. 9. Clouded leopard. 10. Black baza or cuckoo falcon. 11. Crested goshawk. 12. Besra sparrowhawk. 13. Indian black eagle. 14. Hawk eagle. 15. Sumatran rhinoceros. 16. Great Indian rhinoceros. 17. Indian elephant. 18. Gaur. 19. Serow. 20. Axis deer. 21. Sambar. 22. Crested wild boar. 23. Malayan tree rat. 24. Bamboo rat. 25. Flying squirrel. 26. Orang-utan. 27. Gibbon. 28. Langur. 29. Pheasant. 30. Peacock. 31. Flying fox.

Right and facing page : Gibbons are born acrobats and thanks to their long, well developed arms are able to swing rapidly from one branch or liana to another, often leaping 30–40 feet through space. When hanging by one arm, sometimes supported by a foot, the free arm can be used for food gathering.

GIBBONS

Class: Mammalia
Order: Primates
Family: Hylobatidae
Diet: fruit, leaves, insects, small birds, eggs
Gestation: 200–212 days
Number of young: one

LAR GIBBON
(*Hylobates lar*)

Length of head and body: 18–25 inches (45–64 cm)
Weight: 11–18 lb (5–8 kg)

Very long, slender body; no tail; extremely long arms. Much individual colour variation. Long fingers (but not thumbs) and toes.

SIAMANG
(*Symphalangus syndactylus*)

Length of head and body: $29\frac{1}{2}$–$35\frac{1}{2}$ inches (75–90 cm)
Weight: 18–29 lb (8–13 kg)

Similar to gibbon but larger. Black body and limbs. Hair particularly long on legs, shoulders and armpits. Laryngeal sac swells when animal emits calls. Membrane connects second and third digits.

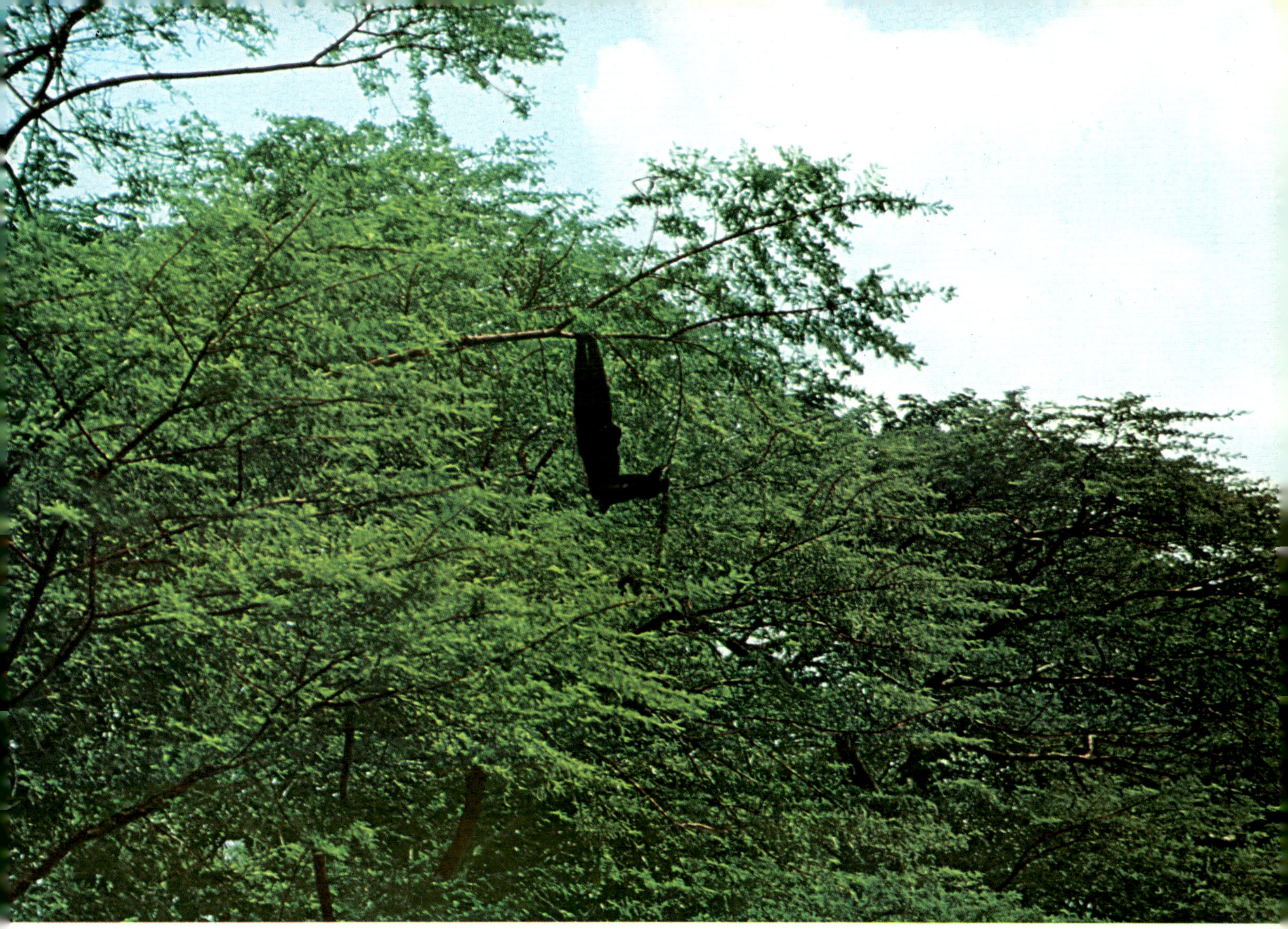

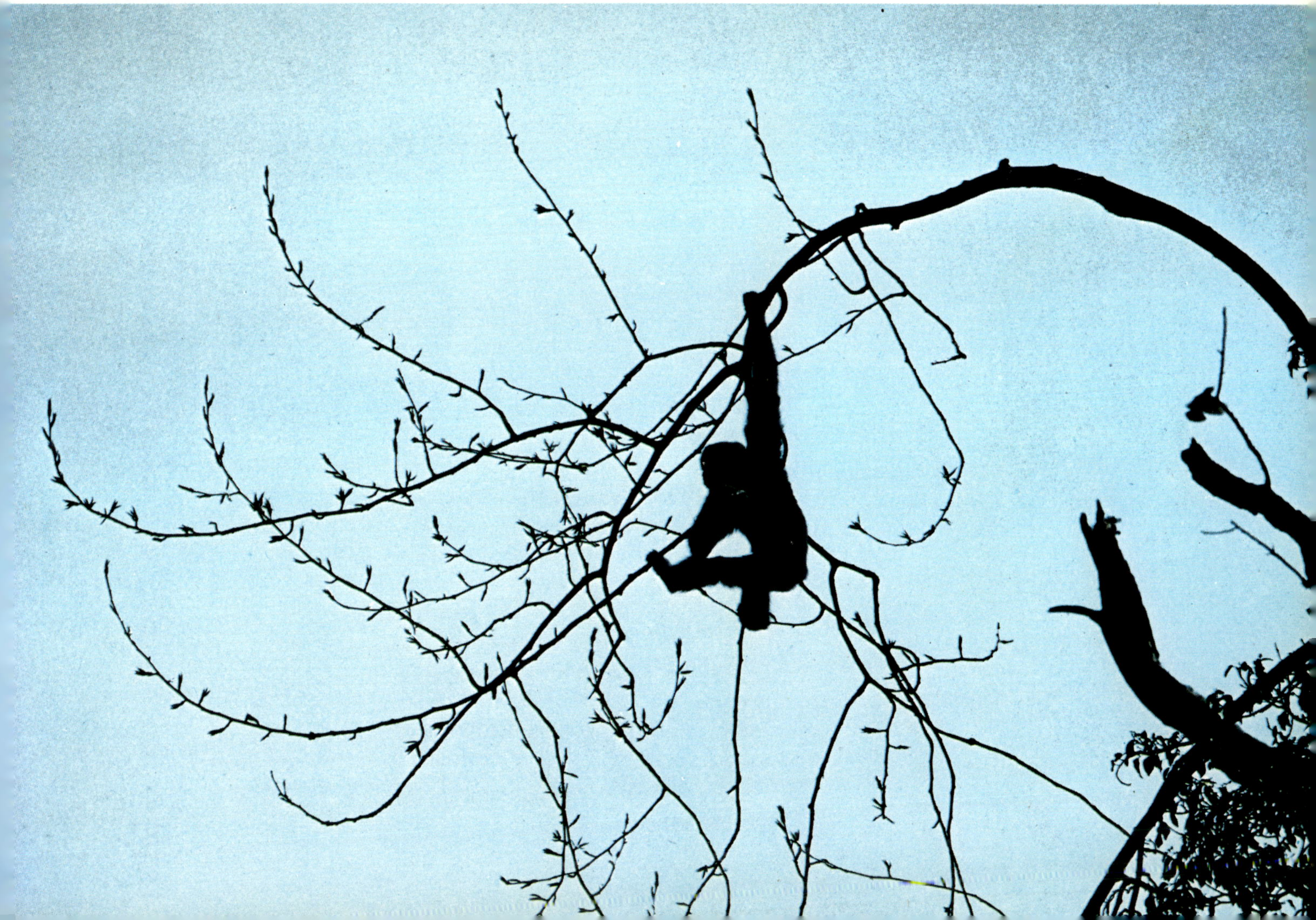

Above and facing page: The gibbon is the only primate apart from man which can walk steadily in an erect position, using only two legs; but although it sometimes descends to ground level it is essentially a tree-dweller. Unlike other anthropoid apes it does not build a nest for sleeping.

will comprise two individuals—couples too young or otherwise unable to have progeny—and the largest family unit is six. Adult females give birth to one baby only once every two or three years and it is unusual for a fifth birth to occur before the eldest of the offspring has asserted its independence (immature animals customarily leaving their parents at about the age of seven years). But in the case of other species, particularly the hoolock, social groups may be much larger, with up to twenty or thirty individuals. Among all species, however, the male tends to be monogamous and the young are normally expelled before they reach sexual maturity.

Even among lar gibbons the patterns and relationships are not necessarily so simple, as in the case of two of the twenty groups studied by Carpenter. In group 10, for example, there were three males, one a dominant and another a very old individual who was almost certainly sexually inactive and therefore of little social importance. The third male was a much younger adult which although it had reached the age of emancipation still lived with the family, apparently freely and without arousing resentment. This arrangement seemed to suggest that in due course the younger gibbon would supplant his father and force the latter to abandon the family group for good.

Group 13 was different, comprising two females, one of them gravid, and her daughter, scarcely mature but to all appearances closely attached to the dominant male, suggesting that she had usurped her mother's place in his affections. But having rounded up the group, Carpenter examined the vaginal mucus of the younger female, the analysis showing no trace of spermatozoa,

indicating that the father had not coupled with the daughter, at any event not recently.

Gibbons do not wander at random through the jungle. Each group establishes itself within a portion of territory which, although comparatively small, is constantly patrolled and defended against all intruders. The territory consists of those trees in which the various members of the family find their daily sustenance. Where conditions are favourable the total area may range from 30 to 100 acres and the frontiers are scrupulously observed. Where the local gibbon population is small and food perhaps in shorter supply this area may be considerably larger, possibly extending over more than 250 acres.

In the jungle territorial bounds are seldom marked by visual means, and in the case of gibbons notice of ownership is announced with loud cries. Similar warning cries are emitted when territory has to be defended but these may not be sufficient to deter rivals, in which case fighting may break out. Each dawn of a new day sees a gathering of the clan on the same spot – normally the most exposed and vulnerable part of the domain – this being followed by a raucous sequence of high-pitched cries, instantly recognisable and comparable either to the barking of dogs or the strident trilling of birds.

This dawn chorus appears to be initiated by the females who are soon joined by their male companions and echoed by members of neighbouring family groups until the entire forest resounds with their clamour. Of all Old World primates the gibbons have a reasonable claim to be the noisiest but there is of course a purpose to their dawn concerts. The forest foliage is so dense

Given the opportunity, the gibbon proves itself to be the most skilful and confident of tightrope walkers.

that only the voice is of any practical use for communication and recognition by others of the species.

The early morning commotion usually lasts only about an hour and is not always without incident. It may happen that the concerted warning fails to intimidate an intrepid young gibbon who deliberately or unwittingly trespasses on alien territory. His intrusion will result in an acrobatic chase through the trees, the outcome invariably being the headlong flight of the stranger. A less frequent occurrence is an outright confrontation and battle between two adults. When such a contest does take place each animal does its best to bite, scratch or hammer the opponent, but care is taken to confine such attacks to those parts of the body adequately protected by thick fur, namely the arms, the back and the chest. Thus it is rare for any serious injury to be inflicted on either side.

The female members of the various family groups are just as fierce and indomitable when it comes to defending their domain against other females. Like other tree-dwelling primates, they are as large and heavy as their male companions.

Aggressiveness is usually most in evidence when intruders of the same species are involved. In the case of a larger and stronger predator there is no sense and little disposition to stand and fight and both male and female gibbons prudently seek safety in flight.

Each species of gibbon gives out its own characteristic cry and experts are able by this means alone to differentiate between them. Thus a subspecies of lar gibbon (*Hylobates lar pileatus*) emits noises that are almost musical, consisting of a rapid sequence of ascending notes followed by an equally swift descending series, tailing off into silence. The cry of the hoolock is clearly bisyllabic (giving rise to its common name). It seems that this species utters its distinctive calls not only in response to neighbouring groups but also on other occasions, as when the sun

rises or when it begins to rain, and sometimes simply to give vent to its feelings. In any event the noise can be heard more than a mile away.

Unlike other man-like apes, gibbons do not normally roam far and wide in search of food, but when conditions are especially bad a clan will set out for new, unexplored territories, trying to avoid, wherever possible, clashing with other family groups bent on similar quests.

The gibbon's day

After their noisy greeting to the dawning day and having satisfied themselves that their territorial frontiers have not been infringed, the gibbons retire to the surrounding trees and devote themselves to the main business of the day—feeding. As they swing their way through the jungle an individual may pause now and then to pluck a fruit or snatch a passing butterfly, using the free hand or even a foot to transfer the morsel to its mouth. Sometimes it will gobble down this titbit while suspended in what appears to be a most uncomfortable position, gripping a branch by one hand and one foot; alternatively it may pop the food into its mouth and settle down in the fork of a tree for a few moments to consume it before resuming its journey. For this is merely a preliminary—a kind of *hors d'oeuvre*—to the meal proper which will only commence when the whole family has reached its regular banqueting site which offers more staple fare.

Favourite types of fruit are undoubtedly figs of the genus *Ficus,* of which there are more than 600 species, followed by grapefruits and mangoes. Apart from these delicacies gibbons also feed on leaves, shoots, flowers, insects, baby birds and eggs, cracking the shells with their teeth. Naturalists have not been able to determine whether they store excess food or whether they simply gather as much as they need, calling a halt when hunger

These pictures show how the gibbon moves by means of brachiation, gripping a branch (or, as here, a rope) with one hand and swinging its whole body round and forward to find a grip with the other hand, legs suspended in space.

Gibbons live in small family groups, each of which occupies its own territory. This is energetically and noisily defended against other gibbons, communication being effected by a variety of calls.

Facing page : The baby gibbon is carried around by its mother, clutching tightly to the hair of her belly or chest, until it is old enough to clasp her around the waist. She then encourages it to play with its siblings and gradually it becomes accustomed to family life, asserting its independence long before it is sexually mature but remaining within the family circle. Maturity comes at about seven years of age and it is only then that male and female pair off and part from the parents.

is appeased. Since at no time of the year is food likely to be in drastically short supply there is probably no incentive for them to lay in reserves.

Because vegetation alone does not offer sufficient liquid content for their daily needs gibbons satisfy their drinking requirements by lapping up any rain water they can find in the hollows of trees or in flower corollas. If still unable to quench their thirst by such methods they may come down to ground level to find a convenient puddle, perhaps hanging by one hand from an overhanging branch or liana, scooping up the water with the other hand and then conveying it to the mouth.

After the morning meal the adult gibbons settle down for a rest, impassively looking on as the youngsters leap about and chase one another playfully from tree to tree, emulating the antics of their elders. Some of their games are astonishingly like those of children, especially a form of 'hide-and-seek' in which a couple·of youngsters close their eyes, giving the others an opportunity to slip away. There then follows a frenzied pursuit through the foliage, all to the accompaniment of excited cries. The only limiting factor to such games is that there are seldom more than four participants, given the size of the family.

Young and old gather together for a second meal during the afternoon and towards dusk resume their strident chorus. When darkness falls they settle down to sleep.

Gibbons are the only anthropoid apes which do not take the trouble to build a proper nest for the night. Each individual makes for a favourite branch and squats down on its haunches, cushioned against the rough bark by the callosities that are in evidence from birth, sleeping in a seated position.

Preparation for adulthood

Feeding sessions, play periods and intermittent disputes over territorial rights make up the gibbons' daily round, all young males and females joining in these activities. But although the youngsters mingle freely with one another from an early age there is no display of sexual interest until they are about seven years old. Male and female then pair off and it is believed that such liaisons normally last a lifetime. There is no clear evidence of either individual dominating the other and consequently when offspring arrive there is no permanently accepted head of the family, nor are there rules and rituals for settling quarrels. Sometimes the male yields to his partner, sometimes she defers to his will. Relationships are on the whole harmonious and responsibilities shared. Thus one partner may direct the daily food expedition while the other asserts authority in social and sexual matters.

Gibbons, like other anthropoid apes, couple at any time of year rather than at an appointed season. The female has a sexual cycle of approximately thirty days' duration and her receptivity is indicated by the swelling of her genital parts. Copulation is preceded by the semblance of an embrace in which both animals, firmly supported by both feet and one hand, draw each other close with the free hand and then adopt the customary mating positions.

■ *Hylobates*

■ *Symphalangus klossi*

▨ *Symphalangus syndactylus*

Geographical distribution of gibbons.

Gestation lasts between 200 and 212 days, the female becoming progressively slower and more deliberate in her movements and going down on all-fours whenever she is on the ground.

No zoologist has yet been fortunate enough to see a birth in the wild but observations of gibbons in captivity indicate that the mother emits a cry prior to the event and that she swallows about two-thirds of the placenta, the remainder being consumed by the male.

The newborn gibbon is completely helpless and supports itself by clutching the hairs of its mother's belly (it is very rare for her to carry it on her back). When she uses the standard arm-over-arm method of swinging from branch to branch she bends her legs upwards so that she can more easily cradle the baby against her belly. After a couple of weeks the youngster is capable of clasping its arms around her waist but still needs the additional support of her legs when she moves from place to place. By this time, however, she no longer needs to carry it on all occasions and can safely leave it to fend for itself among the other members of the family while she watches nearby. In this manner it is encouraged to learn how to play and to get accustomed to hanging from branches and lianas – activities which are interrupted at regular intervals for suckling. For long journeys she will continue to transport her offspring and only needs to take up the familiar, slightly crouched position for the latter to fling itself in her direction to take up its protective clasping posture close to her chest or belly.

At six months the baby gibbon is well on the way to becoming an expert acrobat and is able to walk upright on the ground, although still keeping its knees bent. Two months later it has gained complete confidence and moves around, both in the trees and on the ground, with as much skill as its elders. All it lacks at this stage is the muscular strength to make the particularly long and daring leaps practised by the adults.

The mother-baby relationship is all-important in these early weeks and months for it is by watching and imitating her that the youngster gradually becomes an integral part of the family group. Its dependence on her is such that it will faithfully reproduce her facial expressions and cries. Later it begins to mimic the ritual behaviour of its siblings and father. Carpenter's observations indicate that a significant phase of its initiation is reached when the mother encourages her youngster to wander off by itself when the family is gathering food.

Games play an important role in the growing-up process, as indeed they do for all mammals. Carpenter subdivides immature gibbons into various groups, based on age, ranging from babies to young adults, and passing through a juvenile stage. He breaks these groups down even further in accordance with size and social habit (the principal features of such development being outlined on the opposite page). Play activity intensifies when individuals reach the third juvenile phase (J3) or when they become young adults; it tends to diminish with advancing maturity and is not in evidence in periods of sexual inactivity.

All the members of a family appear to be closely attached to one another and the same harmonious relationship, doubtless

STAGES IN THE DEVELOPMENT OF YOUNG GIBBONS, ACCORDING TO CARPENTER

BABIES

B1   Newborn individuals and sucklings completely dependent on mother are carried everywhere and seldom left alone. They suckle frequently and their movements are poorly coordinated when they attempt to walk. Never participate in games.

B2   Slightly larger than preceding group, they already show signs of independent activity. Though still suckling they also take solid food. Seldom stray far from mother and begin to play with siblings. Utter sounds and cries, especially when endangered. Full set of milk teeth.

JUVENILES

J1   Although moving about independently they are still closely attached to mother who is now capable of giving birth again. Vocal and locomotive capacity well developed but unable to make long leaps. Very playful.

J2   Larger than preceding juveniles; completely independent, very active, always on the move. Shy when observed. Male's genital organ barely visible. Female lacks teats.

J3   Facial expression less infantile than preceding group. Second set of teeth still small, canines not in evidence. Very active, with much time spent in playful fighting.

YOUNG ADULTS

Difficult to distinguish from adults but face retains its juvenile appearance and canines are not completely developed. Teats similar in both sexes. Neither male nor female yet capable of procreation. Play still figures prominently in daily activities.

Situation	Individuals concerned	Types of call
a) Group travelling towards centre of territory. b) Another group nearby. c) Danger, following type III calls.	Adult males and females, usually the latter, normally emitted by single individual.	Series of cries, gradually becoming shriller, followed by two or three deeper notes. Duration 12–22 seconds.
a) At daybreak. b) Sometimes when group moves from place to place.	Adult males.	Isolated cries, sometimes repeated. Similar to first sequence of type I calls.
Group surprised by hunter, observer or predator.	Adult males, adult females and young, sometimes in unison.	Shrill, repeated single cry, not in series.
Member of family straying when group disperses after ritual alarm call.	Adult males, adult females and juveniles.	Single rising note with something of a questioning inflection.
Not determined.	Males and adult females.	Sharp series of calls, each lasting few seconds.
Group in imminent danger.	Males and adult females.	Guttural growl.
Play or friendly approach.	Young animals.	Small sharp cry.
Abnormal situation; captivity.	Young animals.	Impatient cry.
During group journeys.	Dominants.	Sounds resembling castanets or cluckings.

reinforced by these strong family links, characterises community life in the troop. Sometimes, when they have been temporarily separated but frequently for no apparent reason, two gibbons will come together in an embrace similar to that which precedes mating. Facial expressions are also important factors for mutual recognition purposes and when kindly disposed the animals appear to be smiling. Nevertheless the main form of communication has to be vocal. Carpenter has isolated nine types of calls among lar gibbons, some of which clearly relate to the demarcation and defence of territorial boundaries, others being incitements to play, alarm signals, etc. (see chart above and opposite).

Gibbons are social animals and solitary individuals are few, though not unknown. Probably they will be aged animals abandoned by their offspring or having suffered the loss of a partner. They may wander around the fringes of other family groups, avoiding quarrels, and may even succeed in being adopted by this little community, as was the case in Carpenter's group 10.

Causes of death

The gibbon's prospects of survival are considerably brighter than those of other anthropoid apes. Recent population counts put the number of orang-utans at only 2,500–3,500, gorillas at around 25,000 and chimpanzees at below 100,000. Yet the figure

AND THEIR PURPOSES, ACCORDING TO CARPENTER

Answering individuals	Response call or action	Probable purpose	Type
Other groups or members of same family.	In the case of other groups the same sounds are used, sometimes followed by retreat. Within the family the same sounds are used to announce positions.	Exploration. Defensive action. Territorial protection.	I
Members of nearby groups.	Same calls, emitted simultaneously or alternately.	Location of other groups and avoidance of conflict.	II
Mainly members of same family, also other groups.	Similar calls, preparatory to flight.	Alarm. Defensive warning.	III
Members of same group.	Search and assembly.	Maintenance of group unity and alert.	IV
Majority of groups in same area.	Similar cries.	Not known.	V
Members of same group.	Typical aggressive call.	Defensive.	VI
Play companions; mother.	Continuation of game and signs of affection.	Encouragement of play and mutual approach.	VII
All animals in area.		Solicitation.	VIII
Members of same group.	Young follow adults.	Direction of group.	IX

for gibbons is over 200,000, which is certainly encouraging.

One reason why they have fared so much better than their relatives is their modest size; another is the structure of their social life, with small family units that are capable of surviving in comparatively small areas of tropical forest. Thus three of the groups studied by Carpenter lived in a fairly restricted jungle district surrounded on all sides by cultivated land.

Man is without doubt the principal enemy of the species, if only because by modifying his environment he has deprived the animal of natural habitats and food. Hunting is a lesser risk. Buddhists hold the gibbon in high esteem, treating it almost as a sacred animal; not so certain tribes such as the Meo of southern China and Vietnam and the Karens of Burma, who eat it.

Fractures and bruises caused by falls sometimes result in death and so too do a number of parasitic diseases. In this connection Carpenter refers to several types discovered by Hoffman in the excreta of gibbons. They include the protozoa *Entamoeba coli* and *Balantidium*, a tapeworm (*Bertiella studeri*) and a roundworm (*Trichuris trichiura*).

Predators, on the other hand, appear to exert only a minimal influence on the gibbon population. Large snakes certainly kill young gibbons, but although there is a specialised monkey-eating eagle in the Philippines no equivalent raptor exists in the Indo-Malayan region. The leopard, however, is a redoubtable enemy.

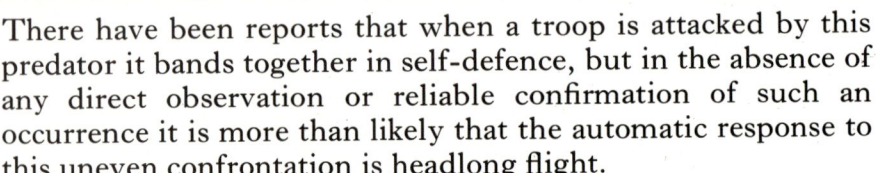

The siamang is the largest of the gibbon family. Although it differs little from the smaller species in social behaviour it is physically distinguished by a membrane linking the second and third fingers and a laryngeal sac which swells enormously when the animal emits its call, serving as a sound-box which amplifies the voice.

There have been reports that when a troop is attacked by this predator it bands together in self-defence, but in the absence of any direct observation or reliable confirmation of such an occurrence it is more than likely that the automatic response to this uneven confrontation is headlong flight.

In captivity, of course, gibbons are protected from all these natural hazards, though the price they often have to pay is boredom and comparative inactivity. Zoo gibbons have been known to live for up to thirty years.

The vociferous siamang

The gibbons of the genus *Hylobates* are often found in mountain forests up to the 8,000-foot mark; but the siamangs of the genus *Symphalangus,* commonly classified with the former, are able to exist at greater heights, in fact up to 10,000 feet above sea level. In the two species concerned, *Symphalangus syndactylus* and *Symphalangus klossi,* evolutionary processes only sketchily outlined in true gibbons have reached fruition.

Of the two species, the former is considerably larger and heavier. Understandably, since it is nearly double the weight of an ordinary gibbon it is not anything like as agile. The distinctive features both of this and the related dwarf siamang are, firstly, an interdigital membrane linking the second and third toes and, secondly, a laryngeal sac which magnifies the voice. This vocal apparatus is dilatable, swelling when the animal howls to the size of the head itself and serving as a voice-box. The chorus of a troop of siamangs at dawn is so powerful that it can be heard for a distance of two or three miles (or, as local people term it, 'an hour's journey away'). Amid this din it is possible to pick out the deep-throated sound of the males as compared with the shriller calls of the females and young. Before giving voice siamangs throw back their head and sometimes cup the hands

over the mouth, which has the effect of further amplifying the sound and varying the register.

Siamangs differ from gibbons in that family groups are somewhat bigger–from four to ten individuals, often more. Other aspects of social behaviour are identical but they are more accomplished at swimming.

Man of the woods

Animal legends circulating throughout the ages have often alluded to curious creatures, half-man, half-monkey, and extremely ferocious by nature. These little man-monkeys have been reported as living in the heart of dense forests in the remotest parts of the world. But such regions are no longer inaccessible and expeditions of exploration have satisfied most people that these creatures do not really exist. In Africa, for example, 'eye-witness' accounts by hunters have invariably proved to be sightings of gorillas or chimpanzees–man-like in certain respects but in no sense the 'missing link' between ape and man.

Such stories have also been current in Asia. Local inhabitants of Sumatra and Borneo have long used the term 'orang-utan'–which, literally translated, means 'man of the woods'–to describe a resident of their own forests, though apparently the reference is not to a man-like ape but to certain members of a savage tribe of the interior. This was confusing for the first western explorers of the islands who assumed that the name applied to the huge anthropoid ape of these jungles. Whatever the original intention and meaning, orang-utan (*Pongo pygmaeus*) is nowadays the name used to refer to the only representative of the Pongidae to be found outside Africa.

Looking at this animal in a zoo (and few of us are likely to see it anywhere else) it is not hard to understand how such a mistake arose; for although it is more closely related to the gorilla and chimpanzee than to man, the orang-utan frequently adopts facial expressions which are similar to our own, albeit in our sadder moments. In their melancholy way these looks seem to reflect the sad history of a species which may be doomed to extinction.

The fate of the peaceful orang-utan has indeed been tragic and man has been almost entirely responsible. The gradual destruction of the primeval jungle to make way for arable and pasture land has affected many mammals apart from the orang-utan but few have been singled out in such a callous, cold-blooded manner for persecution in other ways. Tracts of forest have been deliberately set on fire for the sole purpose of trapping the older males, while females have been hunted and slaughtered so that their offspring can be taken alive. The animals have then been consigned to distant zoos to spend the rest of their life, bored, inactive, frequently deformed, behind bars. Many of them die as a result of contracting lung diseases transmitted by keepers and visitors. Little wonder that this is a vanishing species.

In days gone by the orang-utan's range was considerable, for fossil remains dating from fourteen million years ago have been discovered in the Punjab. Long before that, some five hundred

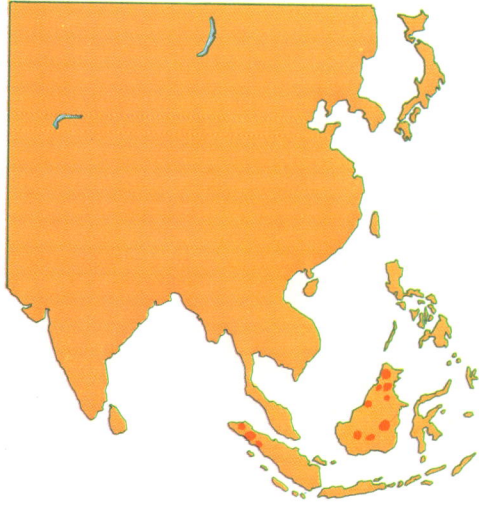

Geographical distribution of the orang-utan.

The siamang is sometimes found in mountain forests up to 10,000 feet, this altitude being beyond the range of gibbons of the genus *Hylobates*. Although not as agile and acrobatic, the siamang is a better swimmer.

The gibbon, the orang-utan, the chimpanzee and the gorilla are known as anthropoid or man-like apes. The first two are tree-dwellers, occasionally coming down to ground level, the other two are terrestrial, although they sleep in trees. Comparative study of these primates shows how they have progressively adapted to walking on the ground, a process perfected by man with his fully erect stance and bipedal gait.

ORANG-UTAN
(*Pongo pygmaeus*)

Class: Mammalia
Order: Primates
Family: Pongidae
Height: 45–59 inches (115–150 cm)
Weight: 88–220 lb (40–100 kg)
Diet: fruit, bark, leaves, eggs, etc.
Gestation: 8–9 months
Number of young: one

Sole non-African member of Pongidae and second in size only to gorilla. Grey skin, sparse, reddish hair, face almost naked. Adult male has beard and moustaches. Subcutaneous layers of fat on cheeks often deform face, expecially in captive animals. Laryngeal sac. Long arms, fairly short legs. No tail.

Facing page : The orang-utan, with its rather melancholy expression, is the rarest and most imminently threatened of all the anthropoid apes.

million years ago, the species roamed freely from China to Java. Today, however, this once-vast distribution is restricted to Borneo, largest and least populated of the Sunda Islands, and to Sumatra, a land so low and flat that it has been picturesquely described in native lore as lying 'just below the wind'.

The only ape exceeding the orang-utan in size is the gorilla. An adult male may stand 5 feet high (one exceptionally large specimen in Borneo measured almost 6 feet) and weigh between 165 and 200 lb. Females are smaller, standing at most 45 inches high and weighing, on average, 90 lb. An orang-utan's skin is greyish and there is a thin covering of reddish hair. The face is almost naked but the adult male is provided with a beard and moustaches. The cheeks of both sexes, particularly in the case of older reproductives, are often deformed by layers of subcutaneous fat. This phenomenon is especially marked in zoo animals (it is aggravated by lack of exercise) so that the face takes on a swollen and somewhat grotesque appearance.

Most orang-utans are nowadays found in low-lying forest regions, including coastal swamp areas. It is rare for them to venture much higher than 2,500 feet but in the mountains of Borneo a few orang-utans have been sighted at around the 6,000-foot level. Tracts of primary forest are preferred but they also inhabit secondary forest which has long been abandoned and virtually restored to its original condition.

Unanswered question

The orang-utan is an animal which spends the major part of its time high up in the trees, seldom descending to ground level. The question which has puzzled scientists is that although this is an arboreal species, males and females show a dramatic contrast in size and weight, a characteristic of species which live principally on the ground.

Of the four types of man-like ape two (gibbons and orang-utans) are tree-dwellers and two (chimpanzees and gorillas) terrestrial.

The tragic fate of the orang-utan seems to be reflected in the large brown eyes of this baby, snatched from the jungle and destined to spend its life behind bars in a zoo.

Facing page : The trustful nature of the orang-utan is detrimental to its interests. Poachers have no difficulty in capturing the peaceful 'man of the woods' as it gazes curiously down from its tree refuge. Experts predict that the species may be extinct by the end of the century.

Baboons, as we have previously noted, exhibit a large measure of sexual dimorphism (difference in appearance and size between male and female). This undoubtedly came about as the species gradually abandoned its arboreal existence to adopt a new mode of life on the ground. The males, in accordance with their defensive function, became larger and stronger than the females.

In the case of primates which continued to live in the trees, such as the gibbons, this sexual contrast is not in evidence, primarily because such species are seldom menaced by large predators and in the event of such confrontations are able to escape fairly easily. Yet orang-utans are curious exceptions to this rule, sexual dimorphism being as pronounced as it is among terrestrial gorillas and chimpanzees. Why this should be is the question which still awaits a satisfactory answer. Theoretically, the mystery could be solved by making a detailed study of the species in the wild—as has been done in the case of the anthropoid apes of Africa. Unfortunately there are so few orang-utans left—and those that survive have proved so elusive—that this is easier said than done.

According to Barbara Harrisson, who has been responsible for interesting attempts to reintroduce illegally captured orang-utans to traditional habitats, it is something of an achievement merely to discover the whereabouts of these apes in the wild; but actually to get close enough to them in the jungle to allow even three hours of observation in the course of a week is an occasion for celebration. One Japanese zoologist who paid a visit to what was reputed to be the largest enclave of wild orang-utans in the world failed to see a single animal during his three-month stay. Little wonder that the behaviour of the species is so poorly documented; and if present trends continue we are unlikely ever to find out much more, for unless stringent protective measures are quickly introduced experts predict that the species will be extinct by the end of the century.

Study is all the more difficult because orang-utans do not form troops and normally live alone. Sometimes a mother and her baby will come together with other mothers or will be joined by a male; but it is far more usual for males to remain solitary for the greater part of the year, while immature animals roam the jungle in groups of, at most, two or three. The reason may be due to the fact that older orang-utans are territorial animals, keeping their distance from congeners and emitting calls to announce their whereabouts. Such cries can be clearly heard for considerable distances, the voice being amplified, as in the siamang, by the dilatable laryngeal sac which grows larger as the animal ages. This swelling voice-box also helps to vary the register of the sound emitted so that a number of contrasted calls can be distinguished, according to changing situations.

The orang-utan is active by day and sleeps in a tree (30–80 feet above ground level) at night. It builds a rudimentary platform of small branches, lined with leaves, in a main fork, using this nest for several successive nights—unlike the gorilla which builds a new one every evening—and also occasionally resting in it during the day. It has been suggested that younger animals learn to build a nest by watching their elders but this does not

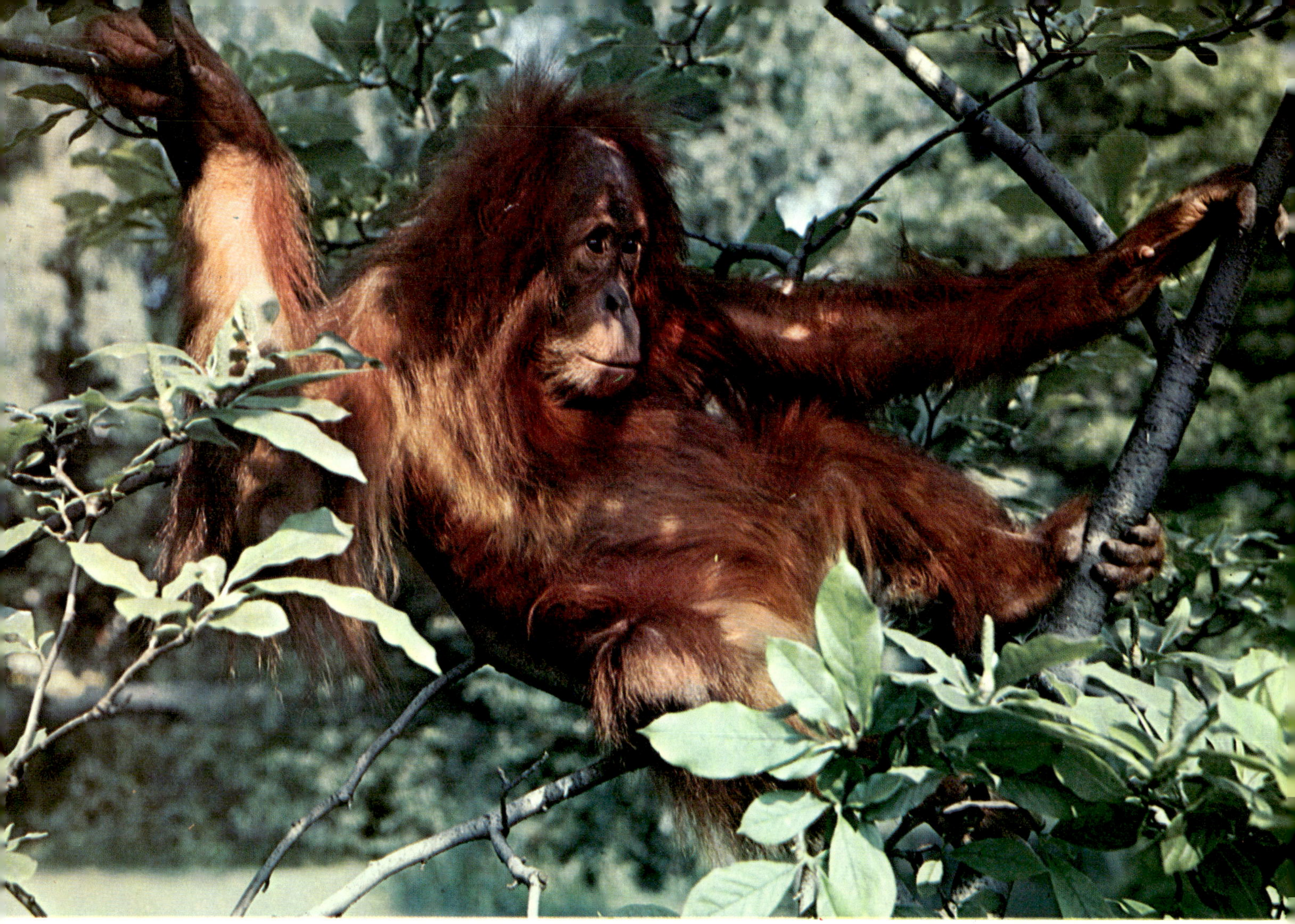
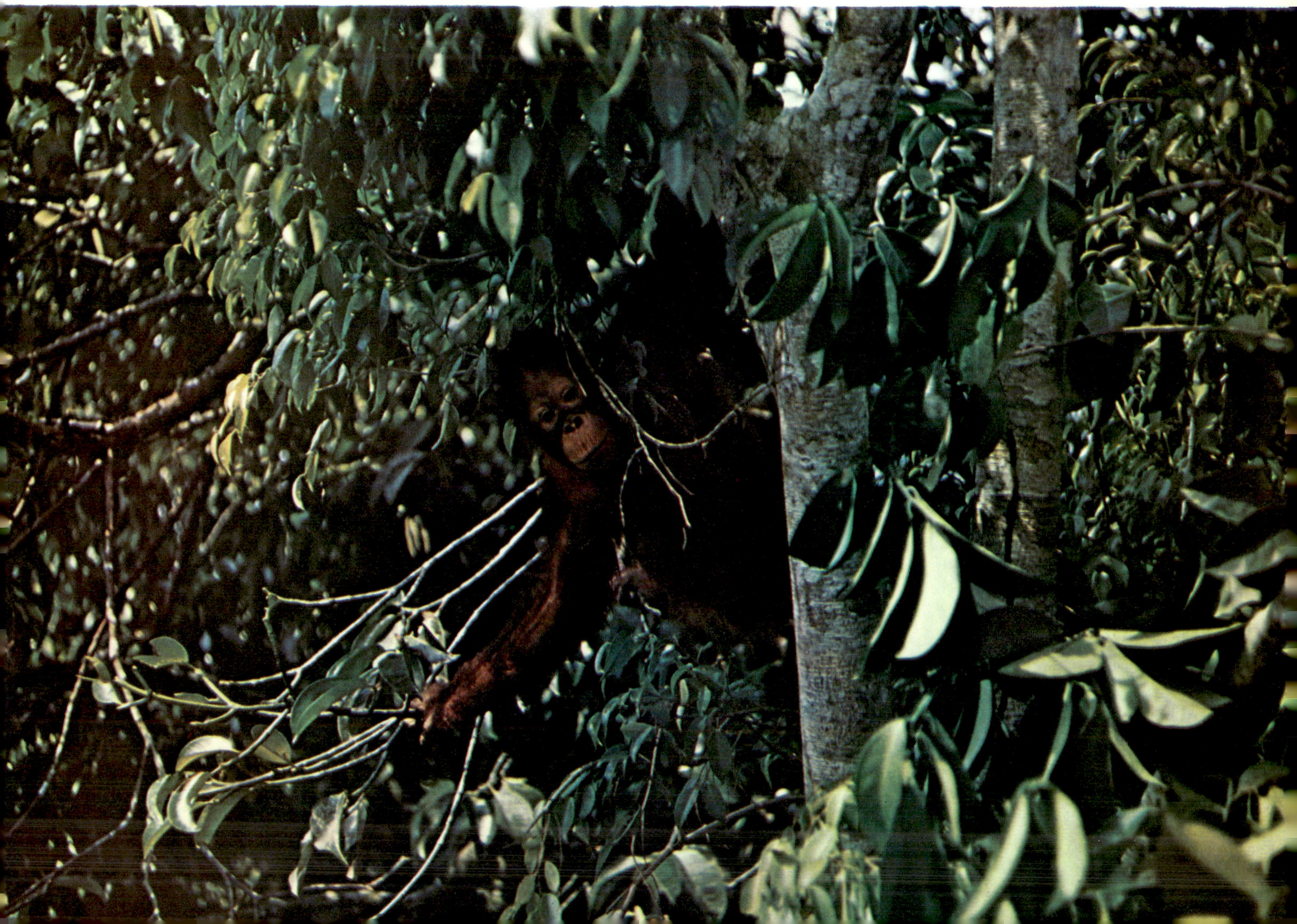

seem to be the case, for orang-utans born in captivity or reared under laboratory conditions instinctively engage in this activity. One animal escaped from London Zoo into neighbouring Regent's Park and was recaptured the following day as it rested comfortably in the topmost branches of an oak. There was certainly no question of it having been taught the habit.

Searching for food in the trees is the orang-utan's principal daytime activity. One type of fruit to which it is especially partial is the durian, about the size of a coconut. The pulp and seeds are edible and the highly unpleasant smell, which makes it unpopular for human consumption, fails to deter the ape. Other food includes leaves, bark, eggs and—now and then—snails.

The orang-utan, like the gibbon, moves about by brachiation, the arms being much longer than the legs; but it is also able to stand upright on a branch, using the hands to retain its balance. In captivity the animal quickly learns to walk unsupported on its hind legs, although with a rather stiff gait. In the wild, when it descends to ground level, it adopts a more natural, four-legged position.

Mating and births occur at any time of the year and rival males are thought to battle fiercely for possession of a female. Such fights have never been witnessed in the wild but evidence of scars on the arms, face, neck and cheeks of some animals have led naturalists to conclude that they are battle injuries. Copulation is preceded by a noisy courtship ritual. The male begins by emitting soft purring sounds which are gradually transformed into loud roars, soon dying away. He then approaches the female and the animals mate, both uttering deep growls.

Gestation lasts eight or nine months, the baby weighing about 3 lb at birth and almost immediately clasping the hair of the mother's chest. Solid food is soon taken in the form of leaves and fruit, carefully chewed by the mother. At the age of one year the youngster weighs 15 lb and is capable of finding its own food. It remains with the mother, however, for four or five years and then wanders off with companions of the same age.

It takes about ten years for an orang-utan to grow to adult size and reach sexual maturity. Theoretically it may then procreate every year but this never happens, for a female will not produce a second baby until the first has left her to lead its independent life; and because the rate of infant mortality is very high (around 40 per cent) a female orang-utan will rarely raise more than two or three offspring in a lifetime.

In the wild an orang-utan may live for thirty or forty years but in captivity life expectancy is much shorter for it is highly susceptible to pulmonary diseases transmitted by man. Even when it escapes such infection difficulty is experienced in maintaining the animal in perfect health.

A number of experiments have been carried out to test the orang-utan's intelligence which is regarded as about on a par with that of the chimpanzee. But research is handicapped by the fact that whereas the chimpanzee is outgoing and exhibitionistic, the orang-utan is shy and disinclined to perform to order.

The man of the woods is not greatly menaced by predators. There are no leopards either in Sumatra or Borneo. A few tigers are to be found in Sumatra but the Dayaks of Borneo claim to have exterminated this feline a thousand years ago. So in fact man is the only serious enemy of the orang-utan. Bones of females and young have come to light in caves once occupied by prehistoric man, indicating hunting activities dating back thousands of years; but today hunting has degenerated into slaughter. The inflated prices offered for the capture of live animals have spurred poachers to adopt a variety of cruel and highly reprehensible procedures, secure in the knowledge that they will go unpunished, for the protection of wildlife has never been a vital concern in these parts. Although much patience is needed to locate an orang-utan, once it is found the problem of catching it is absurdly simple. Apart from being unafraid of man, the animal is extremely inquisitive, so that as the hunter climbs the tree the placid orang-utan gazes down curiously and trustingly from its branch, making no attempt to get away. An adult female will be killed in cold blood and her baby taken alive. An alternative method is to locate the tree in which an animal is perched and then systematically destroy all the trees in the vicinity. Having consumed all food close to hand the ape is compelled to come down to ground level and is promptly caught by means of a net. If the poachers do not have time to spare they will hasten the process by smoking out their victim.

Whichever of these methods is adopted the result is always the same. If taken alive the unfortunate animal is doomed to spend the rest of its days in captivity. Assuming it reaches its destination in sound, healthy condition (and this is far from certain) the chances are that it will not live for more than three or four years behind bars.

Some efforts have been made to save this species from extinction, notably by trying to reintroduce selected individuals to some of the Asiatic habitats where they are no longer found. This operation was organised by the naturalist Barbara Harrisson and met with a modest degree of success. In 1963 she was responsible for a count of the wild orang-utan population in the islands of the Malayan archipelago. The figures were 2,000 in Sabah, 1,000 in Kalimantan, 700 in Sarawak (giving a total of about 3,700 in Borneo), and a further 1,000 in Sumatra. Yet only one year later the number in Sumatra had been drastically reduced to about 100 animals, basically as a consequence of poaching for the benefit of zoos and research institutes.

In order to save the species it is obviously essential to set up an international organisation to control this illicit trade, beginning with the legal prohibition of the sale of live animals. Some zoos have already agreed not to purchase any orang-utan which is not accompanied by a valid export certificate. But these gestures of good intention are mere drops in the ocean.

Little can be expected from the 300 or so captive animals which make up about one-tenth of the world population. Even under the most favourable zoo conditions reproduction is difficult and the rearing of young a hazardous affair for the mothers generally neglect their offspring. Only about five baby orang-utans are born in zoos every year—hardly sufficient to guarantee the future of the species!

Above and facing page : The orang-utan is the largest of the tree-dwelling anthropoid apes, with a life expectancy in the wild of thirty or forty years. Artificial rocks and caves are no substitute for the dense foliage of its jungle habitat. Zoo specimens seldom live long, succumbing to lung disease or at best becoming fat from boredom and lack of exercise.

CHAPTER 85

Ups and downs of forest life: langurs and macaques

In addition to the two species of anthropoid ape described in the previous chapter, there are a large number of monkeys to be found in the Oriental region, occupying a wide diversity of habitats. The two main groups to be discussed here are inhabitants of the most typical biomes of South-east Asia–the tropical rain forests–but some of the species concerned also live in other surroundings.

A rapid survey of the monkey population of Asia shows that the non-anthropoid primates (*Cercopithecidae*) are broadly divided by zoologists into two subfamilies–Colobinae and Cercopithecinae. The former is represented by the langurs or leaf-monkeys, the latter by the macaques or rhesus monkeys.

The langurs are slender, agile, tree-dwelling monkeys subsisting on an exclusively vegetarian diet, mainly leaves. All phyllophages (leaf-eaters) require an enormous amount of food and this specialised diet has necessitated a number of adaptations. In the case of the langurs these include particularly long intestines and a large, lobular stomach.

In India langurs are regarded as sacred; hence they have multiplied unchecked, often causing considerable damage in built-up areas. Because they have lived cheek-by-jowl with humans it has not proved difficult to study their habits in a city environment; but this type of existence has so modified their behaviour that it bears little relationship to that of their original wild habitat. Unfortunately, observations of the species in the jungle are seldom undertaken.

Rhesus monkeys are, on the other hand, terrestrial by habit and much closer, for example, to baboons. Unlike langurs they are omnivores, eating a fair proportion of animal protein. They too are worshipped in some parts of India but their fate has not

Facing page : Although macaques or rhesus monkeys are worshipped in some parts of India they are much in demand in the West for scientific research because of their similarities to humans.

been all that happy because for some time they have been caught and exported for use in scientific laboratories and medical research institutes, primarily due to their physical and mental makeup, which is very similar to that of humans. Thus they are more satisfactory laboratory subjects than rats or dogs. Vivisection has become a controversial subject in recent years, partly for humane reasons, partly because the validity of many such experiments is disputed. Nevertheless the work can be justified by recognising that it has contributed to medical progress and helped to save millions of human lives. It is disquieting to realise, however, that hundreds of thousands of rhesus monkeys are sold for such purposes every year, inevitably resulting in a dramatic decline in their numbers.

The good-humoured leaf-monkeys

Langurs live in small but well organised social groups and display the characteristic behaviour patterns of most arboreal monkey species. Thus their instinctive reaction when faced by a predator is to run away rather than to stand and fight. Because of this the individual langur is notable for a lack of aggressiveness and since there is no call for them to form defensive bands the structure of the troop as a whole is not characterised by rigid class divisions.

The typical social unit consists of from five to fifteen individuals in the forests of central India, increasing to between thirty and one hundred and twenty in the more arid regions from Nagpur to Raipur. The area of territory occupied by such a troop varies with the surroundings, being directly linked with the quality and quantity of available food. In the drier districts the domain of a langur troop may extend to six or seven square miles, but not all the terrain is uniformly exploited. Certain sites are visited more frequently and for longer periods than others. The centres where the majority of animals congregate normally include clumps of trees which provide shelter and food, and a good source of water.

Daily journeys vary in duration and distance but as a general rule langurs do not travel more than a mile or two, so that even the feeblest members of the group can keep pace with the others.

Because of its arboreal habits even an animal on its own can fend for itself quite effectively so that belonging to a troop is not a precondition of survival. Thus solitary individuals are often in evidence, although many of them are not completely independent, rejoining their companions when night falls. All-male groups, consisting of two to ten monkeys, are also not uncommon. Sometimes these small clans succeed in integrating themselves with other troops but it is quite likely that they will encounter hostility from other males.

Lack of aggressiveness and a fairly flexible social hierarchy within the troop are not conducive to a strong sense of solidarity and territoriality. It is not surprising, therefore, that when two strange troops meet their attitudes towards each other are reasonably friendly and that fights are uncommon. Two or more troops may indeed happily settle down to feed in the same trees

Facing page : The female langur devotes the utmost attention to her baby, parting company from the rest of the troop and rejoining it only when the youngster is weaned. At about fifteen months the latter is ready to become an independent member of the group, males and females remaining separate until they are old enough to mate.

COMMON FOREST LANGURS

Class: Mammalia
Order: Primates
Family: Cercopithecidae
Length of head and body: 17–31 inches (43–80 cm)
Length of tail: 20–42 inches (50–107 cm)
Weight: 15½–40 lb (7–18 kg)
Diet: vegetarian
Gestation: 196 days
Number of young: one

Slender body, long tail, delicate hands, small thumb.

HANUMAN OR ENTELLUS MONKEY
(*Presbytis entellus*)

Colour generally greyish with golden or ochre tints; white head, black face.

PURPLE-FACED LANGUR
(*Presbytis senex*)

Black, with scattered greyish hairs, whitish whiskers.

DUSKY LEAF MONKEY
(*Presbytis obscurus*)

Blue-grey, with white rings around eyes. White or flesh-coloured lips.

MAROON LEAF MONKEY
(*Presbytis rubicundus*)

Reddish-brown, blue face. Crest of hair at top of head.

GEE'S LANGUR
(*Presbytis geei*)

Thick golden fur; black muzzle.

DOUC LANGUR
(*Pygathrix nemaeus*)

Chestnut head with light brown stripe under ears. White cheeks and neck, white whiskers. Body marbled, grey with black ticking. Hindquarters and tail white. Underside of arms, legs, hands and feet black. Forearms white and calves of legs yellow.

Above and facing page : The troop as a social unit is not of great importance to the langur, although even solitary individuals join their companions at night to sleep. Regarded as sacred in India these monkeys freely roam towns and villages and often cause damage to fields and gardens.

and to drink from the same pools and streams, so that it is virtually impossible to determine which monkey belongs to which group. At such times the langurs emit appeasing noises; and in certain situations, whether it be feeding or preparing to sleep, they may let out cries ostensibly for the purpose of locating their family group.

Langurs are just as peacefully inclined when they happen to meet other monkey species. They are, for example, on the best of terms with macaques, for although the latter are by nature far more aggressive, they do not compete for the same living space or food. This sometimes leads to the formation of mixed troops—a very interesting phenomenon—in which langurs and macaques live together in perfect harmony. This is in itself unusual enough but what is particularly fascinating is the fact that when such a mixed troop comes across a group consisting wholly of langurs, perhaps encroaching on their feeding grounds, it is the macaques who incite their langur companions to expel the intruders.

Although the hierarchy of a langur troop is not anything like as clear-cut and rigid as that, say, of a colony of baboons, there is, nevertheless, a rudimentary form of social organisation, with dominants and subordinates of either sex. But nobody has determined the precise roles that are played by these individuals nor how they affect relations between the sexes. A dominant male is not necessarily the troop leader on every occasion nor does he command a monopoly of female partners, for even the younger males appear to share their favour; nor do the females invariably recognise the chief's authority. Thus all manner of social and sexual situations may arise and although rituals are strictly observed on some occasions, discipline is relaxed at other times.

In surveying the different types of behaviour patterns of monkeys one should of course avoid making close comparisons with the human situation. The peaceful existence of the langur may seem eminently desirable compared with the rather more eventful, excitable life of the macaque, but the world of the former is not necessarily a paradise. To say that the individual langur is happy because it does not have to knuckle down to authority is to ignore the positive aspects of a more rigidly organised society such as that of the belligerent macaque. Thus only in the latter form of troop do we find examples of self-denial and sacrifice in the interests of group survival. The aggressive macaque is a much more attentive father than the docile langur, who shows no interest in his progeny. In this and similar ways the macaque is the more highly evolved species.

Responsibilities of motherhood

The complex processes of reproduction, culminating in the creation of a new individual are, no matter at what level of animal evolution, wonderful and, in many respects, mysterious. The purely biological considerations are of course of prime importance but equally interesting, particularly for the ethologist, are the inexplicable effects of such processes on the mental

attitudes and behaviour of the mother, both before and after the birth. These are more clearly marked and easier to observe among higher mammals, especially primates.

Maternity brings about profound changes in the behaviour of the langur. The moment a female gives birth to a baby her entire mode of life undergoes a complete alteration. Whereas previously she was accustomed to living in a group she now displays no interest whatsoever in her former companions and wanders off on her own to devote all her time to the newborn baby. It is especially intriguing to note that most of the other members of the troop respect her wish for privacy but that some of the younger females try to follow her, hovering about in the evident desire to share by proxy the duties of motherhood. In fact one of these immature females may be given charge of the youngster for brief intervals, instinctively adopting the maternal gestures suitable for pacifying the baby. The mother will nevertheless remain close at hand, ready to intervene should the temporary guardian display any sign of intolerance or simply when she feels it is time to reclaim her offspring. The young nurse will hand over her charge without dispute but will often wait around hopefully until another opportunity arises. One advantage of this 'exchange' is that if the mother should die the baby will promptly be adopted by another female who will show genuine affection and tenderness towards the orphan.

If danger should threaten, the mother picks up her baby, which stays tightly clasped to her chest, and scampers off to safety. The youngster learns to recognise its mother soon after birth, extending its arms to her whenever she approaches.

As the young langur grows its pelage takes on a progressively lighter hue. Although the mother continues for some time to carry it around in her arms the attitude of the other females of the group gradually changes. They may still rush to its assistance if they see it being punished or tormented by an older play companion but will make no special effort to pick it up.

The reaction of a mother to her baby's death is a sad and moving sight. No more capable than other monkeys of understanding what has happened, she makes frantic attempts to revive her offspring, trying to attach it to her body, catching it as it falls to the ground, warming and caressing it. As time goes on, with no answering expressions and cries, she eventually gives up these attempts. Her interest slackens and she finally abandons the dead baby.

The blissful period of babyhood ends as soon as the young langur is weaned. This is a traumatic experience because the youngster cannot possibly understand why its mother, previously so attentive, not only refuses to let it suckle but pushes it violently away. The result may be a genuine contest of wills in which the young monkey tries to mollify its mother by adopting infantile postures. Initially it may succeed and she will allow it to suckle, but this state of affairs cannot last, and she firmly and finally rejects her offspring. By this time it will be getting on to fifteen months old and the mother, again sexually receptive, is looking for a mate. About six months later she will give birth to another baby and the cycle is completed.

Now the young langur is ready to take its place as an independent member of the troop, mingling with other individuals of the same age and sex. Young females generally remain with the elders of the group, sometimes helping the mothers to look after their babies and training, as it were, for future motherhood. The young males form separate groups, spending most of their time playing and fighting. These games become rougher and more serious as the participants grow, serving to determine the social positions each individual will occupy in adult life. The strongest and most resourceful animals are the future leaders.

The regimented macaques

The life of the macaque is very different from that of the langur. Descent from the trees to the ground has compelled this species to adapt to an environment where survival depends on a show of aggressiveness rather than meekness, with security the responsibility of the group rather than the individual. In the animal kingdom a strict and efficiently organised social structure is a sure indication that each individual member of that community will react aggressively when roused, and so it proves here. The macaque has evolved as a highly belligerent animal and the macaque community is correspondingly regimented, with a clearly defined hierarchy. Such a tightly-knit group

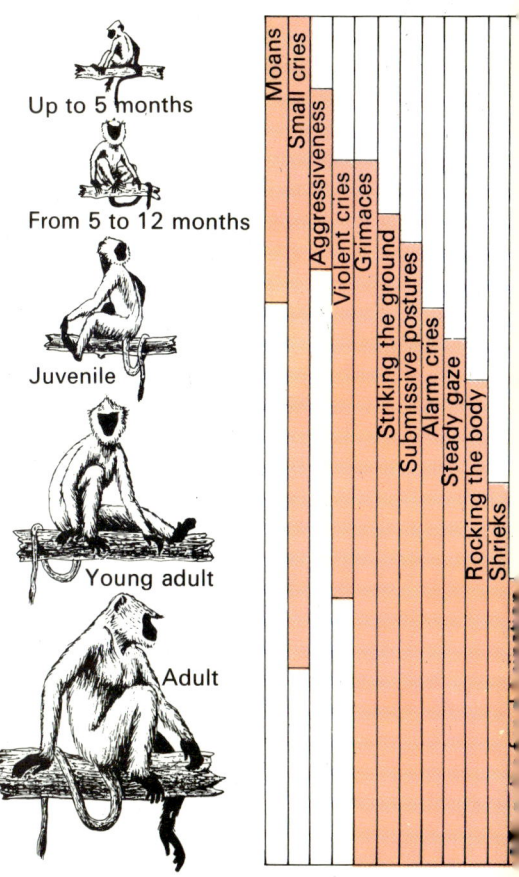

The behaviour of the langur changes as it grows older. This chart shows the various methods of communication used as it develops to adulthood. Although not strictly regimented the langur community has its leaders and its subordinates, though the roles are not clearly defined. Both on the individual and group level communication is achieved primarily by various types of calls.

The female langur is entirely responsible for rearing and educating her baby for the father takes no interest whatsoever in his offspring.

Macaques are popular in zoos all over the world and are valued for their use in medical and biological research. The discovery in their blood of the agglutinogen now known as the rhesus factor (often present in human blood) helped to solve the mystery of blood groups, making it possible for blood transfusions to save millions of lives. Rhesus monkeys have also been used for experiments in manned space flight. The danger is that their numbers will become so depleted in the wild that they may eventually disappear, though, unlike orang-utans, they flourish in captivity.

Facing page: In parts of the Orient the macaque, like the langur, is a sacred animal and is frequently seen in built-up areas. Here a mother and her baby squat on the walls of a Hindu temple.

structure is obviously effective when the troop is threatened by a predator, banding together in self-defence; but its significance extends far beyond day-to-day preservation. In an animal society controlled by a single dominant individual the death of that leader is likely to result in a complete breakdown of order and a period of anarchy during which time the entire group is subjected to all manner of hazards. On the other hand, in a community directed by what one might term an aristocracy rather than an omnipotent leader, the problem of succession in the event of death is speedily resolved by the surviving dominants, so that discipline is maintained at a crucial time.

Once again we must not be tempted to seek analogies with human societies. There is no point or validity in claiming that the individual macaque is less content than its langur counterpart. Although forced to submit to authority it is not under continual stress and indeed life within the troop appears to be pleasant and undemanding, each individual having the opportunity of developing in accordance with its ability and nature. Relations between dominants and subordinates are controlled by a sequence of rituals designed to inhibit aggressiveness and it is rare for rivalry to spill over into serious conflict.

Marching orders

The ritualised relationships between dominants and subordinates within a macaque troop make a fascinating subject for study. A representative of the 'higher orders' will begin by threatening a lower-ranking animal with gestures that become increasingly belligerent but which never go beyond mutually acceptable bounds. Thus the dominant will retract its lips and bare its

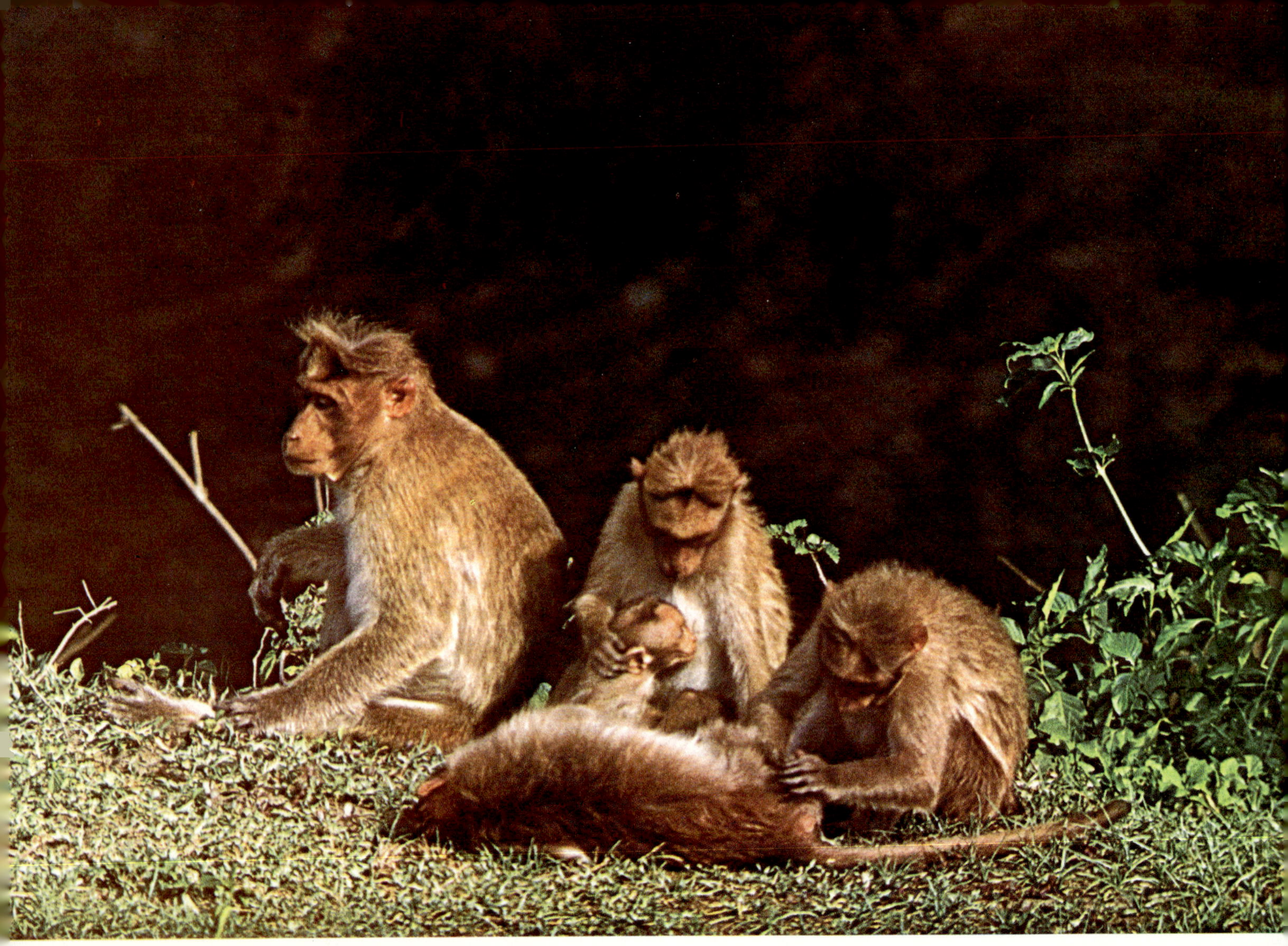

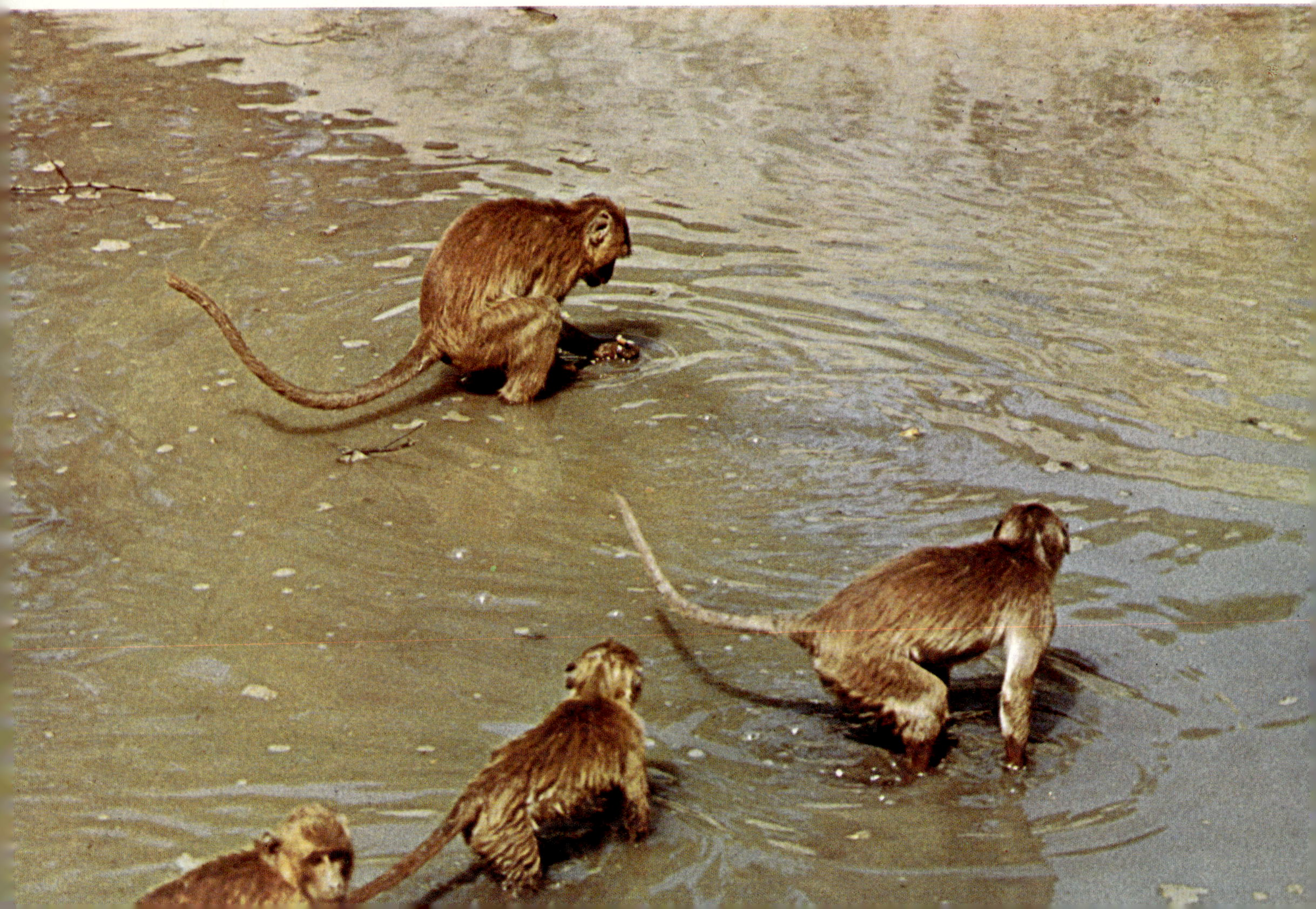

canines – obviously aggressive actions designed to intimidate the underling – but will not under any circumstances bite the other animal. It is tantamount to the reaction of an infuriated man who shakes his fist at someone without actually striking him.

If the inferior animal which has been the cause of the dominant's wrath consents to accept the latter's superior status, it will adopt a suitably submissive posture, reminiscent of the female prior to mating (likewise intended to ward off an aggressive approach). In this situation, however, which is devoid of sexual significance, the dominant simply simulates the act of copulation, thus proving symbolically that he is the master. In this manner he makes a demonstration of power without trying to inflict any punishing injury on the audacious animal which has offered him defiance. By ritualising his actions he also conserves energy which might be better expended in the defence of the weaker members of the troop. The male dominants frequently simulate the sexual act to underline their authority.

There are various other ways in which the different monkeys in the community establish mutual links. Cleaning and grooming, for example, which are primarily hygienic functions pure and simple, are here transformed into ritual gestures of pacification when performed by lower-ranking macaques and placidly accepted by their superiors. It is a way of maintaining group unity and cohesion. Although as a rule it is the dominants who receive the closest attentions of this nature, any two individuals, whatever their rank, may engage in mutual grooming, so that the procedures frequently cut right across class barriers.

Problems of experimental research

Animal behaviour is an extremely interesting but dangerous field of study – dangerous not in the sense of physical risk but in being laid open so easily to error and misinterpretation. In the case of social animals such as monkeys it is particularly important to carry out double checks, with observations in the wild (conditions permitting) as verifications of tests in the laboratory, and vice-versa. A naturalist may publish a report based on eye-witness observation in the wild, but it is nevertheless essential to examine such findings scientifically so as to rule out all possibilities of error. Conversely, it is not enough to describe the behaviour of an animal or group of animals under test-tube conditions as being characteristic of that species if one entirely ignores their possible reactions to similar situations in the wild. Many mistakes have been made in this field, either because few ways and means have been available for the study of a species in its natural habitat (travel and transport problems have of course been overcome to some extent in recent years) or because funds and facilities have been lacking for proper scientific research (a situation which is unlikely to improve much).

Thus false conclusions have arisen in the case of monkeys born in captivity and subsequently restored to the wild, a traumatic experience resulting in aberrant social behaviour. Care must also be taken in evaluating the behaviour of different groups of animals in captivity which have never experienced

Faced by a predator, the meek langur (*above*) instinctively seeks safety by climbing up into the trees. The belligerent macaque (*below*) will, however, confront the foe and often defend itself with unbelievable courage.

Facing page : Because macaques live at ground level their social organisation, like that of baboons, is based on a rigid class division. This is common to all species. Those shown here are (*above*) bonnet monkeys and (*below*) crab-eating macaques.

Unlike langurs, macaques are seldom found on their own for their survival depends on being a member of a well organised troop. This Japanese macaque, which has become temporarily separated from its companions, will be compelled to rejoin them sooner or later; alternatively it will have to try to become a member of a strange troop, if need be tracking it from afar for several months until socially accepted.

the kind of normal inter-group contact found in the wild.

No stable animal society can come about purely as a result of improvisation; there are certain recognised laws and procedures that have to be observed by each individual and an animal growing up naturally alongside others soon learns to become an integrated member of the group. Imagine a situation—which could well arise in the not distant future—in which a party of humans, each from a different country, decided to colonise a new planet. The difficulties which might predictably arise would not be due principally to the strange environment (one assumes that technology would ensure suitable adaptation and survival) but to the diversity of individual natures and backgrounds. Linguistic and cultural problems would hamper communication and make it impossible to exchange views as to how best to improve living conditions. Although some form of social structure would probably emerge it would already be carrying the seeds of its own decay and destruction, for nobody would be capable of understanding anyone else. Certainly it would bear little resemblance to a normal, civilised human society on earth; and although it might be instructive to study the behaviour of our imaginary space settlers, both as individuals and as a group, it would be ludicrous to suggest that it could provide any clue to their normal social behaviour at home.

The same remarks apply to the study of animal behaviour in the wild and in captivity. It is comparatively easy and convenient to carry out experiments with a caged animal; and even when the subject is allowed to roam freely for the occasion it will, because it is accustomed to human presence, stay where it is placed, well within range of camera or tape-recorder. The value of such studies is strictly limited and any general conclusions about the behaviour of the species as a whole are likely to be based on false premises, since the observations have been carried out under abnormal conditions. To have any validity, a broad survey of animal behaviour must be based on tests, conducted over an extended period, on individuals born and reared in surroundings which do not change in the course of investigation.

If we possess more detailed information about the life patterns and behaviour of the macaque than we do of many other monkeys, it is because due care has been taken to avoid these pitfalls. By comparing and correlating separate experiments in natural and artificial surroundings, based on observations of a number of successive generations in both environments, naturalists have been able to build up a comprehensive picture of the species, applicable to varying conditions and situations.

The effect of captivity on the social and reproductive habits of macaques is well illustrated by a series of experiments conducted by the naturalist William A. Mason. It was immediately obvious to him that the day-to-day behaviour of the laboratory monkeys under observation bore very little similarity to that of the species in the wild. The sexual behaviour of the males, for example, was completely upset, affecting not only the frequency and duration of couplings with the females but also the nature of the act itself. The animals no longer adopted the normal mating position of the species and consequently found it difficult, sometimes impossible, to copulate at all. The artificial environment also inhibited the macaques' grooming activities and other hygienic functions, these being carried out rapidly, automatically and evidently without any sense of enthusiasm. Another interesting point was that the macaques, away from their natural surroundings, were even more belligerent than in the wild, and much less gregarious. Standard rituals designed to inhibit aggressive behaviour now failed to have the desired effect, so that chance encounters which might have been expected to have a peaceful outcome frequently resulted in bitter feuds. In fact the whole social hierarchy seemed to be turned topsy-turvy, the relationships of dominants and subordinates being ill-defined and the lower-ranking individuals often proving to be the worst-tempered. One of Mason's conclusions, confirmed by other naturalists pursuing different enquiries, was that among social species such as primates the normal processes of development and maturation are not inborn, but that each individual inherits certain tendencies and potentialities that are fulfilled as it grows up and shares in the life of the community.

The Japanese macaque

The most remarkable studies so far carried out on macaques under natural conditions have been those of Japanese zoologists on a local species, *Macaca fuscata*. Although Japan is not part of the Oriental region, the scientific interest of these experiments and their application to other macaque species are sufficient excuses for mentioning them in this chapter.

Macaques display their aggressive feelings in several ways. Increasing anger is shown in the three characteristic attitudes illustrated above, beginning with a steady, fixed glare, changing to a snarl with mouth open and teeth bared, and culminating in an up-and-down swaying of the head. Should the adversary not be intimidated, fighting will be the consequence.

Whether on the move or at rest macaques are well disciplined and each member of the troop takes up an allotted position, depending on its rank. When on the march the male dominants stay in the middle to protect the females and babies; when they stop to feed the leaders likewise take up a central position with the weaker members of the group, posting the younger males around the perimeter.

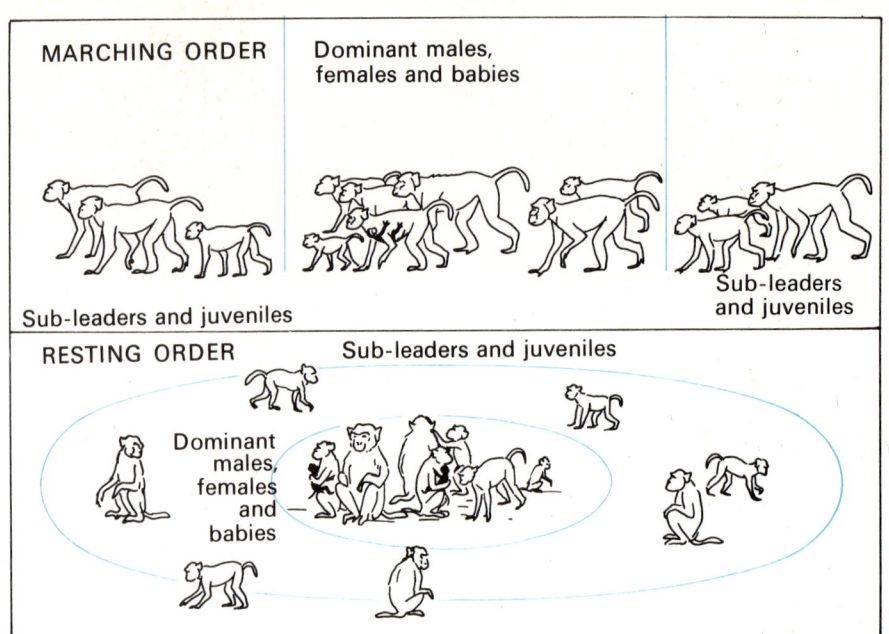

MARCHING ORDER | Dominant males, females and babies

Sub-leaders and juveniles

Sub-leaders and juveniles

RESTING ORDER | Sub-leaders and juveniles

Dominant males, females and babies

Facing page : A small family of Japanese macaques evidently enjoying a sun-bathing session in the lower branches of a tree.

Japanese macaques are gregarious animals but because the species is on the decline troops tend to be widely separated from one another. Observations have shown that these groups vary considerably, not only in respect of total numbers and the proportion of males to females, but also with regard to individual personalities and social behaviour.

Each group or *oikia* (the name applied by Professor Imanishi) leads a nomadic existence, roaming its own piece of territory, the area of which ranges from one to five square miles, according to the size of the troop and the availability of food. Although the domains of different troops may overlap there is seldom overt rivalry and conflict, for the class structure which determines the position of each individual macaque within the group appears to be extended to the troop at large. Thus the approach of a dominant troop will automatically cause a subordinate group to vacate that particular area.

Calls play an important role in communication, especially in dense forest where the monkeys cannot see one another. Broadly speaking, these calls are of two kinds. Cries which are accompanied by varied facial expressions and gestures are usually indicative of a high degree of nervous tension and are designed to communicate with a particular individual not very far away; other calls are intended to carry over a longer distance, mainly to locate the positions of different groups.

The Japanese scientists were, however, able to break down this pattern of intraspecific calls into six distinct categories: 1) calls from one individual to another, without overtones of alarm; 2) warning cries; 3) defensive, rallying calls; 4) aggressive, challenging cries; 5) calls of young; 6) cries of females on heat.

When a troop is moving from place to place in the forest the first type of cry is frequently used, whereas those of the fourth type are generally emitted when danger threatens, to be followed immediately by withdrawal of the entire troop to a safer area. During the retreat there is complete silence. When the alarm signal is given by a number of immature males it is quickly taken up by their older and more experienced companions.

More than two-thirds of these characteristic cries are used specifically for maintaining contact between the different members of the group.

Japanese macaques, which have to contend with bitter winter temperatures, have a reproductive cycle adapted to the climatic conditions of their habitat. Births usually take place from mid-May to mid-August so that by next winter the babies are big enough to survive. The infant mortality rate is thus much lower than it would be if births were to occur in a random manner at any season.

Leaders and sub-leaders

Within the macaque troop discipline is the order of the day and when the group is on the move its progress has all the efficiency of a well planned military manoeuvre. The monkeys advance in single file, the younger males being positioned at the head and tail of the platoon, the females and babies collecting in the centre and being escorted by several dominant males. Order is maintained even when the troop settles down to feed on open ground but now the monkeys fan out in concentric circles. Once again the females and young, accompanied by one or more of the male leaders, make up the inner circle while their sub-ordinates take up positions around the outside, ready, if necessary, for defensive action.

The Japanese naturalists who made such a detailed study of the macaques of their country remarked that some of the younger males would normally not join their companions until the time came for the troop to set off on its travels. At that point they would place themselves either at the front or rear of the file, as if they had no other function than to form an escort and no other contact with the group save on these occasions. No satisfactory explanation has so far been found for this type of behaviour.

In one island group, referred to as the Takasakiyama troop, there were about two hundred macaques led by six dominant males. Just below them in rank were ten slightly smaller males who in certain clearly defined situations took the place of the dominants. An indeterminate number of lower-ranking and immature males circulated about these two top groups. There is a similar pattern of leaders and sub-leaders in most macaque troops, their numbers varying according to the size of the local population, as a rule in direct proportion to the total. But in the case of the Takasakiyama troop this law did not seem to apply, for although the total numbers of the macaque community increased steadily to about four hundred between 1953 and 1956, the zoologists noted that there were no additional leaders and sub-leaders during the same period.

Simple observation makes it possible to recognise the relationships between the dominants and their deputies. From time to time, especially when feeding, the leaders jump on the backs of their subordinates as if to underline their authority, and in this way observers can determine any changes in social rank that may occur. But the class differences among the females and the younger males are much less clearly defined.

Privileges of birth

In contrast to the majority of primates macaques mate at a precise season and not indiscriminately at any time of the year. In the case of the Japanese species this begins toward the end of December and may continue until the last days of March.

Females in heat are easily identified by the fact that both the face and the skin surrounding the genital parts turn bright pink. A dominant male then mates with a number of receptive females in succession, engaging in sexual activity with each partner for

three consecutive days. If a particularly large number of females happen to be in heat simultaneously some of the sub-leaders on the outer edges of the group will perform substitute roles.

Gestation lasts approximately five months, the babies being born, depending upon when they were conceived, between mid-May and mid-August. Within a week they are able to walk and a month later they are already on solid food. The close relationship between mother and baby lasts for some ten months, during which time the latter learns to feed itself and to perform all the other activities necessary for adult life within the troop.

In a macaque colony there are two principal types of social contact – one between males and females and mothers and young, the other between animals of the same sex and similar age. It is in this latter form of contact that the pattern of dominant and subordinate assumes paramount importance.

The young macaque's introduction to the second type of social contact begins when it is getting on to eight months old, at which point it is big enough to start mingling with other individuals of about the same age. At that stage it does not matter greatly whether the play companion is a male or female;

Female macaques with babies tend to form small groups, providing the opportunity for the youngsters to play with one another at an early age. Play activity is of course important for developing mind and body in preparation for adult life as a member of a well organised, disciplined society.

but about a year later, when it can be classified as a juvenile, it seeks out other individuals only of the same sex. At two years of age the young male still lives with its mother but strays more and more frequently towards the outer edges of the group. A female of this age will always remain near the centre of the group. At three and a half years she is sexually mature but it is another year before she is ready to procreate. The males mature somewhat later and will mate for the first time at five years of age.

Social ambition is a feature of this as of most other highly organised animal communities. The first step on the path which may lead to dominant status is for a juvenile on the periphery to rise gradually in rank to become a sub-leader; but only a few will reach this status and an even smaller proportion will succeed in getting to the very top, or rather to the centre, where the dominant leader class holds sway.

Some macaques are undoubtedly more fortunate than others, due to the pure accident of birth. For the offspring of two dominants a position of leadership, though not inherited, is a strong possibility. Class distinction perhaps, but in the macaque community such an individual is likely to be better equipped for the leadership than most of its companions. A dominant male will normally choose a dominant female for a partner, in preference to a subordinate, and since both animals will have reached this status partly by virtue of superior weight and strength, these qualities will probably be passed on to their offspring. So for these privileged individuals there may be a short-cut to advancement.

Above and facing page: The baby macaque spends the first eight months or so with its mother, learning how to play, find food and perform all the activities which will be required when it leaves her to lead an independent life. As experiments with different troops of Japanese macaques have shown, each group has its own behaviour patterns and food preferences and one of the things the baby learns is to distinguish between what is edible and what is not.

Father takes a hand

Naturalists do not yet know very much about the true relationship that exists between adult males and young in monkey societies. But it seems to be well confirmed that in many communities youngsters recognise and acknowledge their fathers, and vice-versa, especially in ground-dwelling species.

It was the Japanese zoologist Furuya who first suspected that this might be the case in the course of studying the Takasakiyama macaques, remarking that certain members of the troop would use the daily ritual grooming sessions to seek out particular companions. Such contacts evidently had no sexual significance (even when an adult male and a young female were involved) but appeared to be founded on family ties.

Later surveys were to confirm this theory and it is now known that among certain groups of Japanese macaques there are close and affectionate links between dominant males and young. Nor is it an exaggeration to describe these adult attitudes as paternal – a revealing discovery, in fact, challenging the view of many anthropologists who have always claimed that paternal feelings are the singular prerogative of humans and that this is something which distinguishes us from other primates.

The Japanese naturalists noted that whereas the mothers accepted full responsibility for the rearing of their offspring during the months of babyhood, the male leaders and sub-leaders would extend their protection for a year or two to certain immature individuals when the latter could no longer be cared for by their mothers. That these attitudes sprang from true paternal instincts was indicated by the fact that such behaviour was distinct from normal activity, being directed specifically towards one individual rather than indiscriminately towards several members of the group. It could not, for example, be mistaken for standard protective procedures in the event of danger, which are in any case rare in the Japanese macaque.

What happened here was that a particular male would single out the same youngster time and time again, establishing a firm personal bond with it, and lavishing upon it all the attentions, apart of course from suckling, that had previously characterised the mother-baby relationship. The father (for he it must have been) was seen to fondle, groom and carry the youngster about. At other times he would walk beside it and be ready to protect it in moments of crisis. In the company of the younger macaque the male's behaviour was marked by unaccustomed docility, but it seemed to provoke other adult males to greater belligerence.

This was no isolated occurrence. Among the Takasakiyama macaques five dominant males and nine or ten sub-leaders were seen to act in this manner. When individuals of under a year old were concerned the adults showed little discrimination between male and female; but things were different in the second year. Out of twenty-five two-year-old macaques selected for such treatment, twenty proved to be females, the reason being that by this time most of the males had wandered off towards the boundaries of the group territory, leaving a nucleus of females in the centre.

Studies of the Japanese macaque population showed that play activity was an important part of the growing-up process and this has been reinforced by laboratory experiments. Denial of opportunities for play, which rarely happens in the wild, may lead to abnormalities of adult behaviour.

Any individual beyond the age of two years which continued to be pampered by a dominant male tended be delicate, sometimes puny. This was another significant finding, the implication being that protection of the weak (which is contrary to the laws of natural selection) is not, as commonly believed, exclusive to human societies. It raises the interesting question as to whether this form of behaviour may not be more widespread among animal communities than has generally been assumed.

A more detailed examination of this phenomenon in the Takasakiyama group showed that the sub-leaders were even more attentive than the leaders. Although one dominant went through the protective ritual 35 times in the course of a day, another sub-leader repeated the performance no less than 60 times. Both were exceptional, however, the average daily figure being 5·8 for the leaders and 6 for the sub-leaders. Among the subordinate males on the outer fringes of the group the maximum was 4 and the average only 0·4, proving indisputably that this type of paternal behaviour was essentially symptomatic of the two higher-ranking orders. Yet beyond that it was impossible to detect a fixed pattern, for among the leaders and sub-leaders themselves it seemed to be the intermediate ranks who most frequently displayed their fatherly feelings.

The zoologist Junichiro Itani subsequently proposed that the various members of the two upper-class groups might be classified according to three criteria—sociability, aggressiveness and degree of interest in the nucleus of the group. In this way he identified six distinct types of adult personality, measuring in each instance the frequency of paternal activity. He found that the individuals displaying the closest care for their offspring were notable for a high degree of sociability, a comparatively gentle temperament and a keen concern for all that was going on in the centre of group territory—in short, those who most aspired to positions of dominance. There was thus a clear and direct link between paternal responsibility and social ambition. It was noticeable that on a number of occasions a sub-leader would not be repulsed by the adult females and dominant males once he had staked out a central position and was actively engaged in caressing a younger animal, apparently as a mark of social consideration which would not necessarily have been accorded under different circumstances.

Paternal behaviour among the subordinate males of the troop, though seldom observed, was likewise most marked in individuals who showed particular interest in the activities of the central members of the group, suggesting that they too were aspirants to a higher social rank.

One very interesting conclusion to be drawn from all these observations is that in a highly bellicose society such as that of the macaque, aggressiveness is not the only factor which determines social status and that other individual qualities are equally important.

Another revealing aspect of the father-offspring relationship is that a youngster which is regularly cared for and protected by an adult male enjoys certain privileges, especially at feeding time, and is in fact entitled to the first pickings. This individual

generally displays a superior attitude towards many of the adult females and companions of its own age, so that the signs of future dominance are already there. Although it has not so far been ascertained whether this pattern is permanent, there are strong grounds for thinking that paternal care in early life plays an influential role in determining the later social position of a young macaque. But it is not necessarily a one-way relationship for, as will be described in due course, the adult may acquire certain habits and customs from the youngster – a case of the father learning from his child. Further studies will doubtless resolve some of these questions which extend beyond the confines of social status alone. It has already been remarked, for example, that the adult males most involved in this type of paternal behaviour are sexually active all the year round, not just during the customary breeding season.

Caution is also needed in applying these findings to the species at large. Thus among eighteen troops of Japanese macaques paternal behaviour was strongly in evidence in three groups, spasmodic in six others and completely absent in the remaining eight. Yet neither geographical distribution nor the numbers of

Among several groups of Japanese macaques it was evident that the agile, inquisitive youngsters would experiment more freely than the hidebound adults, especially when it came to savouring new types of food. By the process of imitation the younger animals were responsible for introducing new customs to the troop, reversing the usual patterns of acquired behaviour whereby the adults pass on their experience to the younger generation.

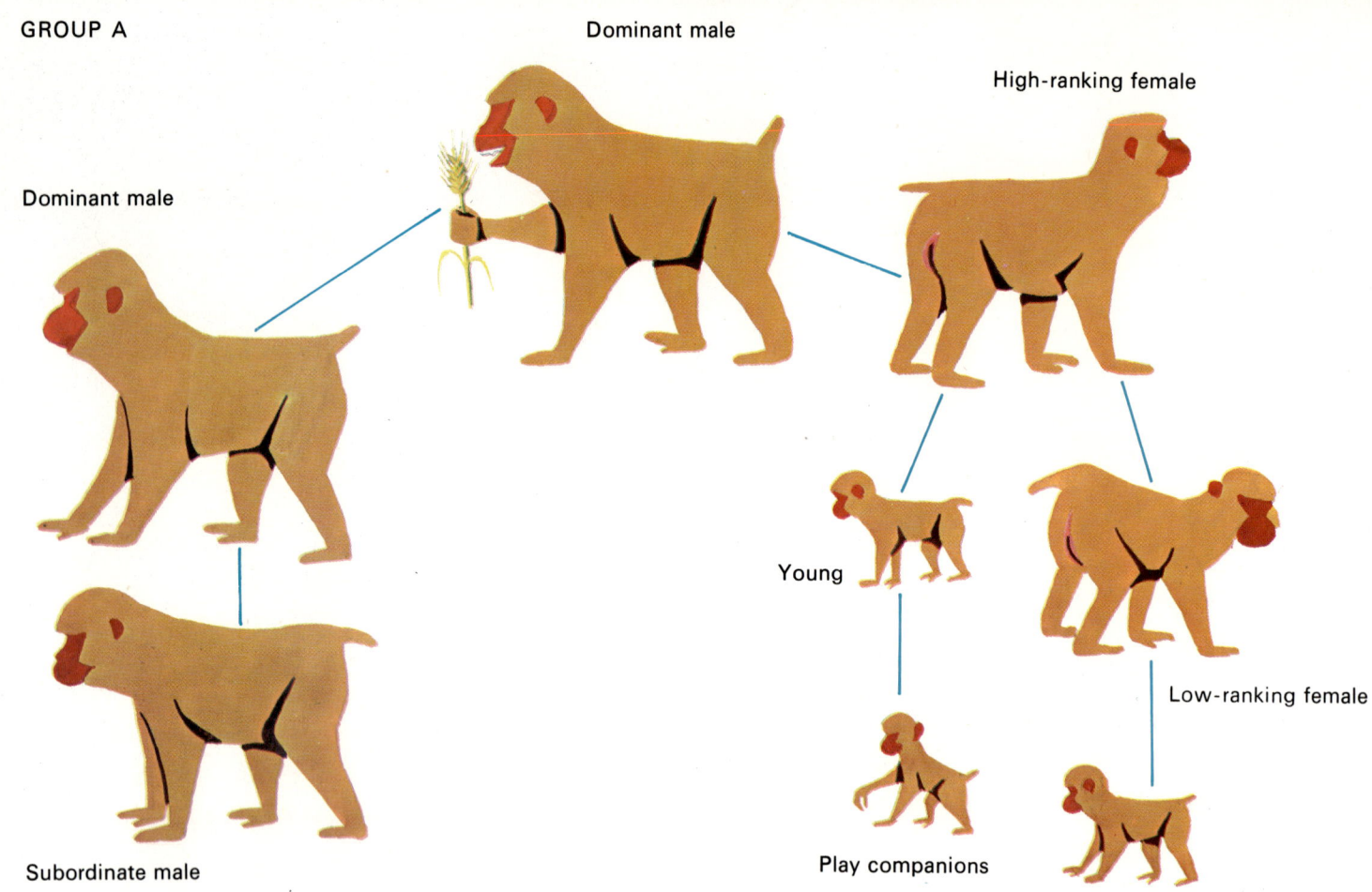

Dominant male

Dominant male

High-ranking female

Young

Low-ranking female

Subordinate male

Play companions

Young

Above and facing page : The most interesting information to come out of the study of the macaque troops of the Japanese islands related to the acquisition of new habits and the transmission of new patterns of behaviour from one individual to another. In some cases this proceeded from adults to young but in others in the opposite direction. Thus in group A from Minoo, a dominant male discovered that a blade of corn, never previously experienced, could be eaten; within four hours the entire group had followed his example, beginning with other dominant males, high-ranking females and their offspring, then continuing with lower-ranking animals and their progeny. In group B from Takasakiyama, however, a young female who decided to wash a sweet potato before eating it was imitated by her play companions and in due course, via the mother, by other adults of the troop. But it took more than a year for the custom to be widely adopted and even then only 51·2 per cent of the animals engaged in this activity. Whatever the circumstances of such an innovation it is only when acquired by dominant males and females and subsequently adopted by their offspring that it can become a permanent feature of community culture.

monkeys in a troop appear to have any significant bearing on the problem, the conclusion being that this, in similarity to other aspects of social behaviour, is an example of acquisition by imitation and experience.

Discovery and progress

Naturalists specialising in the subject of animal behaviour, who have had both the time and the facilities to conduct experiments with several successive generations belonging to the same group, have concluded that whereas many actions are purely instinctive, others are definitely acquired by learning. Baby and young animals, by watching and imitating adults, gradually perfect complex actions upon which future survival may depend. Among the more important of these activities are hunting, finding food and building nests or other forms of shelter and refuge.

One of the most extraordinary examples of acquired behaviour to be found in the animal world is that of the great tit and the related blue tit in the British Isles, to which brief reference has already been made. Taking advantage of the fact that foil caps are used to seal milk bottles and that milk deliveries are left outside on doorsteps, these ingenious birds have acquired the habit of pecking through the foil and sipping the contents. This behaviour was first noted among great tits in one part of the British Isles but spread progressively to other districts, being adopted by blue tits as well. Doubtless the discovery was accidental to start with but certainly it was subsequently copied and used for the particular purpose of obtaining an additional supply of food.

Just as there were stubborn souls during the 19th century who

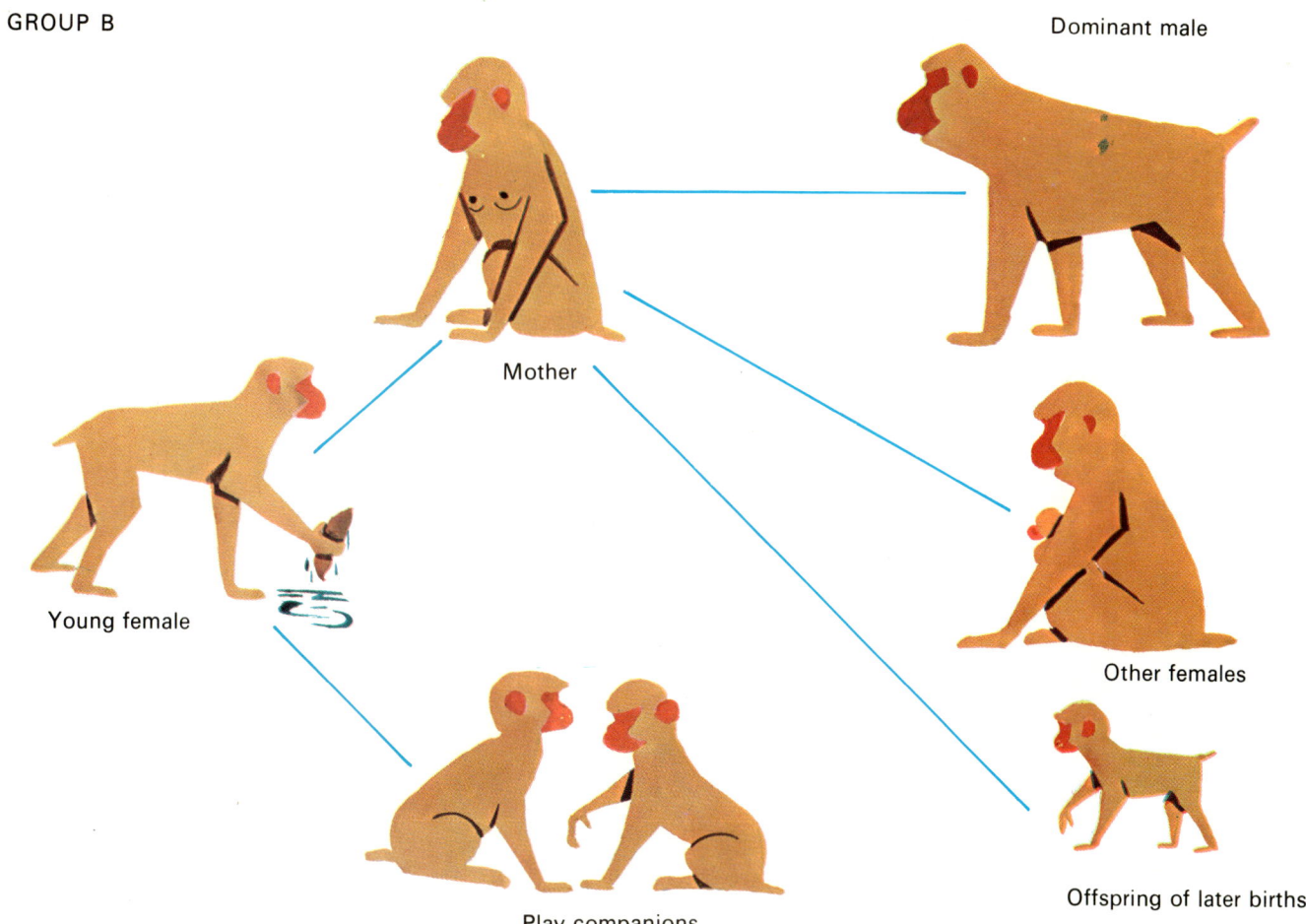

Dominant male

Mother

Young female

Other females

Play companions

Offspring of later births

refused to acknowledge the validity of Charles Darwin's theories of evolution, there are still some traditionalists who cannot accept the evidence now pouring in that animals, particularly higher vertebrates, are capable of learning – an activity which had previously appeared to be exclusive to man. The conclusion, uncomfortable though it may be to human vanity, is that the barriers dividing us from other primates may be far more flimsy than we had been led to suppose. How closely our individual and social behaviour can be compared with that of animals is a controversial but highly fascinating field of study.

Claims for animal intelligence should, nevertheless, not be pitched too high, for it is all too easy – and common – to confuse activities which are the direct consequence of example and imitation with those connected with the normal processes of growth and maturation (such as the evolutionary phases through which a fledgling passes before it becomes capable of flying). In many bird species, including eagles, the organs necessary for flight are present and can theoretically be used long before the fledgling embarks on its first flying attempts. Only when all its movements are properly co-ordinated will the bird begin flapping its wings. So what we loosely call 'learning' to fly is simply part of the normal growing-up process.

Zoologists have carried out a large number of tests in their efforts to distinguish between innate and acquired patterns of behaviour, both at an individual and a group level. The study of the Japanese macaque population has provided valuable material in this field, particularly concerning the transmission of what has come to be called 'animal culture'.

The most detailed findings – mainly because they proved to be the easiest to observe – related to the acquisition of new feeding

As they approach the age of two years young males, though still dependent on their mothers, stray off from time to time with others of their own sex. Until then they will have mingled contentedly with females.

Facing page : Jigoku Dani, a valley celebrated for its thermal springs, is situated in the Japanese province of Nagano. It is a meeting point in winter for a large troop of macaques who, when the temperature drops below 0°C, plunge into the warm water.

habits. It was immediately evident that each troop had its individual food preferences and standard diet. Thus the Takasakiyama macaques were seen feeding, at various times, on eighteen types of plants, together with a few miscellaneous insects, whereas a group from Minoo showed a partiality, among other items, for birds' eggs. Each young macaque adopted the feeding habits of its particular troop, a baby accepting what its mother gave it. In captivity, however, it has proved difficult to interest a macaque in the type of vegetation which its counterpart in the wild might consume as a matter of course. The only food which a baby macaque in a zoo or laboratory will take instinctively is milk.

There is one very important difference between human and animal culture. In the former language is used to convey information from an individual to the group at large; but in the latter it is a question of direct imitation, one at a time. In the Japanese experiment the naturalists were especially intrigued to note the effect which age had on cultural transmission. The adults often adopted new habits in the slowest and most painstaking fashion, whereas the younger animals often proved to be much more alert and flexible, making discoveries and quickly putting them to practical use. Thus, contrary to expectation, it was sometimes the younger members of the community who dictated changes of behaviour to their elders. As soon as one of them settled down to an unusual activity this would be taken up in turn by its play

Facing page : The expressions on the faces of these Japanese macaques as they take their ease in the warm waters of Jigoku Dani are surely as eloquent as words. The monkeys are good swimmers and often play about in the water for several hours.

COMMON FOREST MACAQUES

Class: Mammalia
Order: Primates
Family: Cercopithecidae
Length of head and body: 15–30 inches
   (38–76 cm)
Length of tail: 0–24 inches (0–61 cm)
Weight: up to 28½ lb (13 kg)
Diet: omnivorous
Gestation: 5–7 months
Number of young: one
Longevity: upwards of 30 years in captivity

COMMON RHESUS OR INDIAN MACAQUE
(Macaca mulatta)

Overall colour brown; pink face.

BONNET MONKEY
(Macaca radiata)

Grey-brown pelage; hair of head en brosse, giving appearance of black cap.

TOQUE MONKEY
(Macaca sinica)

Gold or reddish-brown pelage, pink face, long hair on head.

HIMALAYAN MACAQUE
(Macaca assamensis)

Large animal, yellowish-brown to dark brown, living at high altitudes.

BROWN STUMP-TAILED MACAQUE
(Macaca arctoides)

Large animal, dark chestnut-brown pelage; tail a mere stump.

JAPANESE MACAQUE
((Macaca fuscata)

Large animal, yellowish-brown pelage, pink face.

TIBETAN MACAQUE
(Macaca thibetana)

Large, hairy and bearded; chestnut muzzle.

LION-TAILED MACAQUE
(Macaca silenus)

Robust build; black pelage with thick white ruff around face; tail terminates in long tuft of hair.

companions and they would subsequently teach it to their mother although in other respects they continued to receive instruction from her.

Yet such behaviour was not consistent. Thus a three-year-old female belonging to the Takasakiyama troop one day proceeded to wash a sweet potato before eating it, passing on the custom to her young companions and then to her mother. Three years later the naturalists watched ten adults doing the same thing.

In the Minoo group, on the other hand, the procedure was reversed. A dominant male suddenly began eating an ear of corn and was imitated by other leaders, then by high-ranking females and their young. Some time later the habit spread to lower-ranking females and then to their offspring.

It is clear that individual social relationships, whether they are between mother and baby, two play companions or juvenile and adult, are of considerable importance in animal culture, this depending above all on the nature and frequency of contact. Social progress, however, comes about not only by means of discovery and transmission but by diffusion. Although new habits may be transmitted in a sideways direction from one young animal to another, they will only take on real social significance if passed down from one generation to another. This change in direction (vertical instead of horizontal) occurs when the innovating youngsters themselves reach maturity and produce offspring, educating the latter in activities which will be progressively adopted by the entire community.

The behaviour of certain animals in the group may well delay or prevent the propagation of new habits. Thus a mother, simply by keeping her youngster on too tight a rein (protecting it from danger at all times, for example) will give the latter no chance to discover anything for itself. Moreover, the natural instinct of a dominant is to discourage individual experimentation. So whereas strict control and rigid discipline may be necessary for the survival of the species, too much authority may in some circumstances be a hindrance to cultural growth.

Reactions to novel situations vary greatly from group to group and even within a single troop. Thus when jam was offered to the Takasakiyama macaques it took them a year to become accustomed to the new food and then only about half of them would touch it. Yet when the Minoo macaques were handed ears of corn, a cereal hitherto unknown, it took only four hours for almost all the animals in the troop to start experimenting with this novel type of food. There are doubtless a number of reasons for this difference in attitude. In the former instance the discovery was made by a young animal and in the latter case by an adult. Two examples are of course insufficient for establishing a general theory and it is impossible to decide on this basis which is the rule and which the exception. It may be that just as important as the direction of cultural transmission (adult to young or young to adult) is the personality of the innovator, so that general application of a discovery may be a pure matter of chance, perhaps encouraged, perhaps checked.

Broadly speaking, those members of a community who adapt to changing habits and cultural patterns have a better chance of

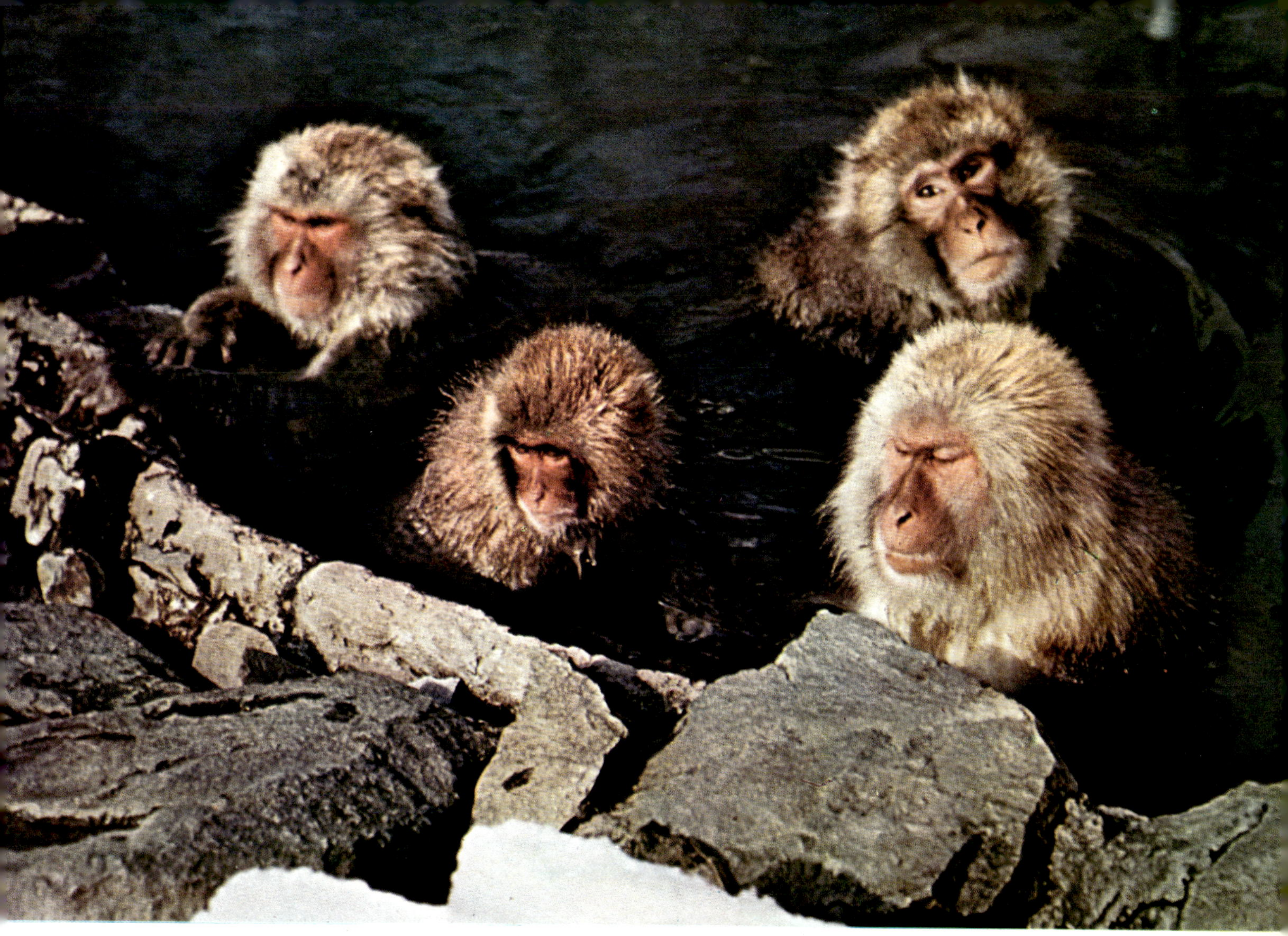

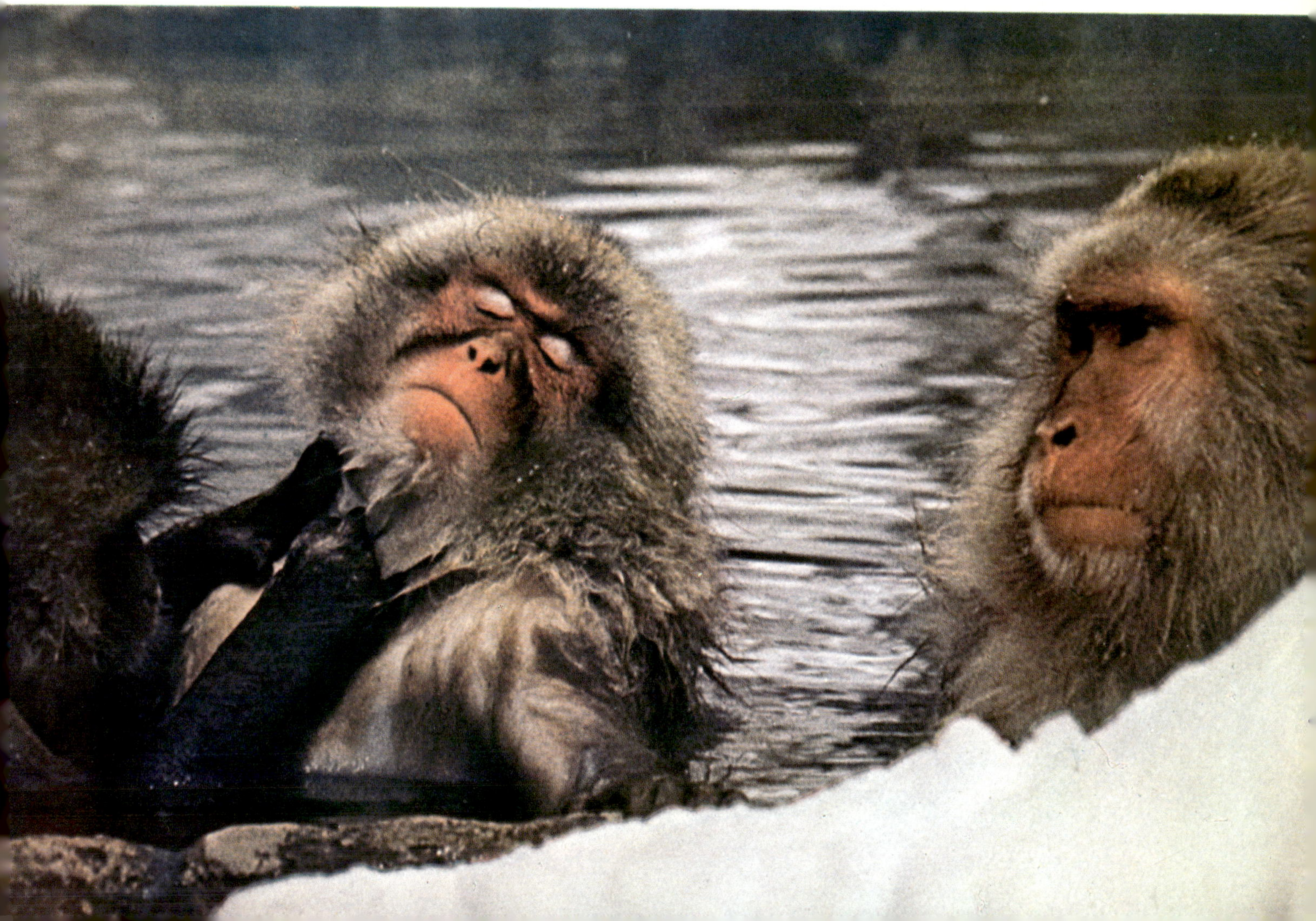

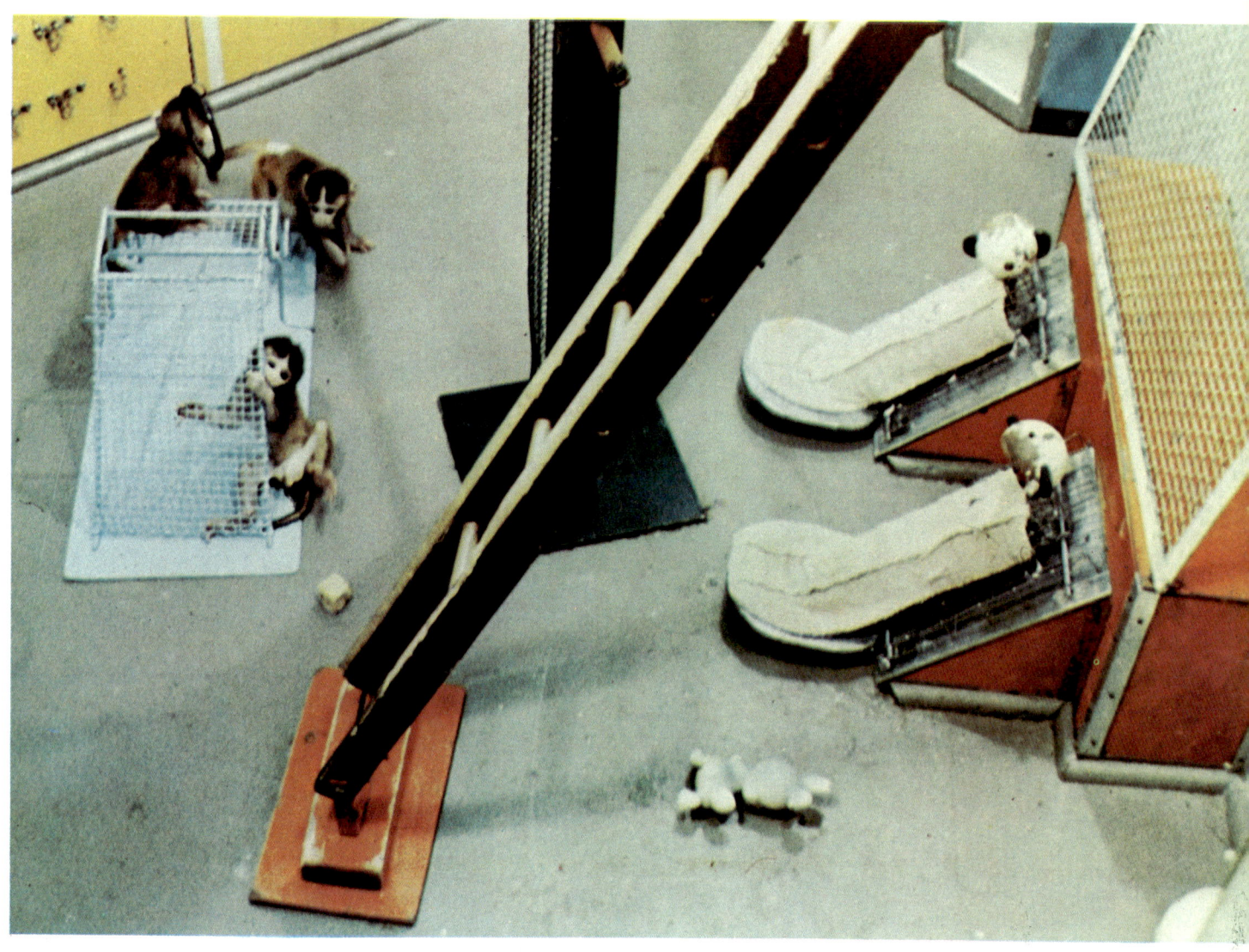

survival than those who stubbornly resist such novelties. The same applies to the group as a whole. In the case of the Japanese macaque the naturalists made a clear distinction between troops which could be described as 'progressive' and others whose social structure appeared to be completely stagnant.

The macaque and human neurosis

The branch of science concerned with animal psychology, which at the outset was dismissed as of minor importance, is rapidly becoming recognised as a valuable aid to the better understanding of the causes of human emotional disorders.

Dr Harry F. Harlow of the University of Wisconsin is one of many eminent scientists who maintain that the identification of abnormal behaviour in monkeys can throw much light on the human situation, especially psychiatric disturbances resulting, for example, in juvenile delinquency and other forms of crime.

Harlow's initial experiments were based on the mother-child relationship, recognised as being of prime importance to individual development. Yet Harlow's laboratory tests with a group of macaques suggested that the almost sacrosanct assumption

Dr Harry F. Harlow of the University of Wisconsin carried out a series of fascinating experiments with macaques which revealed interesting information about their habits in captivity. In one set of tests two metal doll-like objects served as artificial mothers, one to feed the baby, the other to provide comfort and security.

Facing page : The custom of washing sweet potatoes in water, accidentally discovered by one young female macaque, later became a normal activity for more than half the members of the troop.

that there could be no substitute for the nursing mother and the maternal breast might have to be amended. He used macaques because their development (in the sense of co-ordinated movements) is both more advanced and more rapid than that of human babies, so that their behaviour can be interpreted more objectively.

In one experiment a baby macaque was placed, a few hours after birth, in a room alongside a couple of metal doll figures, one equipped with a feeding bottle, the other covered in a soft, furry fabric. Harlow was intrigued to note that the baby ventured near the former only when it wanted to suckle and that it spent the rest of its time huddled against the soft body of the other doll. This seemed to underline the importance of sheer physical contact in the mother-baby relationship, providing the latter with a sense of security, and this was confirmed by later tests. Confronted by mechanical toys and deprived of the presence of its artificial mothers, the baby macaque recoiled in terror, but as soon as the dolls reappeared in the cage it hurled itself at the one which offered comfort, never the one which supplied food. The contact eventually pacified the baby sufficiently to allow it to venture out to play with the toys. Other experiments showed that signs and movements also played a significant role in forging babyhood links and that the actual warmth of the maternal body was not an influential factor.

Harlow showed too that contacts between a mother and baby are established early in life and that if the two are separated during this time the later relationship may be quite different. Babies deprived of all maternal contact in the first eight months of life and then placed near an artificial mother tended to remain close to the latter twice as long as babies that had been reared in this way from the beginning.

Other tests revealed the serious psychoses and neuroses that might arise from isolation. Monkeys brought up in isolation, away from mother and companions, showed severe psychological disturbance in adult life, characterised by abnormal sexual and social behaviour.

The broad conclusions to be drawn from such tests are that some form of maternal care and early companionship are necessary for stable individual development. Companionship may compensate for the absence of maternal affection, however, for orphans reared with other babies will develop quite normally.

That nervous disorders are at the root of certain physical ailments is nowadays generally accepted. Gastric ulcers are a case in point, and one remarkable experiment with two macaques confirmed the mysterious influence of mental stress on physical functioning. The monkeys were placed in chairs which gave off a mild electric shock. This could be cut off by depressing a lever and both animals learned to control the current. But when placed together one macaque soon discovered that it could get its companion to do all the work and the latter became increasingly irritable. Medical examination also showed that the monkey exposed to the continual stress of manipulating the lever had developed a serious stomach disorder, whereas the inactive one was still in perfect health.

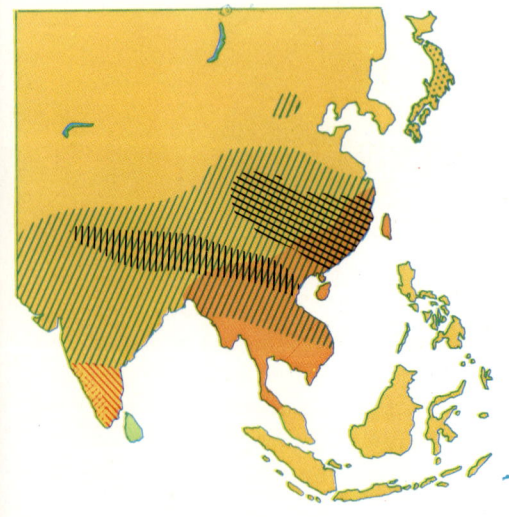

Geographical distribution of common forest macaques.

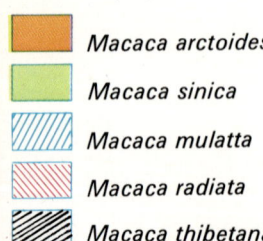

Macaca arctoides

Macaca sinica

Macaca mulatta

Macaca radiata

Macaca thibetana

Macaca assamensis

Macaca fuscata

Macaca silenus

In Dr Harlow's laboratory experiments the baby macaque spent much more time huddling against the soft, furry covering of the object which spelt security than with the figure equipped only with a feeding bottle. Although initially frightened by the presence of mechanical toys, contact with the soft material restored confidence. These reactions indicated that the protective function of the mother was of greater significance than her nursing function.

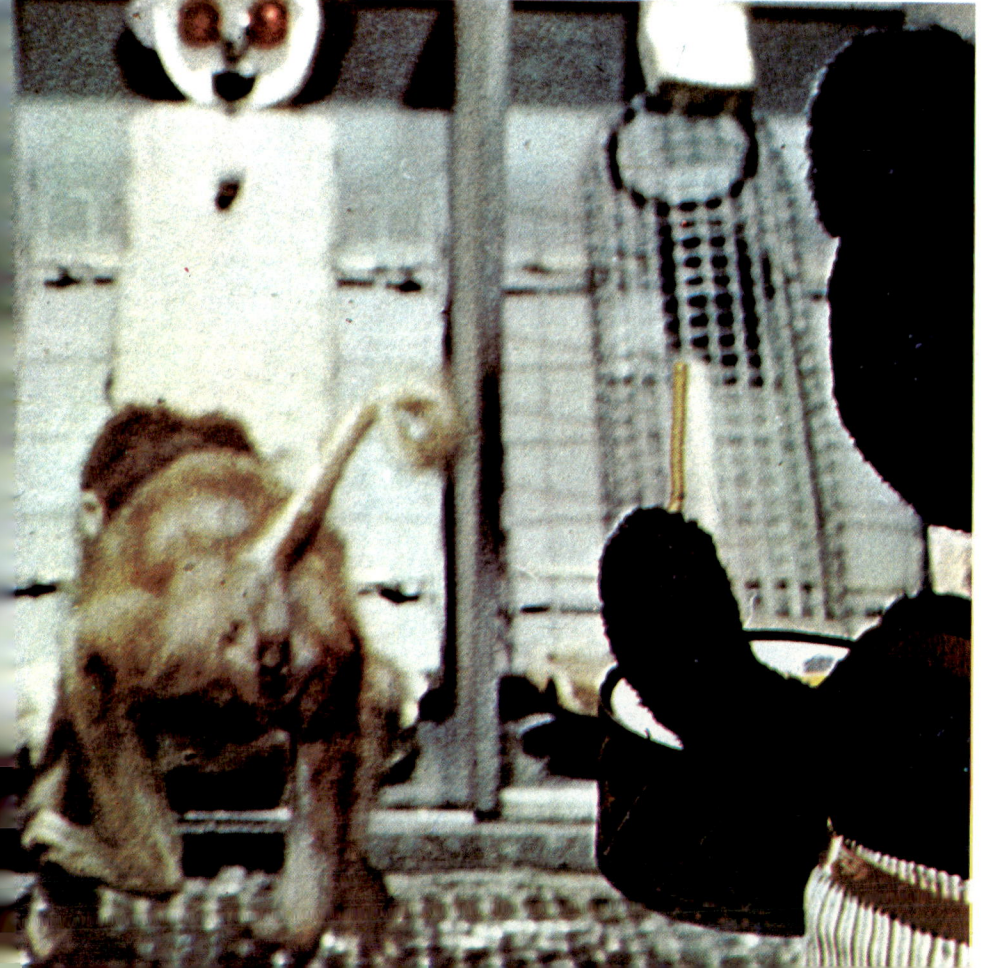

CHAPTER 86

The jungle: a paradise for plant-eaters

Plant growth in the Asiatic tropical forest is extraordinarily rapid and prolific. A path that has been hacked out with considerable difficulty will be overgrown with vegetation in a matter of days, completely obscuring all traces. The dense tangle of coiling, trailing lianas constitutes a nightmare for the forest traveller but the profusion of plant life represents a veritable paradise for a multitude of herbivorous mammals, especially in those parts where the secondary forest, with its luxuriant undergrowth, has replaced the primeval jungle and its floor of mud and leaves. A rich variety of fruit, flowers, twigs, foliage and vegetational debris of all kinds sustains a huge army of forest animals, ranging in size from the diminutive tree rat, barely 4 inches long, to the giant of the jungle, the enormous Asiatic elephant. Here too are massive rhinoceroses, majestic wild cattle and goats, ferocious wild boars, deer with magnificent antlers, and an unparalleled number of terrestrial and tree-dwelling rodents.

The hordes of invertebrates, amphibians, reptiles and plant-eating birds which compete with these mammals for food do not threaten a way of life which has changed little for millions of years. Unhappily, it is man who is the gravest menace to the well-being of this delicately balanced community. Because of the population explosion in Asia which has forced back the forest frontiers and destroyed the natural haunts of the jungle inhabitants, the larger mammals of the Oriental region are immediately endangered and a few species may become extinct before scientists even have the opportunity of studying them in detail. This may be the lot of the three species of Asiatic rhinoceros.

The tropical rain forest provides refuge for some of the rarest animals in the world, notably the gaur (*Bos (Bibos) gaurus*), the

Facing page : Although its range of distribution is nowadays much more restricted than in the past, the great Indian rhinoceros, one of the three surviving Asiatic species, is protected in a number of reserves in Nepal, Bengal and Assam.

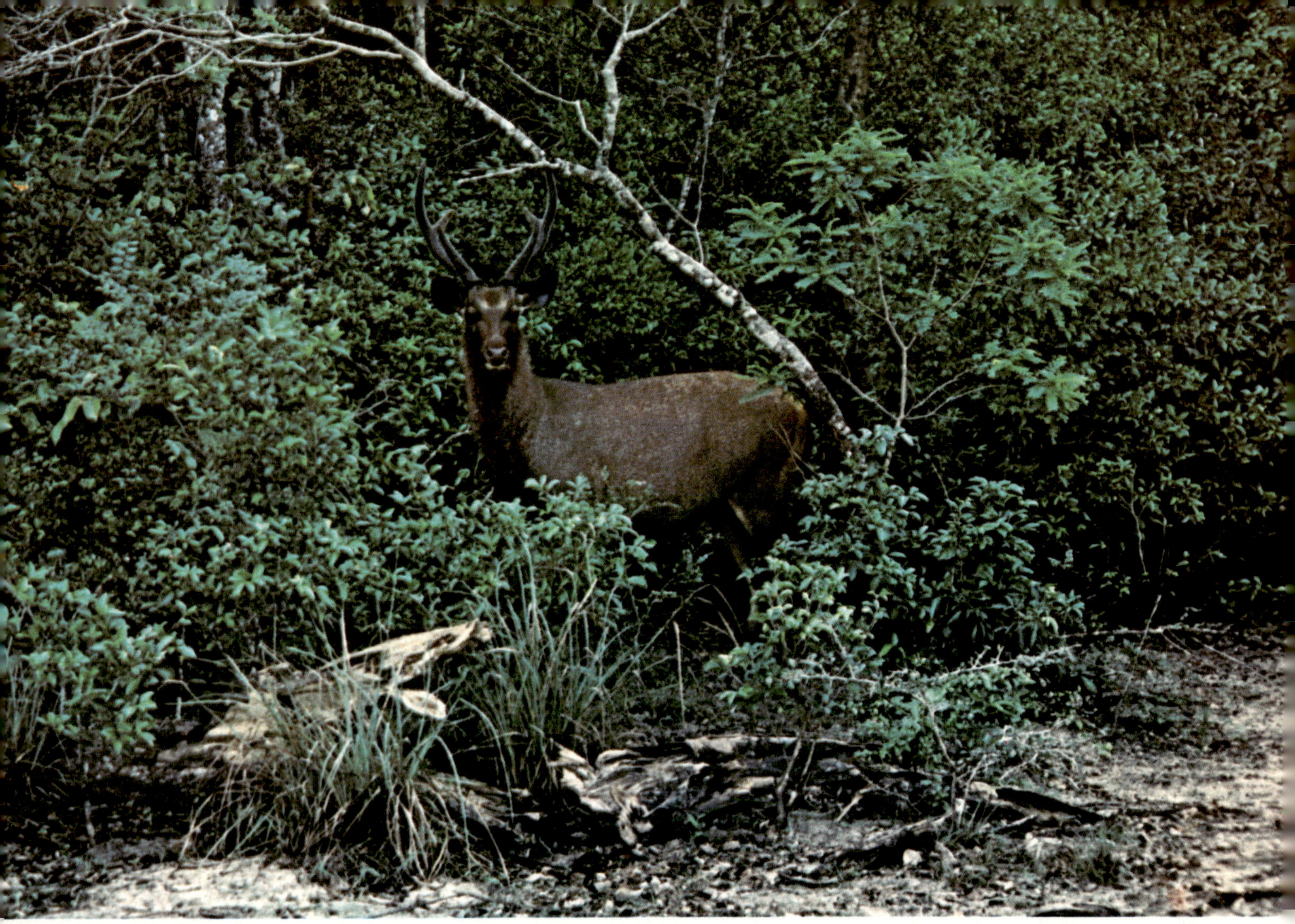

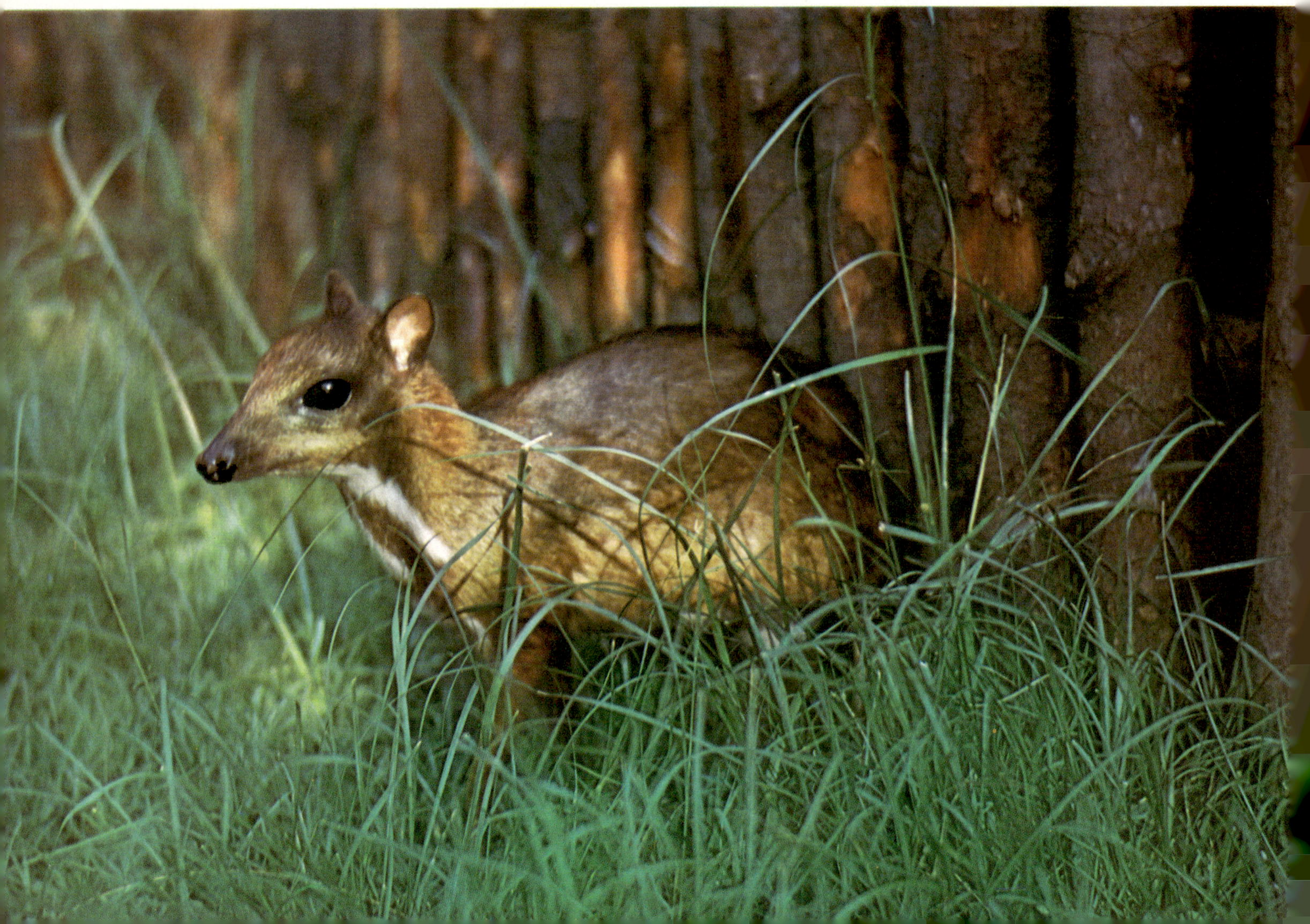

Indian or crested wild boar
(*Sus scrofa cristatus*)

Bornean wild boar
(*Sus barbatus*)

axis or Indian spotted deer (*Axis axis*) and the Malayan tapir (*Tapirus indicus*) whose only living relative is the tapir of the South American jungle.

Other animals are comparatively familiar, strongly resembling related species from Europe. Thus the Bornean wild boar (*Sus barbatus*), found only in Asia, is similar to the common European wild boar but is notable for its size and for the swollen area of long white hairs surrounding the snout.

There are many species of deer. In addition to the axis with its white spotted russet coat, there are the Indian sambar (*Cervus unicolor*), the Timor deer (*Cervus timorensis*) and a number of muntjacs or barking deer (*Muntiacus muntjak, M. pleiharicus, M. feae, M. crinifrons* and *M. reevesi*).

Representatives of the Bovidae include the gaur (largest of the genus *Bibos*), the banteng (*Bos (Bibos) banteng*), the kouprey (*Bos (Bibos) sauveli*)–in danger of extinction–and the Sumatran serow (*Capricornis sumatrensis*), a hardy animal with a wide range extending from sea level to the foothills of the Himalayas.

But the most spectacular plant-eating mammals of the Oriental region are certainly the rhinoceroses. There are three species in southern Asia, the great Indian rhinoceros (*Rhinoceros unicornis*), the Javan rhinoceros (*Rhinoceros sondaicus*) and the Sumatran rhinoceros (*Dicerorhinus sumatrensis*). They all have the appearance of animals of a prehistoric age, the first two with huge folds of skin on the back which are suggestive of armour-plating, the third (and smallest) being hairy. Of the three, the Javan species is the most seriously imperilled.

In addition to these large and medium-sized phytophages there are, as in every other part of the world, a great number of rodents. The majority of these species are arboreal, such as the pencil-tailed tree mouse (*Chiropodomys gliroides*) which spends the day in its nest between the nodes of a bamboo stalk. Communication with the outside is by way of two holes an inch or so in diameter, one of which is only used for escape in the event of danger.

Another arboreal rodent is the red tree rat (*Pithecheir melanurus*) which lives in a circular nest of leaves situated some 6–12 feet above the ground. The Sumatran bamboo rat (*Rhizomys sumatrensis*), on the other hand, leads a subterranean existence, excavating and exploring long galleries under the roots of the bamboo which constitutes its principal type of food. At night the rodent climbs up the stalks, stripping off slivers to carry back to its burrow. Higher up in the trees live the Oriental squirrel (*Callosciurus notatus*) and the giant Malabar squirrel (*Ratufa bicolor*) as well as a number of flying squirrels. It is important

Similar to their counterparts in Europe are the Indian or crested wild boar and the Bornean wild boar.

The pencil-tailed tree mouse spends the day concealed in a nest burrowed in a bamboo stalk.

Facing page : The largest deer of the Oriental region, with thick antlers, is the sambar (*above*). The mouse-deer or chevrotain (*below*) is not a true deer. Standing less than a foot high it has no horns. At the slightest sign of danger it flees into the underbrush.

Most of the familiar tree-dwelling rodents are represented in the Oriental region. The tropical forest is the home of the giant Malabar squirrel (*left*), which may grow to a total length of about 3 feet, and the striped palm squirrels (*right*), which are among the smallest species.

to point out that these rodents are not necessarily found in every part of the Oriental forest region, because the jungle itself is discontinuous, especially in the south where the islands of the Malay archipelago are the homes of subspecies found nowhere else in Asia.

The Asiatic elephant

A test was recently carried out in a cross-section of British schools in which children of different ages were asked to list their favourite and least popular animals. There were twelve thousand answers, some of them ('the coca-cola animal', 'the otamus', 'the jumping worm') showing a high degree of imagination, others being quite unidentifiable. The varied responses not only threw a great deal of light on the comparative educational effects of books and television, but revealed points of particular interest concerning the child's vision of the animal kingdom. It was noticeable, for example, that there was a direct relationship between the age of the child and the size of the favourite animal. The younger the child, the larger the animal selected. Thus about 15 per cent of the four-year-olds chose the elephant as against only 3 per cent of the fourteen-year-olds. Other smaller children picked the giraffe while many of the older children chose the bush-baby.

Why did the younger boys and girls express such a clear preference for animals so much larger than themselves? Subconsciously, no doubt, because they are symbols of power and strength, and to some extent because they appear to possess human-type qualities of sociability and intelligence. We will

Giant brown squirrel
(*Ratufa affinis*)

Prevost's squirrel
(*Callosciurus prevosti*)

Giant Malabar squirrel
(*Ratufa bicolor*)

Temminck's flying squirrel
(*Petinomys setosus*)

Sumatran bamboo rat
(*Rhizomys sumatrensis*)

return to the latter point shortly.

To the children who put the elephant at the top of their list, the species was unimportant although many of them may have been aware of the principal distinctions between the African and Asiatic animals. We have already dealt with the former in some detail, making the point that it is very difficult to tame. So the animal most of the children had in mind was doubtless the one that provided rides at the zoo, namely the Indian elephant. If we refer to this species here as the Asiatic elephant it is only because, in addition to India, its habitats include Ceylon, Burma, Thailand, Malaya and Sumatra—lands where it has been successfully domesticated for many centuries.

Although there are obvious outward resemblances, the Asiatic elephant (*Elephas maximus*) is smaller than its African relative. Its forehead is domed, the back is arched rather than hollowed, the ears and teeth are relatively small and the trunk is smooth, terminating in only one prehensile 'lip'. Four nails are visible on the hind feet, as against three in the African species.

The food habits of the two species vary considerably. The African elephant feeds principally on stringy vegetation such as bark and leaves, tough and rich in silica, but the Asiatic elephant's diet consists mainly of tender grass. Such food preferences are reflected in the contrasting surface structure of their molars. Both have enamel ridges on the crowns, surrounded by cement, but whereas in the African species these are few and lozenge-shaped, suitable for grinding, in the Asiatic species there are more ridges which are transverse and almost parallel to one another. The mastodons, primitive proboscideans which in various shapes and sizes roamed over the entire globe, except

Rodents, as in other parts of the world, make up by far the majority of the mammals of the Oriental region. Here are five characteristic species – the Sumatran bamboo rat and four types of squirrel.

Above and facing page : Like their African relatives, Asiatic elephants make daily journeys to streams and rivers in order to drink and bathe.

ASIATIC ELEPHANT
(Elephas maximus)

Class: Mammalia
Order: Proboscidea
Family: Elephantidae
Length of head and body: 216–252 inches (550–
640 cm)
Length of tail: 47–59 inches (120–150 cm)
Height to shoulder: up to 118 inches (300 cm)
Weight: up to 4–5 tons
Diet: herbivorous
Gestation: about 21 months
Number of young: one, rarely two

Distinguished from African elephant, which it superficially resembles, by domed forehead, arched back, smaller ears, less developed teeth and almost smooth trunk ending in one prehensile 'lip'. The forefeet and hind feet have five and four nails respectively.

for Australia, and became extinct in the Quaternary era, also had ridged molars, but the difference between these and the teeth of their descendants was that the separating areas were not filled with cement so that they were only suitable for chewing soft types of vegetation.

As a consequence of traditional habitats being systematically destroyed, the Asiatic elephant population is today much reduced as compared with former times. Even in Assam, the region lying between Bangla Desh and Burma, the forest mass which once extended continuously from Bengal to the foothills of the Himalayas has been drastically cut back. Nevertheless, this is the most densely wooded province in India and about one-quarter of its total area is covered by jungle. This then is the true home of the Indian elephant and in these parts it is still to be found in comparatively healthy numbers.

It was in Assam that man first discovered how to capture a wild elephant and to train it in such a way that it could be used for transport and various simple forms of heavy work. The rules for domesticating elephants were first laid down in an ancient Assamese book called the *Hastividyarnava* which is still used today as the most detailed procedural guide for those engaged in this skilful and often dangerous activity.

The method of domesticating these elephants has in fact remained virtually unchanged for centuries. A team of beaters first herds the animals into a large, extremely strong wooden stockade. Inside this enclosure a group of tame elephants will be waiting, their function being to act as a calming influence on

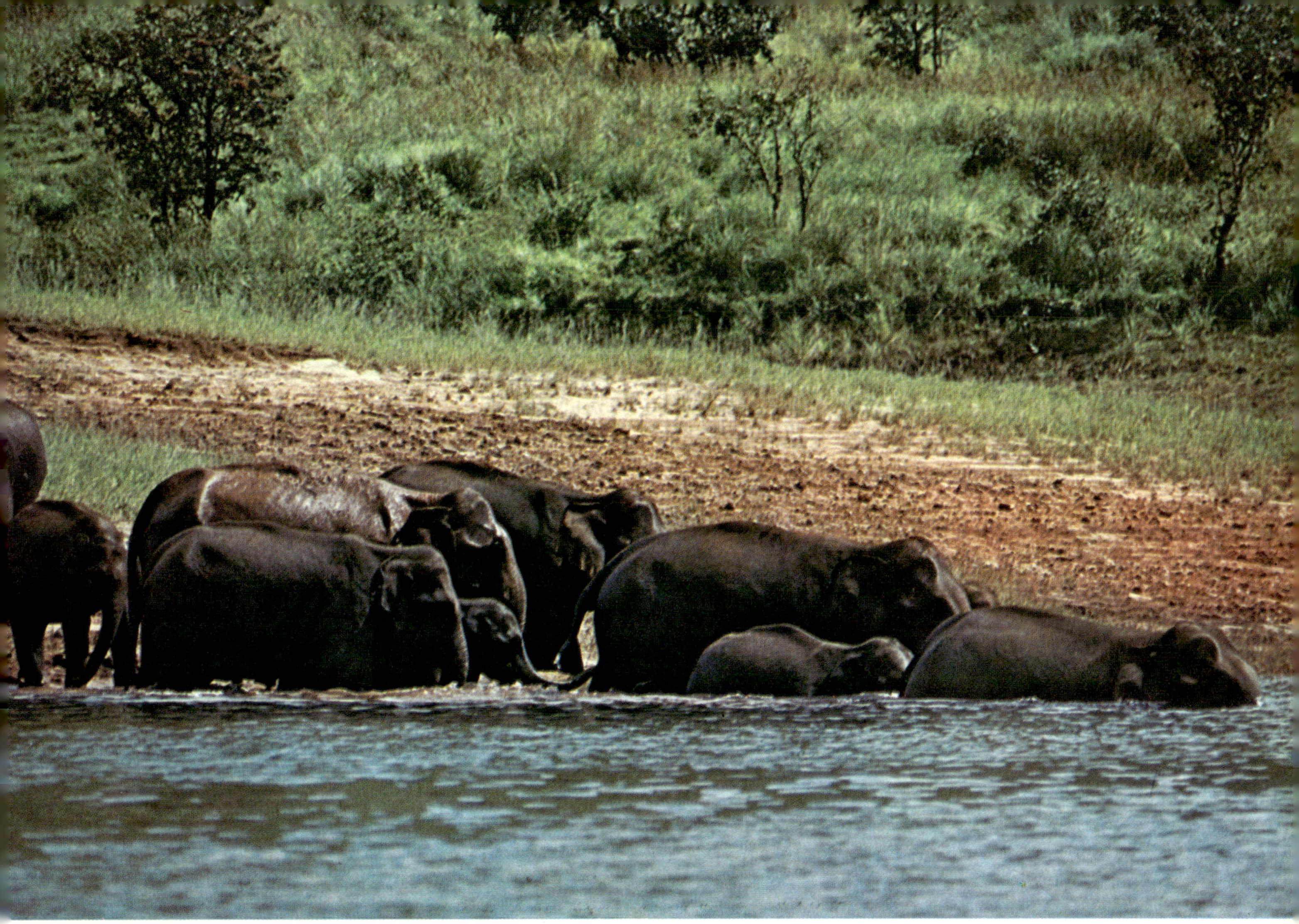

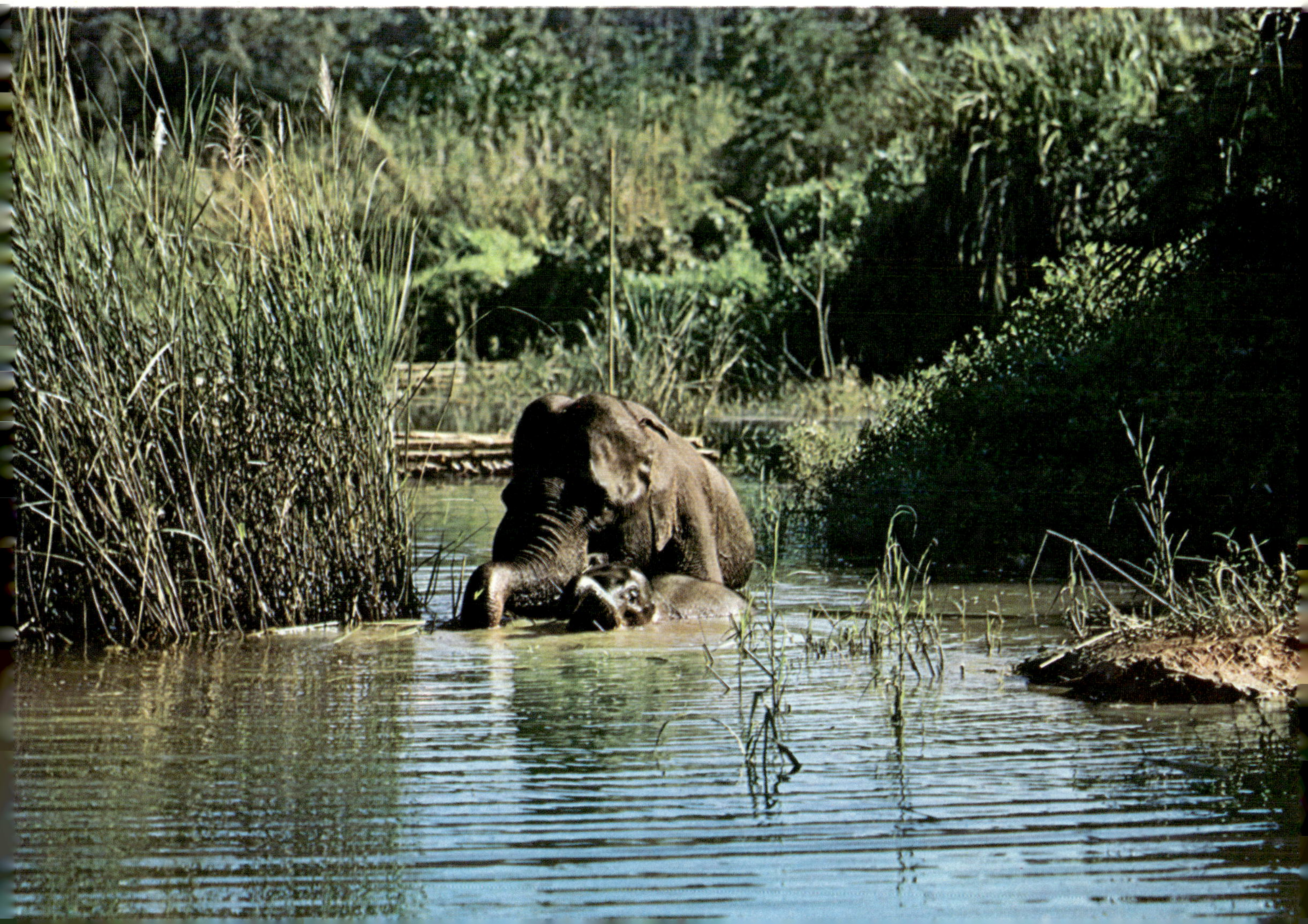

There are a number of anatomical differences between the African and the Asiatic elephant. The latter is smaller, its back is arched rather than hollowed, the trunk is smooth instead of ringed and ends in a single, not double, prehensile 'lip'. The Asiatic species has four nails on the hind feet as against three in the African species. Since it eats grass and soft vegetation the enamel ridges of the Asiatic elephant's molars are transverse, whereas those of the African elephant are lozenge-shaped, suitable for masticating tougher, fibrous vegetation.

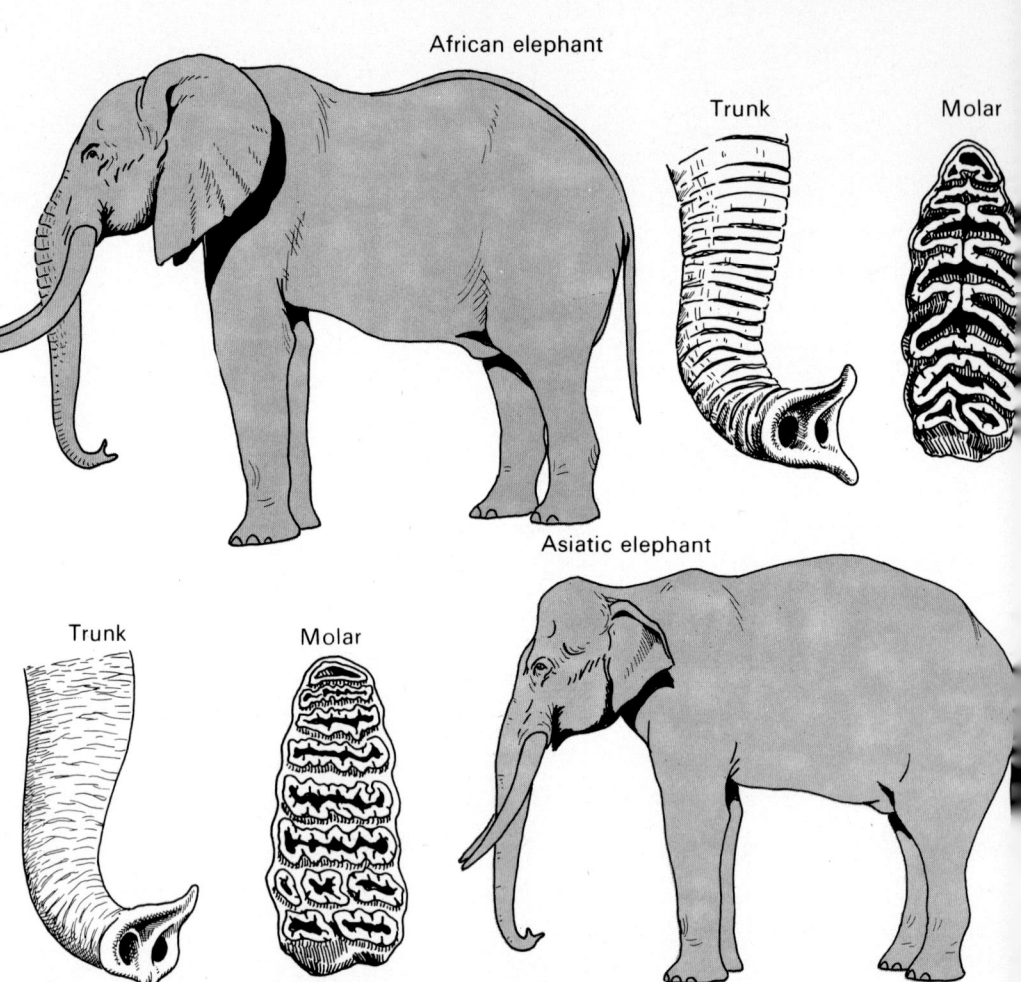

African elephant

Trunk   Molar

Asiatic elephant

Trunk   Molar

the newly imprisoned beasts. The presence of the domesticated animals gradually pacifies the wild elephants. When their angry trumpeting dies down it is safe enough for coolies to dash in and shackle them with stout ropes.

When it has abandoned its attempts to break free and has quietened down sufficiently, each elephant is given into the care of a driver or *mahout* who from now on will be inseparable from his charge. During the first couple of days of captivity the animal is kept tightly bound and deprived of food and water. After that the *mahout* approaches, gradually loosens the shackles, drives off the hordes of tormenting insects and offers food to the weakened and much more docile elephant. All his actions are gentle and deliberate and sometimes he will croon a simple repeated melody so that the animal gets accustomed to his presence and voice. It will not be long before the elephant permits the *mahout* to come close enough to stroke it and rub its back with handfuls of grass. Once it has accepted such treatment with good grace, the time has come for proper training to begin.

In the first instance the new pupil is accompanied wherever it goes by a couple of domesticated elephants who are already at their master's beck and call. Patient observation of their responses to the *mahout's* verbal instructions soon has the desired effect and by the end of two or three weeks the trainee is capable of understanding commands such as 'move', 'get up' and 'kneel down'. But it is a slow process which cannot be hurried. Several years may pass before the elephant is ready to become part of a work force.

Eventually the newly domesticated animal will obey about two dozen commands, the most important of which are to move forwards, stop, move backwards, turn, lift a foot, lie down, roll

over on one side, lift the trunk, hand over an object, enter the water, drink, spray its back and belly, push with the foot, push with the head, and go round or step over an obstacle.

Any animal that manages to assimilate this list of simple commands is capable of performing a variety of tasks and with experience will even do them without being asked.

Whereas the African elephant is essentially a creature of the savannah, the Asiatic species is a jungle-dweller. It too lives in herds whose social structure is similar to that of its African relative.

A matter of intelligence

The elephant has always been regarded as an animal of high intelligence. It is reputed to possess a remarkable memory and local folk traditions abound with stories of individual and group activities that, if true, can only be explained by the fact that the animal's mental powers are indeed of a superior order. Those who have lived and worked with Asiatic elephants both in the wild and in captivity need no convincing. The positive response of the animal to training procedures and the type of work it is capable of undertaking, often on its own initiative, would seem to support such claims. But scientists are cautious people and, very properly, hard to satisfy. Since memory and intelligence can nowadays be measured and evaluated in the laboratory, it is possible to call on science to confirm or refute the traditional opinions. The results of such experiments have to a large extent vindicated the earlier claims. The Asiatic elephant is in fact a pretty intelligent animal.

In order to test the legendary intellectual capacity of the elephant, a group of West German research scientists from the Münster Institute of Zoology, under the direction of Professor Bernhard Rensch, travelled to India to study a domesticated herd in its natural surroundings. Having completed their observations

One of the rarest animals in the world is the Javan rhinoceros, the last few dozen of which are protected in the beautiful Udjung Kulon nature reserve in the western part of Java. Because of its shy habits it has proved difficult to study the behaviour of this imperilled species.

Geographical distribution of the Javan rhinoceros (*Rhinoceros sondaicus*).

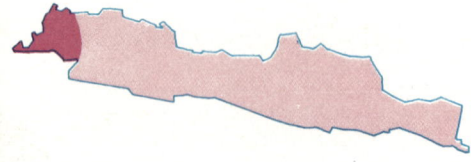

ASIATIC RHINOCEROSES

Class: Mammalia
Order: Perissodactyla
Family: Rhinocerotidae
Diet: herbivorous
Number of young: one

GREAT INDIAN RHINOCEROS
(*Rhinoceros unicornis*)

Length of head and body: up to 165 inches
(420 cm)
Height to shoulder: up to 83 inches (210 cm)
Weight: up to 4 tons

Largest of three Asiatic species, with one horn, at most 2 feet long. Skin deeply folded, with rounded knobs on flanks and upper legs; hair only on tips of ears and tail.

JAVAN RHINOCEROS
(*Rhinoceros sondaicus*)

Same general appearance as Indian rhinoceros but smaller. Skin smooth with deep folds.

SUMATRAN RHINOCEROS
(*Dicerorhinus sumatrensis*)

Length of head and body: 98–110 inches (250–280 cm)
Height to shoulder: 59 inches (150 cm)
Weight: up to 1 ton

Smallest living rhinoceros and only Asiatic species with two horns, rear one larger. No skin folds; short, hairy coat.

112

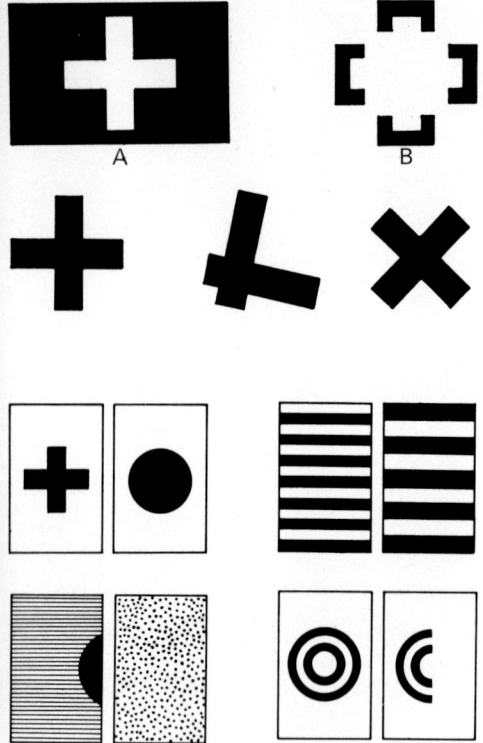

To test the intelligence of the Asiatic elephant a group of West German scientists conducted a number of tests with a five-year-old Indian female in the Münster Zoo. Two boxes marked respectively with a black cross and a black circle were placed in front of the animal, the former containing a titbit, the latter being empty. It took 330 attempts for the elephant to pick the correct box but later experiments in which the symbols were changed in various ways (as shown above) required much less time. Only when a white cross (A, B) on a black background was substituted for the original black cross on white did the animal have any difficulty in making the right choice. Tests were successfully repeated a year later, proving that claims concerning the elephant's remarkable memory are not exaggerated.

Facing page : Although there are still wild elephants in the jungle, particularly in Assam, most Indian elephants have been domesticated. Thanks to their prodigious strength and their high intelligence these animals are valuable aids in such activities as tree-felling and constructional work.

they then decided to try to verify their findings under artificial conditions–no simple task in view of the size of the animal and the restricted facilities offered even by the most fully equipped laboratory. Eventually they compromised by setting up their experiments in the Münster Zoo, using a five-year-old female Indian elephant as their test subject.

In one of these intelligence tests the elephant was offered the choice of two wooden boxes with cardboard lids, one of which had a black cross and the other a black circle painted on it. Inside the former was a reward. The test had to be repeated 330 times before the animal managed to recognise the box containing the prize, but once having done so she made no further mistakes. When the symbols were changed for the second experiment she got the right answer much more rapidly; and by the fourth test she needed only ten attempts to pick the correct box. After several weeks the animal was able to identify accurately twenty pairs of symbols, arranged in thirty different ways. Then the signs themselves were altered, though the basic outlines were retained. The elephant had little difficulty in picking out the box containing the titbit but was clearly perplexed when colours were reversed and a white cross appeared on a black ground. When auditory stimuli were substituted for visual signals the tests were equally conclusive.

The German scientists also tested the elephant's memory. One year after she had been given the final test with thirty patterns the subject scored 73 per cent of successes in 520 attempts, although the proportion of correct results went down to 67 per cent in one especially difficult test. There was no doubt, however, that this was evidence of an extraordinary capacity for remembering; and once again, when sounds were substituted for symbols the results reinforced this conclusion.

Java's Udjung Kulon nature reserve

The largest wildlife sanctuary in South-east Asia is the Udjung Kulon-Panailan nature reserve at the western tip of Java. More than 100,000 acres are almost completely covered by primary and secondary forest. The reserve is remarkable enough for its magnificent flora but its principal importance is that among the various animals protected in it are the Javan rhinoceros, a now-rare species that formerly roamed the island and the Asian mainland through Indo-China and probably as far as Assam, and the tiger. Both animals appear in the IUCN Red Data Book of endangered mammal species.

In addition to the massive rhino there are in the reserve bantengs, Timor deer, muntjacs, two races of wild boar, and chevrotains; and among the large population of carnivores, apart from the handful of tigers which are found nowhere else on the island, are panthers, wild dogs (dholes), fishing cats, civets, otters and mongooses. Primates include gibbons and langurs. Moreover, although ornithologists have hardly begun to count the multitudes of birds, they have already identified 230 different species.

The flora and fauna of Java–of such variety and interest–

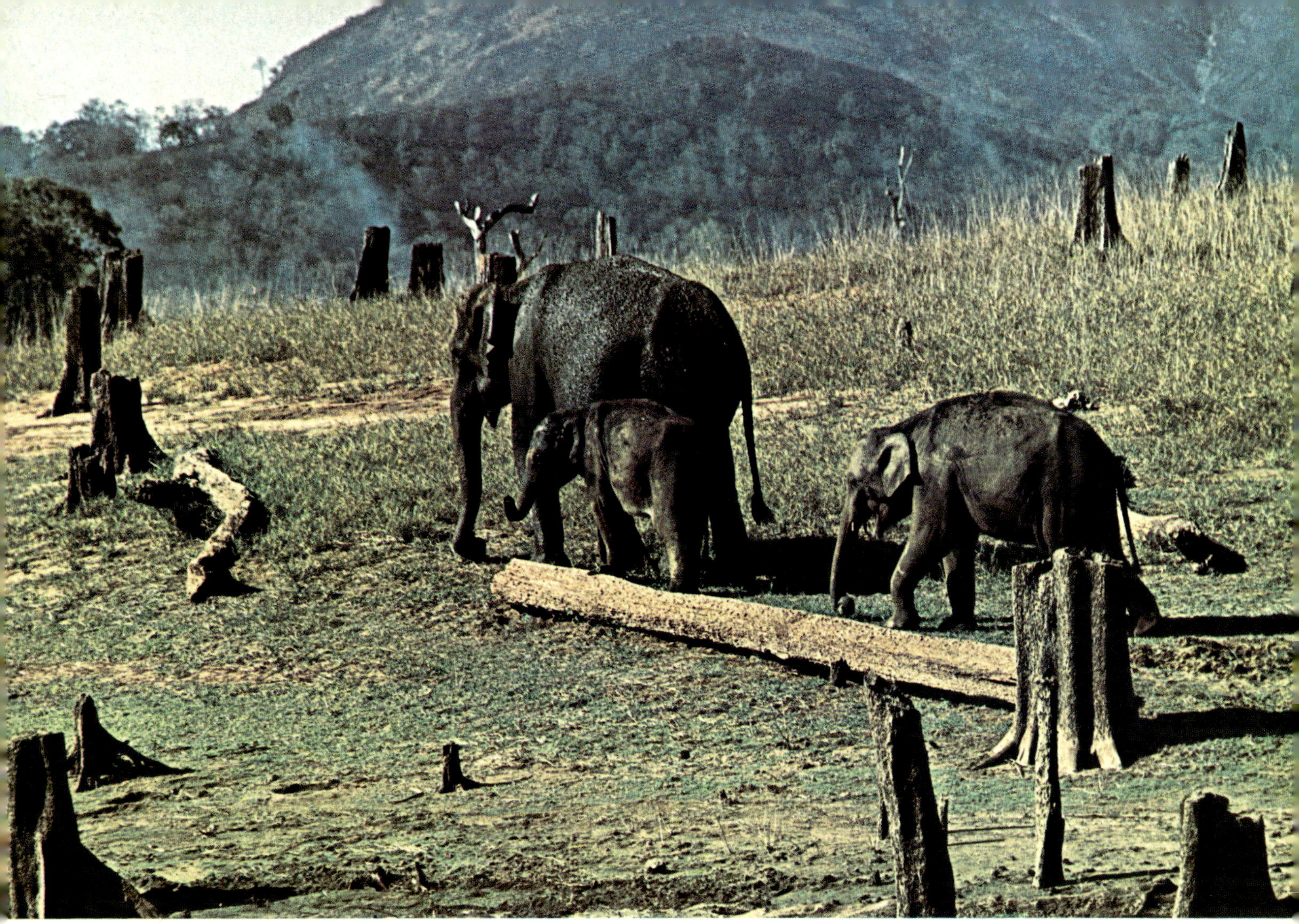

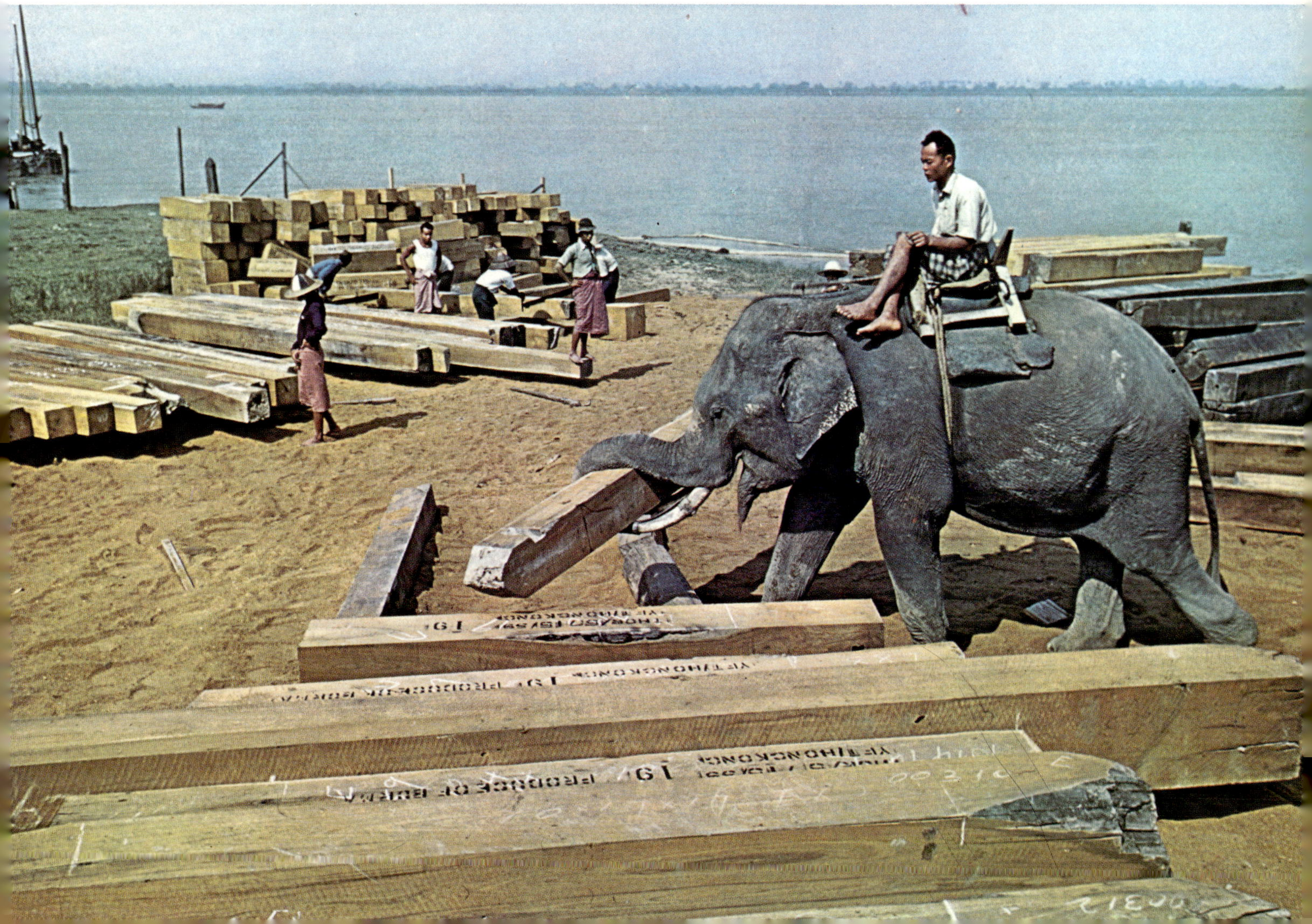

The footprints of the Javan rhinoceros are often the only means of determining its whereabouts and providing a rough guide to its present numbers.

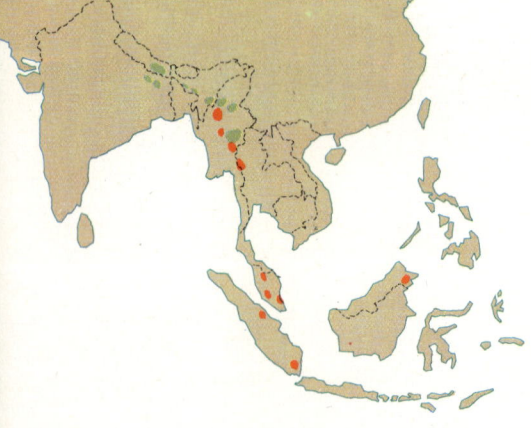

🟩 *Rhinoceros unicornis*
🟥 *Dicerorhinus sumatrensis*

Geographical distribution of the great Indian rhinoceros (*Rhinoceros unicornis*) and the Sumatran rhinoceros (*Dicerorhinus sumatrensis*).

Facing page : The fact that the great Indian rhinoceros and its two allied Asiatic species are today threatened with extinction is in large measure due to centuries of superstition. All have been hunted for their horns, hide and blood.

have been officially protected in this wonderful nature reserve since 1921. In its early years the regulations were not rigidly enforced but today the park is efficiently administered by a staff of forty. Botanists and zoologists have journeyed from all over the world and have published many articles on the plants and animals which make the reserve justly famous.

Rarest of the world's pachyderms

The most valuable animal in the Udjung Kulon reserve is indisputably the Javan rhinoceros. Two centuries ago the animal ranged freely across the island but today this is its last refuge and the only barrier between it and extinction.

The other species of Asian rhinoceros also give rise for concern although both are protected. The principal refuge of the great Indian rhinoceros, largest of the three species, is the Kaziranga reserve in Assam. The Sumatran rhinoceros, distinguished by its shaggy coat and two horns, is protected in that island's two largest wildlife reserves, the Gunung Loser and Sumatra-Selatan nature parks. Because it is so rare and so timid, however, the Javan rhinoceros is the least known. Unlike the related species the female either has only a very tiny horn or lacks one altogether.

The discovery of fossil remains dating from the Pleistocene indicates that the range of the Javan rhinoceros was originally vast, extending to the Punjab and Ceylon. It was reasonably common throughout South-east Asia a century and a half ago and at the beginning of the 1914–18 War it was still found in the Mekong river valley. By 1940 its range was confined to Sumatra and southern Java but later it vanished from the former island. Reports that there are still some Javan rhinoceroses in Thailand lack confirmation. The Udjung Kulon reserve is the only place in the world where it definitely exists today.

Like all wild animals, the Javan rhinoceros has suffered grievously as a result of man's invasion of its habitats. The rapid increase of the island's population over the past hundred years has been one cause of the species' steady decline, with agricultural land encroaching upon traditional forest domains. But a more important reason, which has equally affected the other two Asiatic species, is hunting and poaching–stimulated by the ancient, and utterly baseless, belief that the horn of a rhinoceros has aphrodisiac properties. Other parts of the animal's body have also been considered to have curative powers so that the hide, the blood and the dried bones are crushed into a powdery mixture to be sold by purveyors of drugs and medicines. Naturally these preparations have traditionally commanded very high prices and once formed part of the tribute annually offered to emperors or of wedding dowries in wealthy families.

Many of the surviving rhinoceroses sought refuge in the swamps of western Java and also on the slopes of volcanoes. Although neither of these areas could be described as typical habitats (it is essentially an animal of low-lying regions) tracks indicated that some individuals regularly made their way up to a considerable altitude; and they may have been attracted to the

Great Indian rhinoceros
(*Rhinoceros unicornis*)

Javan rhinoceros
(*Rhinoceros sondaicus*)

Sumatran rhinoceros
(*Dicerorhinus sumatrensis*)

Of the three species of Asiatic rhinoceros, the Javan is the most gravely imperilled, with fewer than 50 animals now alive. The smaller Sumatran species is only a little better off, with less than 170 survivors.

Facing page : The largest herd of Indian rhinoceroses is today protected in the Kaziranga reserve in Assam. This region, flooded annually by the waters of the Brahmaputra, abounds in bogs, marshes and high grass—essential features of the animals' welfare.

swampy coastal districts by the sulphurous nature of the soil. The mountain population was the first to vanish, its escape routes cut off when the land below was gradually given over to farming.

Today the remnants of the island population are usually to be found in remote, well forested lowland regions, abandoning the interior only occasionally to feed on shrubs and shoots on the fringes of the jungle.

How many survivors?

It is almost impossible to determine the precise number of Javan rhinoceroses in the Udjung Kulon reserve and recent estimates can only be regarded as approximate. In 1964 Lee Talbot put the total at between 36 and 56 animals; three years later Schenkel counted 24 or 25 and the following year 25 or 26, of which more than a dozen were under two years old. The principal difficulty is that the plant cover in those areas frequented by the animals is more or less impenetrable so that counts cannot be based on actual first-hand sightings. A zoologist therefore has to rely on tracks, prints, droppings and noises—indirect evidence which is bound to result in a fair margin of error.

If it is hard enough to arrive at a near-accurate figure for the total population, it is practically impossible to break the numbers down by age and sex—information which is essential for a prognosis of the species' future prospects. One age survey, however, has been attempted by the naturalist A. Hoogerwerf, who based his report on a comparison of footprints in the reserve. His conclusion was that the majority of rhinoceroses in Udjung Kulon were mature adults, including a small number of very old animals. The large proportion of individuals of reproductive age seemed to indicate that the population was healthy and thriving but the relatively small number of young was rather disturbing when projected into the future. The causes of this imbalance are still being investigated.

Although tracks may be indicative of age they cannot identify sex and the proportion of males to females can only be roughly gauged by direct observation. According to the records of the reserve's wardens there were, between 1940 and 1944, 78 sightings of males and only 22 of females. Previously, between 1937 and 1942, Hoogerwerf had reported a similar ratio, with 22 sightings of males as against only eight of females. The same naturalist, in a two-year survey ending in 1954, reported far fewer appearances but still in more or less the same proportions—eight to two—these being the cases in which the sex could be definitely determined. These statistics clearly indicate a prevalence of males in a ratio of three or four to one, but the reason is unknown. Certainly it cannot be a consequence of hunting because males are killed more frequently than females. Dr Groves believes that the imbalance may spring from consanguinity within a small group, leading to a special type of genetic structure, but this has not been tested or confirmed.

Very little is in fact known about the reproductive behaviour of the species. The rhinos appear to couple at any time of year,

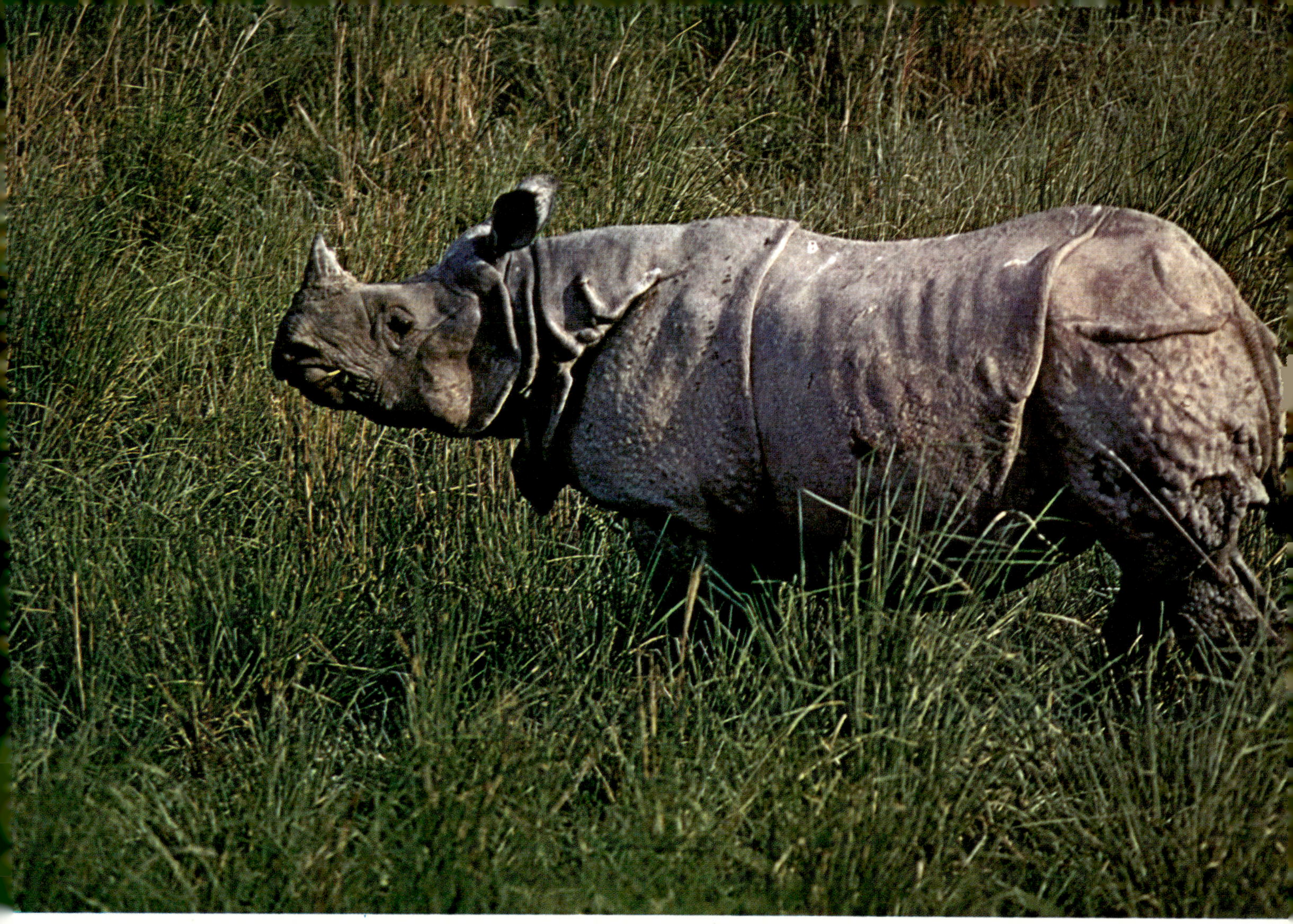

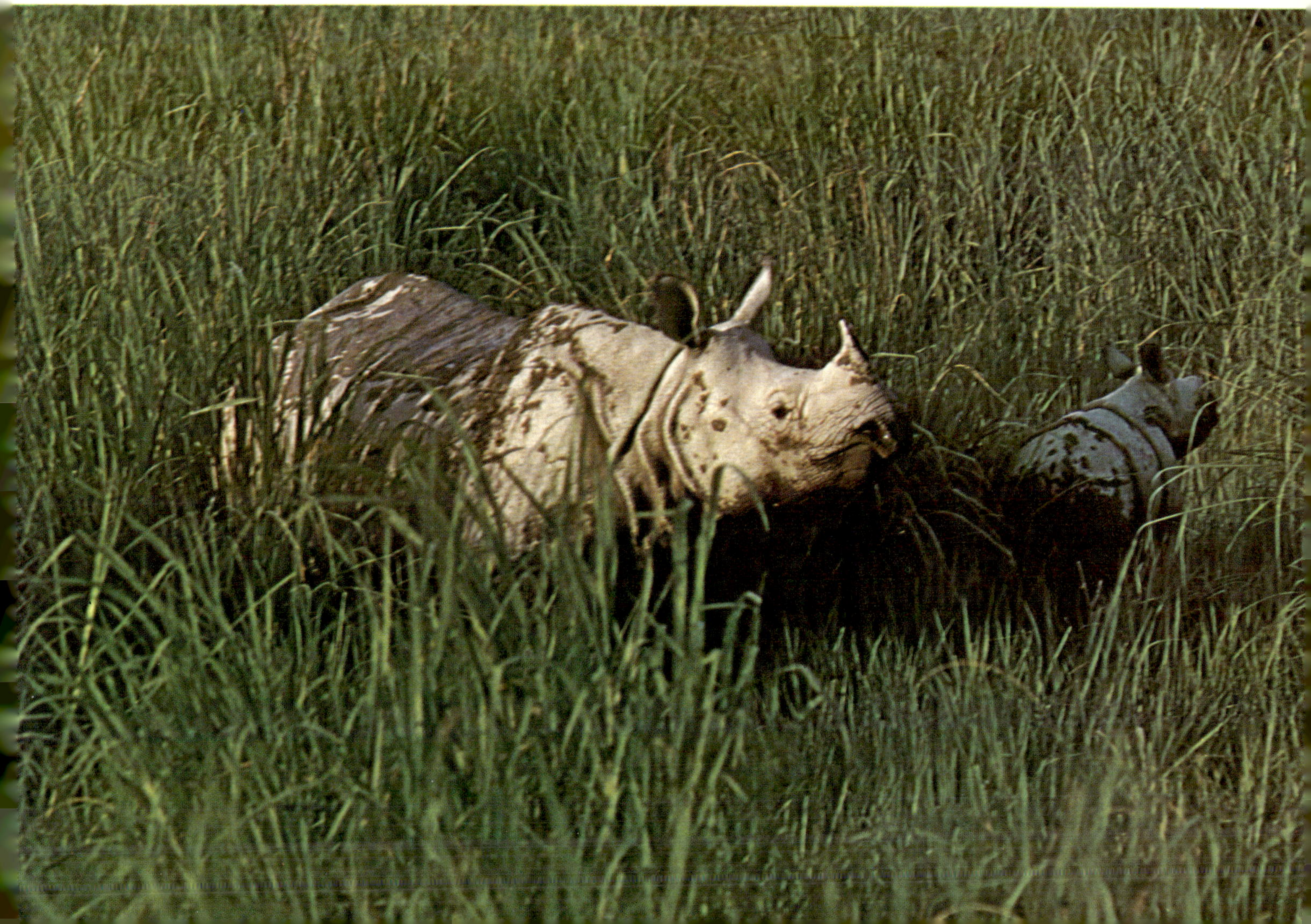

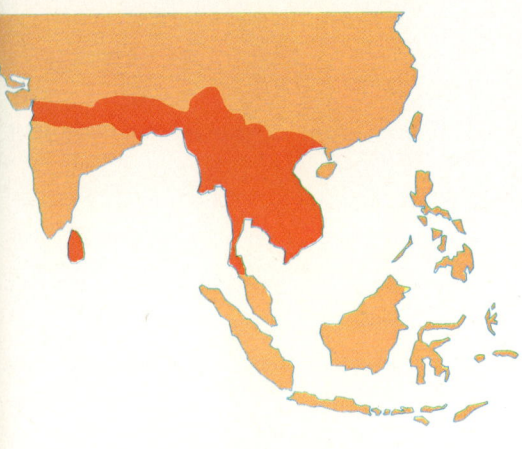

Geographical distribution of the axis or Indian spotted deer (*Axis axis*).

Facing page : With its russet, white-spotted coat, the axis is the most handsome and also the commonest deer of the Oriental region.

and despite the solitary habits of the species pairs of animals may be seen at any season. It is probable that, prior to mating, they pursue each other through the forest and that aggressive action may be the prelude to more amicable relations. Gestation lasts seventeen months and the mother suckles her single baby for two years. This may be the reason why females give birth only once every four or five years.

To live in peace

The Javan rhinoceros is active both by night and day but clearly prefers to feed under cover of darkness. According to Hoogerwerf, it eats more than seventy plant species but his list is probably not exhaustive. The principal plants are *Laportea stimulans, Laucaena leucocephala, Caripa papaya* and various plants of the genus *Musa*. One of the plants for which the animal has a particular craving is a parasitic species which can only be eaten when growing on the lower branches of a tree. Under no circumstances will the rhinoceros eat grass.

All the sites regularly frequented by the animal have one feature in common – puddles where it can enjoy a mud bath. In Udjung Kulon these are usually about 20 feet long and from 10–15 feet wide, their depths seldom exceeding 3 feet, the bottom two-thirds of which consists of slime and the rest of rain water. In dry weather the water, of course, evaporates quickly and the puddles are abandoned for neighbouring river banks. Yet despite their abundance these bathing sites are difficult to locate for they are usually situated in the remotest depths of the forest, concealed by thickets of bamboo and other plants. The surest guides to their position are the tracks leading to them as well as the traces of mud and dung on the trunks of nearby trees. From a short distance away one can smell these sites for even if the rhino has gone it leaves behind strong effluvia.

If their bathing places are effectively concealed by vegetation the animals may visit them either during the day or at night. Some pools may be deserted for long periods whereas others will be used perhaps by two animals at the same time. Although eye-witness accounts of such episodes are few and far between, it seems that the animals share their mud bath quite amicably, either standing up, resting on one side or rolling about.

Other indications of the presence of these rare animals are the heaps of dung that are often found along a path, in an open clearing or at the top of a hill, sometimes close to one another, sometimes a good distance apart.

The Javan rhinoceros has almost nothing to fear from predators. Only the tiger is occasionally likely to kill a calf, but the carnivore has itself become so rare that it has enough trouble finding food without risking a confrontation with an angry adult rhino which will fight fiercely in self-defence. Poaching is the last real threat to the species, but the government of Indonesia, with the backing of the World Wildlife Fund and the support of eminent scientists, is tackling this problem energetically, well aware of its responsibility of preserving in Udjung Kulon one of the world's most precious animals.

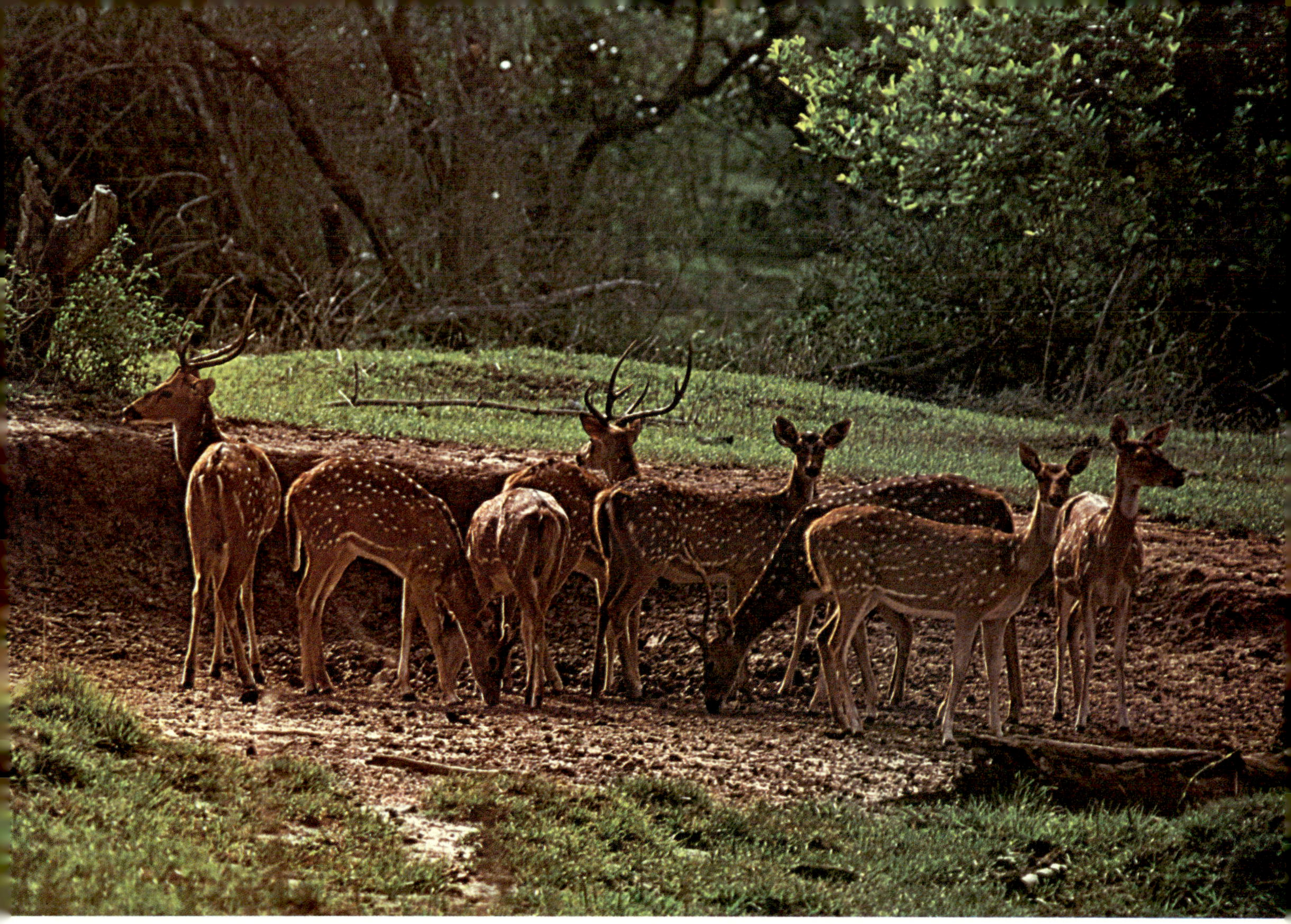

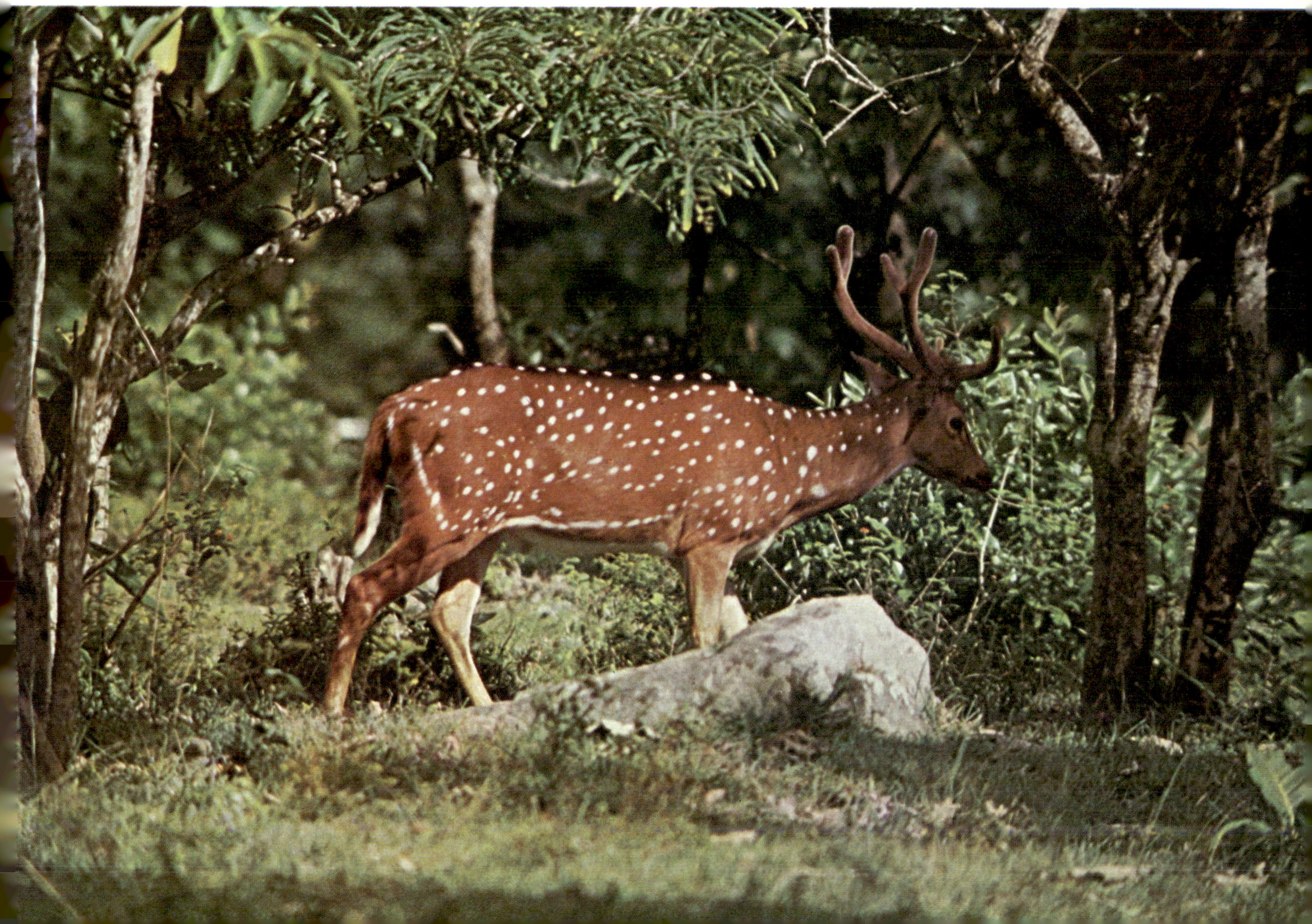

Other imperilled rhinos

In its heyday the great Indian rhinoceros ranged freely over northern India through Nepal and the Himalayan region southward to Burma. Today, because of continued persecution, it is protected in a number of wildlife reserves.

In 1966 zoologists estimated that there were 740 animals still living, of which 175 were to be found in Nepal and the rest in India. Of the latter population 400 were cared for in the 100,000-acre Kaziranga nature reserve in Assam which is situated on the left bank of the Brahmaputra and bounded to the south by the Mikir hills. The reserve is fringed by a 40,000-acre security belt where hunting is prohibited. Every year the plain is flooded, forming an enormous area of swampland, above which only the summits of tall trees emerge. The sole means of transport through the region is by elephant.

Although little frequented by man, this marshy expanse with its tall grass is the ideal habitat for the rhinoceros. This animal, suspicious by nature and so fierce when aroused that it is fully a match for an adult elephant, was in days gone by domesticated quite easily and used by eastern potentates for military purposes. The enormous beasts would be placed in the front ranks of an army and spurred on to charge the enemy with three-pronged spears affixed to their horns. But their value was strictly limited and they subsequently returned to the wild.

The Sumatran rhinoceros is the smallest of the five living species, seldom measuring more than 5 feet tall and weighing up to one ton. Other species of the same genus, *Dicerorhinus*, once had a wide distribution and in western Europe were hunted by prehistoric man, as shown by cave paintings at Lascaux and other early sites. Its population too has suffered a rapid decline as a result of hunting and poaching. Today, although its range includes Thailand, Cambodia, Burma, Malaysia, Borneo and Sumatra, there are only between 100 and 170 animals left. The only hope for the future of the species is in reserves, for experience has shown that they do not breed in captivity. Two such sanctuaries have been established in Sumatra.

The false yeti

In 1953 a team of Indian mountaineers visited the Pangboche monastery in Nepal and examined a much-prized exhibit—a cap made of skin purporting to be the scalp of the legendary yeti or 'abominable snowman'. After photographing it they were allowed to remove a single hair which they sent, together with the photographs, to an American zoologist for identification. Comparison with the hairs of a langur, a bear and a takin all proved negative. That same year the British naturalist Charles Stonor, examining the same evidence, pronounced it to be from the head of an anthropoid ape. The mystery was only cleared up in 1957 when a Nepalese sherpa received permission to bring the whole cap to the British Museum, where it was found to be nothing but a handful of hair from one of the several subspecies of the serow (genus *Capricornis*)—a goat-antelope ranging from

The axis or Indian spotted deer is hunted by the wild dog and to a lesser extent by the tiger and the leopard. It is one of the most beautiful of deer.

Facing page : Muntjacs, or barking deer, are solitary animals, frequently found in forest regions of the Himalayas, at altitudes of up to 10,000 feet.

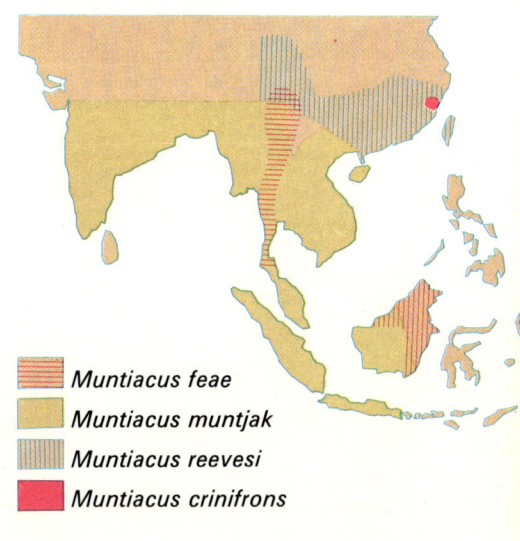

Muntiacus feae
Muntiacus muntjak
Muntiacus reevesi
Muntiacus crinifrons

Geographical distribution of muntjacs.

Above and facing page : The banteng, smaller than the gaur, has been domesticated but is still found wild in the Asiatic jungle.

The antlers of the muntjac are short, with few branches, assuming a characteristic V-shape as they sprout from bony, hair-covered growths on the forehead. The name rib-faced deer is sometimes applied to the species, referring to the deep facial glands under the eyes.

the Himalayas to South-east Asia, with another species in Japan.

Serows are characterised by long, drooping ears, short horns and (in some forms) a mane on the neck. Their habitats are either precipitous rocks close to streams bordered by bushes and bamboos, or the depths of the tropical forest. In summer they tend to wander up to higher ground. Solitary by habit, with a strong territorial instinct, these animals feed at dawn and at dusk, spending most of the day hidden among vegetation. Although once hunted by tigers and leopards, they now have little to fear from these comparatively rare predators; but their numbers are to some extent controlled by wild dogs and birds of prey which attack young, sick and aged individuals.

The Indian spotted deer

Of the many deer of the Oriental region the species that is most abundant is the axis or Indian spotted deer (*Axis axis*), incontestably one of the most beautiful of the world's deer. Not that the antlers are particularly imposing (as is the case with various species of European deer) for in fact these are slender, with not more than three tines. The striking features of the axis are the elegant lines of the body and the handsome white-spotted brown coat. The spots are in the form of discontinuous lines arranged horizontally along the back, the flanks and the upper part of the hind legs. There is a lighter patch around the throat and the underparts and insides of the legs are white.

This animal is an inhabitant of plains, low wooded hills, bamboo thickets and forest areas within easy reach of streams and rivers, for like other deer it is a good swimmer and plunging into water is an effective way of escaping from predators.

Axis herds are mixed, except during the season when the males lose their antlers. Less nocturnal than other members of the deer family, the animals often feed in broad daylight. Their diet is extremely varied and includes leaves, shoots and grass.

Tigers and leopards are no longer the principal enemies of the axis and today they are chiefly attacked by wild dogs. But their numbers cannot be effectively regulated in this way. Thus in the Andaman Islands, where there are few carnivores, the authorities decided to reintroduce the leopard in an attempt to keep down the axis population. But bearing in mind the risk of an introduced species upsetting the local balance of nature, only female leopards were imported. Once they had brought numbers down to an acceptable level they were returned to their former haunts.

The barking deer

Among other deer of the Oriental region are the muntjacs (genus *Muntiacus*) alternatively known as barking deer, because of the sharp call that they emit when alarmed, or as rib-faced deer, from their prominent slit-like facial glands. The unusual feature of the muntjacs is that in addition to short antlers they also have tusk-like canines protruding from the upper jaw. The animals are comparatively small—seldom standing more than 20 inches high—and the hindquarters are slightly raised.

Facing page : The gaur, largest of Asiatic cattle, has little to fear from predators; weighing up to one ton and standing more than 6 feet high, it is a match even for a tiger or a leopard. These animals live for almost the whole year in separate herds, males in one group, females and calves in the other.

JUNGLE CATTLE

Class: Mammalia
Order: Artiodactyla
Family: Bovidae

GAUR
(Bos (Bibos) gaurus)

Length of head and body: 102–130 inches (260–330 cm)
Length of tail: up to 33½ inches (85 cm)
Height to shoulder: 71–79 inches (180–200 cm)
Weight: up to one ton
Gestation: 270–280 days
Number of young: one, rarely two

Largest of Asiatic wild cattle. Black pelage, except for whitish lower legs. Male's horns measure up to 24 inches.

BANTENG
(Bos (Bibos) banteng)

Length of head and body: 71–79 inches (180–200 cm)
Height to shoulder: 51–67 inches (130–170 cm)
Weight: 1100–1980 lb (500–900 kg)
Gestation: 270–280 days
Number of young: one

Male's coat varies from russet to black; lower legs and patch on rump white.

KOUPREY
(Bos (Bibos) sauveli)

Length of head and body: up to 87 inches (220 cm)
Height to shoulder: 75 inches (190 cm)
Weight: 1980 lb (900 kg)
Gestation: about 270 days
Number of young: one

Last large mammal to be discovered by zoologists. Grey pelage with white marks on hindquarters, feet and shoulders.

Habitats range from tropical forest to highland areas with dense tree cover. In some parts of the Himalayan region muntjacs are frequently found browsing in woodland districts as high as 10,000 feet.

Solitary by habit, muntjacs feed only at dawn and at dusk. Although the half-light affords a measure of security the animals display extreme caution when venturing into the open. Food consists of grass, leaves and branches.

Despite the fact that they normally live alone, muntjacs show a sense of solidarity when danger looms, warning one another with their characteristic barking cries.

Cattle of the tropical forest

Three large species of cattle are to be found in the jungles of the Oriental region. Most massive of them is the gaur (*Bos (Bibos) gaurus*) which may stand up to 6 feet at the shoulder and weigh almost a ton. Apart from the lower legs, which are whitish, the pelage is black and the head is adorned with a pair of magnificent, inward-curving horns. As a consequence of its size and strength the gaur has few natural enemies and is a match for any tiger or leopard.

These heavily built animals live for most of the year in small segregated herds, but in spring the bulls seek out the cows, mating with those that are receptive and then taking their leave. These continual comings and goings make it difficult for any stable system of social ranking to be established and there are frequent fights among the bulls for dominance. Although such confrontations are often spectacular, they are not all that violent, for as with other Bovidae there is a recognised dissuasive ritual in which the protagonists stand side by side, display their horns and lash out with their hooves, without any serious injuries being inflicted. The weaker animal eventually yields ground to his conqueror.

Towards the end of May or during the first half of June the bulls gradually part company with the cows, although mating sometimes takes place later in the year. Whether solitary or as part of the herd, the gaur has moderate territorial instincts, although not so jealously protective as other species, animals from different herds often being seen grazing together.

The banteng (*Bos (Bibos) banteng*) is found in Burma, Vietnam, Java and Borneo. It is smaller than the gaur, with a dark coat, except for patches of white on the lower legs and rump. Its behaviour is similar to that of the gaur and like the latter it has been successfully domesticated.

The kouprey (*Bos (Bibos) sauveli*) was discovered in 1937 by a French veterinarian, Dr Sauvel, during an expedition to Cambodia. At first the animal was assumed to be a cross between a gaur and a banteng but examination of a number of individuals killed by hunters showed sufficient differences in appearance, horn structure, coat colour, and other anatomical features to justify its classification as a separate species. Although there is not universal agreement on this, the kouprey may be the ancestor of the zebu or Indian domestic cattle.

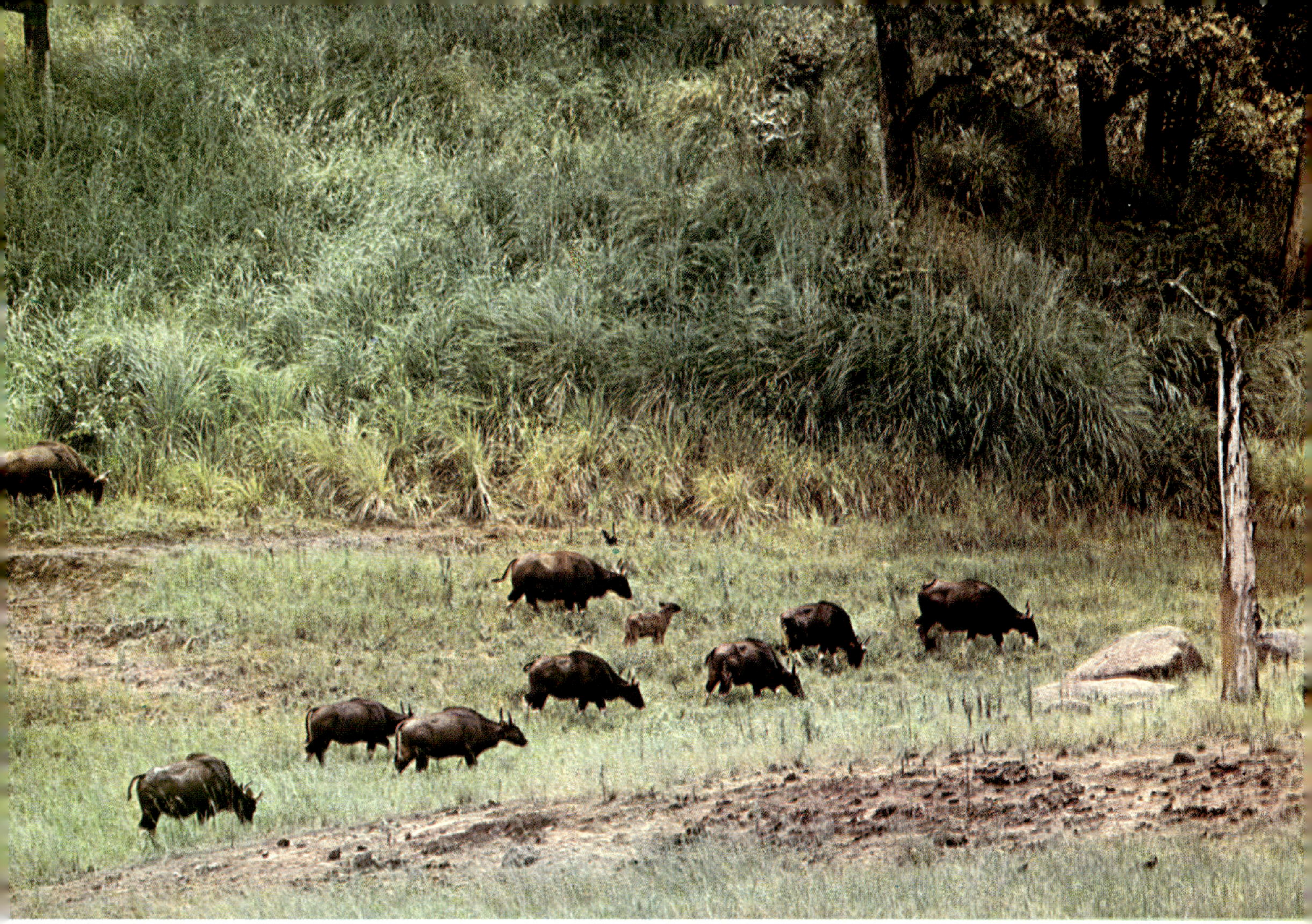

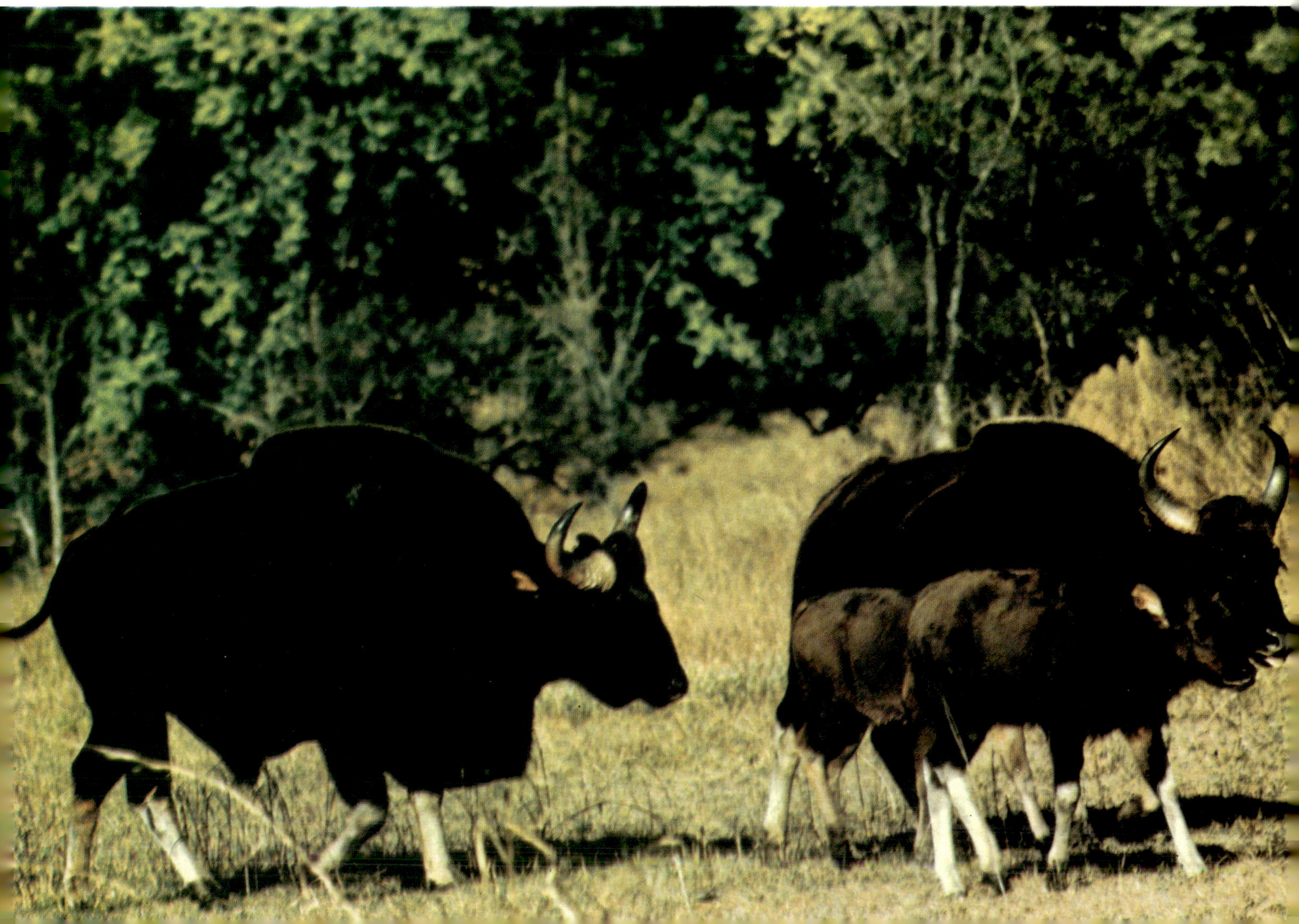

CHAPTER 87

Silent killers of the trees and underbrush

In the primary tropical forest most of the predators spend their time in the trees, for the simple reason that this is where they can most conveniently find their food. Thus the sun bear, one of the two characteristic species of the Asiatic jungle, finds life among the branches much more rewarding than on the ground below; and the other, the sloth bear, although rather less arboreal by habit, feeds exclusively on fruit at certain times of the year. The trees of the tropical forest offer an enormous range of food possibilities. It is not surprising, therefore, that a wide diet is another notable feature of most jungle carnivores. It is true that many hunt rodents, nestlings and insects, but in addition— like the small mammals and birds on which they prey—they feed liberally on eggs, fruit, honey, shoots, leaves and even roots.

One exception to this rule is the tiger which is not, in fact, a typical forest hunter but essentially an animal of the wood-land glades, the savannahs and the steppes—exposed terrain where it stalks and kills large herbivores. The leopard, on the other hand, is much more at home in the jungle, frequently including among its prey tree-dwellers such as monkeys and birds. Deer, however, are the favourite victims of the leopard. In some parts of India a leopard's territory may occasionally overlap that of a tiger but the two felines will normally concentrate on different species, the leopard going after axis, the tiger pre-ferring to track the sambar. The spotted coat of the leopard is of course an advantage for forest hunting, but patterns vary considerably, even within the same region. Some animals are completely black, others so pale as to appear almost white.

Most of the jungle's predators, however, are much more modest in size. The Asiatic pangolins or scaly anteaters, for

Facing page : Wild cats roam freely through the jungles of the Oriental region, although naturalists have had few opportunities to study their behaviour. Some are restricted to particular mainland or island habitats but the leopard cat, seen here, is one of the commoner species, with a fairly broad range.

Like the majority of jungle predators, the Indian python has arboreal habits, feeding on both mammals and birds.

example, look much like their African relatives and have similar habits. The Chinese have traditionally attributed medicinal virtues to the scales and for that reason the animals have always been relentlessly hunted. Incidentally, pangolins are amazingly powerful, as is attested by one reliable story of a peasant in Ceylon who, thinking he had killed one of these animals, wound it round his neck to carry it home as a trophy. The pangolin, however, was merely stunned and on regaining consciousness instinctively assumed its curled-up defensive posture, thereby strangling the unfortunate peasant.

Most of the snakes of the Asiatic tropical forest are likewise tree-dwellers. They have almost the same hunting techniques and ways of life as their African relatives, except for the paradise tree snake, which has no counterpart. The reticulated python, largest in the world, lives in India and the Malay peninsula, but the smaller Indian python, which is under 20 feet long, is better known. Both snakes feed mainly on birds and small or medium-sized mammals, but are capable of consuming larger prey. The stomach of one 17-foot Indian python, for example, was found to contain the remains of a leopard.

The tree-climbing sun bear

Apart from the polar bear, all members of the family Ursidae are omnivores, their varied diet consisting of vertebrates, insects, plants and other substances. Although the Malayan or sun bear (*Helarctos malayanus*) and the sloth bear (*Melursus ursinus*) are no exceptions to the rule, both have their own specialities. Thus the sun bear has acquired tree-climbing skills, enabling it to feed on delicacies such as wasps, insect larvae and honey; and the sloth bear, operating more frequently at ground level, displays anatomical modifications which facilitate its quest for ants and termites.

The sun bear is the smallest representative of the bear family. Its alternative common name is derived from the orange-yellow, crescent-shaped mark on its chest. An inhabitant of southern China, the South-east Asian mainland, Sumatra and Borneo, the bear spends the greater part of the day in trees, climbing with extraordinary agility, thanks to its enormous feet with their long, curved, pointed claws. At dusk it sets out on a methodical exploration of its hunting grounds, searching the vicinity for rodents, lizards and other small reptiles, eggs, fledglings, fruit, honey, snails and termites (the last being mopped up with its long extensible tongue).

The sun bear is, for various reasons, an elusive animal and consequently difficult to study in the wild. Apart from the fact that it is a comparatively rare species, inhabiting remote and inaccessible areas of forest, it is extremely cautious by nature, passing the day concealed and almost invisible in the foliage. It is hardly surprising that very little is known of its normal behaviour apart from the fact that, unlike its relatives of cold and temperate climes, it does not need to reduce the level of its activities in winter.

The naturalist James Alexander Hislop once adopted a sun bear cub, aged approximately two weeks, which had for some reason been abandoned by its mother. Bertie, as he came to be called, was at that stage still unable to see, moving about extremely clumsily. But after a while he learned how to take food from a bottle and in due course accepted such titbits as honey cakes and jam tarts from his master. Bertie slept on his stomach, hands over his head. Before he went to sleep he would lick one paw, just as a baby sucks its thumb – a habit, incidentally, which is common to the brown bear as well. The cub was given back his liberty when he was about six months old but was never confident enough to go for more than forty-eight hours consecutively without wandering back to visit the man who had saved his life. By this time he was capable of building a rudimentary nest in a tree and had discovered how to root out ants and larvae from rotted trunks, spending hours ripping the decaying wood with his sharp claws. At the age of two years Bertie weighed about 120 lb but it was quite clear that he would never accommodate properly to a fully independent life in the wild, for even now he would regularly seek out Hislop's company. Because the bear was by this time much too large to be cared for by one person with many responsibilities, the naturalist decided that he had no other alternative but to place Bertie in a zoo.

Hardly anything is known about the reproductive habits of this species. The female apparently gives birth to two cubs at more or less any time of the year, choosing a suitably well hidden site amongst thick ground vegetation. Adults are thought to pair for life but this has not been definitely confirmed.

JUNGLE BEARS

Class: Mammalia
Order: Carnivora
Family: Ursidae
Diet: ants and termites, small vertebrates, honey, fruit, etc
Gestation: 7 months
Number of young: 2

MALAYAN OR SUN BEAR
(*Helarctos malayanus*)

Total length: 49 inches (125 cm)
Height to shoulder: 23½ inches (60 cm)
Weight: 155–220 lb (70–100 kg)

Smallest of all bears. Black coat, shorter than that of sloth bear; yellow or orange patch on chest. Small, rounded ears; short muzzle; long, curved, pointed claws.

SLOTH BEAR
(*Melursus ursinus*)

Total length: 71 inches (180 cm)
Height to shoulder: 35½ inches (90 cm)
Weight: up to 265 lb (120 kg)

Black or dark brown coat with white V-shaped patch on chest. Very long hair on neck and between shoulders. Large muzzle. Powerful claws up to 3 inches long. Female smaller but with longer hair.

The pangolin or scaly anteater devours ants, termites and other insects, scraping away at bark and trapping tiny victims with its long, sticky tongue. If alarmed it curls up into a ball.

The sun bear (*top*) is the smallest member of its family, climbing trees and rocks with the aid of powerful claws. The sloth bear (*bottom and facing page*) feeds greedily on ants and termites, sucking them up with its mobile lips. Although reputed to be savage, it will only charge if surrounded.

The termite-eating sloth bear

The sloth bear, an inhabitant of lowland forest regions from the southern foothills of the Himalayas through India and across to Ceylon, has the local reputation of being as dangerous, when alarmed, as an enraged elephant. Yet it is traditionally said to flee whenever confronted by a man. Both versions contain a measure of truth for if the animal finds itself trapped it will have no hesitation in charging at its tormentor. The main reason why a sloth bear will not take the easy way out and try, when possible, to avoid a confrontation with an enemy is that its senses of hearing and smell are comparatively mediocre. As often as not, therefore, it has no time to beat a safe retreat and has no choice but to face its foe.

As a rule, however, the sloth bear keeps well away from human habitations and because of its retiring habits naturalists have had few opportunities to study it. As is the case with so many little known species, a wealth of legend has sprung up concerning the animal. It is worth pointing out, by the way, that its very name arose from an error of observation. The 18th-century zoologist George Shaw had heard repeated rumours that the animal was in the habit of hanging upside-down from branches by its claws, often for hours on end; after examining its long, 3-inch nails as well as a skin, he elected to classify it as one of the Bradypodidae (sloths) in the order Edentata. Although subsequent investigation showed that this was a misnomer, the name stuck. In fact, the animal is not significantly lazier than any other nocturnal predator. It is true that the bear's movements are normally slow and measured, but if circumstances so dictate it can run at astonishingly high speed.

The sloth bear is markedly less arboreal by habit than the sun bear, finding most of its food on the ground, generally after dusk and before dawn. Eating habits vary according to the season, but ants and termites are always special favourites. Unlike the sun bear, which licks up these creatures with the tongue, the sloth bear uses its large lips for the purpose, sucking its tiny victims up as if through a tube. The noise which it makes in the process can be heard hundreds of yards away. After a fall of rain, when the soil comprising a termitarium is sodden and the bark of rotting trunks suitably soft, the bear will scrape away diligently with its claws and feast on such insects, together with beetles, snails and rodents; but when the ground is hard and parched it will climb trees to feed on fruit.

Although in Ceylon the bears have been observed mating at all times of the year, in India this seems to be confined to the month of June. Two cubs are born after seven months' gestation and if circumstances permit the mother retires for the purpose to a cavity among rocks. Failing this, she conceals herself under branches which she shapes into a form of nest.

For the first two or three months the cubs do not leave the mother's side, clambering on her back and gripping the long hairs of her shoulder whenever she decides to go hunting. These family groups do not normally break up until the cubs are two or three years old.

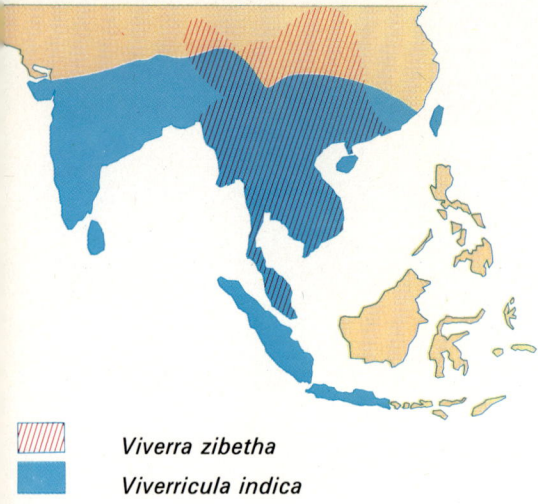

Geographical distribution of Asiatic civets.

ASIATIC CIVETS

Class: Mammalia
Order: Carnivora
Family: Viverridae
Diet: flesh and insects, fruit and other vegetable
   matter
Number of young: 2–3

LARGE INDIAN CIVET
(*Viverra zibetha*)

Total length: 45½–49 inches (115–125 cm)
Length of tail: 16–19 inches (40–48 cm)
Weight: 15½–24 lb (7–11 kg)

Light or dark grey pelage, with black marks
and stripes. Long hair forms a kind of mane
along the back. Neck and throat have black and
white stripes. Tail very thick, broadening at base.
Feet short and slender.

SMALL INDIAN CIVET
(*Viverricula indica*)

Total length: 29½–42 inches (75–106 cm)
Length of tail: 12–16 inches (30–40 cm)
Weight: 4½–8¾ lb (2–4 kg)

Similar to large Indian civet but much smaller.
Brown or greyish pelage with small black marks.
Seven or eight broad black rings on tail. Prominent
black and white lines on throat. Eyes dark, head
rounded. Feet black.

Facing page : The palm civets are among
the most secretive and little-known
hunters of the tropical forest. The striped
face of certain species may serve to warn
enemies of the unpleasantly smelling
secretions they can emit from the anal
glands.

Civets of South-east Asia

The various civets of Africa and Asia have been hunted for
centuries for the sake of the strong-smelling, whitish-yellow
substance (simply called civet) which is secreted in the perineal
glands of the genital region. This glandular secretion is similar to
musk and is used for the same purposes in the manufacture of
perfumes. When emitted it is syrupy, with an extremely unpleasant odour, and it can be collected from the same animal two
or three times a week. Nobody knows what function it serves but
it is generally agreed that it has nothing to do with self-defence.

Civets are slender, long-tailed, fairly long-legged animals,
the body spotted or striped over a general ground colour. Like
most other jungle predators they are nocturnal by habit, spending
the day curled up in a tree hollow or rock cleft. At dusk they set
out to find food, climbing with agility and scampering swiftly
along branches, nosing through the foliage for eggs, fledglings
and insects as well as fruit, flowers and green shoots. At ground
level they are voracious hunters of rodents and invertebrates,
and since they are very good swimmers they often plunge
into water to catch fishes, frogs, crabs and the like. By dawn
they have satisfied their hunger and are back in their refuges.

Alongside the African civet (*Civettictis civetta*) the most
common species are the large Indian civet (*Viverra zibetha*) and
the small Indian civet (*Viverricula indica*). Though little is
known of their life habits it would appear that both Asiatic
species give birth to two or three young, either in a burrow or in
a hiding place under foliage. Civets are not difficult to capture
and thrive well enough in captivity, although they do not breed
readily under such conditions. Nevertheless, they have since
antiquity been artificially reared in parts of Asia and there is
mention of the practice in the reign of King Solomon.

The mysterious palm civets

The rich store of folklore and legend which is part of Asia's
cultural tradition abounds with tales of mysterious, often
terrifying animals (the yeti is the most celebrated example) so
that fact and fiction are sometimes indistinguishable. The
'abominable snowman' is probably the figment of an over-vivid
imagination; but the small jungle predators known as palm civets
undoubtedly exist, even if the veil of hazy information surrounding them makes it hard for zoologists to number and classify
them. There are several genera, none of them comprising more
than two or three species. They are all long-tailed, tree-climbing
animals with nocturnal habits and varying food preferences.

Owston's banded civet (*Chrotogale owstoni*) is very rare and
known only from fifteen specimens discovered in Indo-China and
since distributed among several museums. In two cases autopsy
revealed earthworms in the stomach. Similar to this species, but
with a differently shaped skull, characterised by a very long
muzzle, is the banded palm civet (*Hemigalus derbyanus*). In
common with the Indian palm civet (*Paradoxurus hermaphroditus*)
and the female only of the small-toothed palm civet (*Arctogalidia*

trivirgata), it ejects a nauseous substance from its anal glands, used, as in the skunk, for self-protection. The fact that the secretion is employed for this purpose may explain the black and white striped pattern of the head which shows up clearly in half-light, warning other animals of the civet's ability to strike back most effectively if attacked. The Indian palm civet is also known as the toddy cat because of its habit of sipping palm wine from tapped trees. The Latin name *Paradoxurus,* which literally means 'strange tail', is another example of scientific misunderstanding. The naturalist Cuvier gave the animal this name in 1821 after examining one specimen in the Vincennes Zoo. Because of having been kept in captivity, this animal's tail was deformed and curled round its body, which is not the case in the wild.

Palm civets are adept climbers, leaping skilfully from branch to branch, but a large part of their omnivorous diet is found on the ground. In some areas they are known to invade houses, orchards and gardens, feeding on refuse, domestic poultry, fruit and flowers. Because of this unspecialised diet the canines and carnassials are not especially well developed and are certainly weaker than those of genets and mongooses – other members of the same family Viverridae.

Reliable information about the reproductive behaviour of the palm civets is fragmentary. Most species would appear to have litters of from two to four babies twice a year (after a gestation

The masked palm civet (*Paguma larvata*) is a nocturnal, tree-dwelling animal that feeds on small mammals, birds, insects, worms and fruit.

of approximately four weeks). At about three months the young are already fully grown and able to lead an independent existence.

Linsangs and binturong

As little known as the various species of palm civet are the linsangs, without any doubt the most graceful and acrobatic of all Viverridae. These lithe, slender animals spend their day alone in a tree hollow, emerging only at night for hunting. Although related to the palm civets they are markedly more carnivorous, moving like sinuous shadows over the ground or through the foliage, sliding their streamlined bodies into quite small holes and cracks. Because of their rapid movements they are no easy prey for larger carnivores nor are they accommodating subjects for scientists.

The two species are the banded linsang (*Prionodon linsang*), which lives in Malaya, Java, Sumatra and Borneo, and the spotted linsang (*Prionodon pardicolor*), an inhabitant of Nepal, Assam, Burma and Indo-China.

Both species feed on lizards, frogs, insects, eggs, fledglings, rodents and, from time to time, fishes (these last are much favoured by linsangs in captivity). Very little is known of their reproductive behaviour but, as in the case of palm civets, they probably have two litters a year, each litter comprising two or three babies. One mother with three newly born youngsters was captured under the roots of a large tree where she had constructed a rudimentary bed of leaves, confirming both the number of babies in a typical litter and the type of refuge commonly used for birth and upbringing. At other times linsangs construct a similar nest in trees, high above ground.

The name of the animal, incidentally, is derived from a local Malayan word meaning 'banded' or 'striped'.

The binturong (*Arctictis binturong*) is another cat-like carnivore but it lacks the graceful, elegant lines of the linsangs. Large, heavy and covered with long black hair, this ponderous animal is, despite its weight of 30 lb or thereabouts, almost exclusively arboreal, moving with slow, deliberate steps along branches but never jumping across a gap.

One rather strange feature of the binturong is its prehensile tail which can be used as a support only when the animal is young and still comparatively lightweight. If a fully grown adult were to try to hang by its tail from a branch it would almost certainly tumble to the ground. Among other carnivores only the kinkajou (*Potos flavus*) of Central and South America has the same type of prehensile tail.

The binturong is yet another nocturnal hunter living in the densest parts of the tropical forests of South-east Asia, including the islands of the Malay archipelago. But despite the fact that its distribution is comparatively wide, its numbers are few, and for that reason the animal has been inadequately studied. It is known, however, to feed on various plants and fruits, as well as on carrion. Given this form of unspecialised diet, its teeth are nothing like as powerful as those of most carnivores, although if attacked it will defend itself by biting. Furthermore, as it snarls at its foe, it adopts dissuasive attitudes by making its long hair wave to and fro. This gives the animal a very strange appearance and usually keeps attackers at bay.

Breeding behaviour in the wild is almost a complete mystery but in captivity births have occurred in March, July and November. The gestation period is about 90 days and a litter consists of two or three young. Although the male keeps well away from his offspring as a rule, the female does not seem particularly resentful of his presence when he does venture near. The young leave the nest at ten weeks, about a fortnight after they have been weaned.

Mustelids of the rain forest

Although the most characteristic predators of the South-east Asian jungles are undoubtedly the Felidae (tiger, leopard and various types of wild cats) and the Viverridae (civets, linsangs and binturong), the small and savage hunters of the family Mustelidae are well represented here, many of them being very similar, sometimes identical, to their relatives of the Holarctic region. Some, however, such as the Malay badger and the hog badger, are only found in the forests of the Oriental region.

High up in the trees, the Himalayan marten (*Martes flavigula*), with sleek brown coat and yellow-orange throat, hunts squirrels.

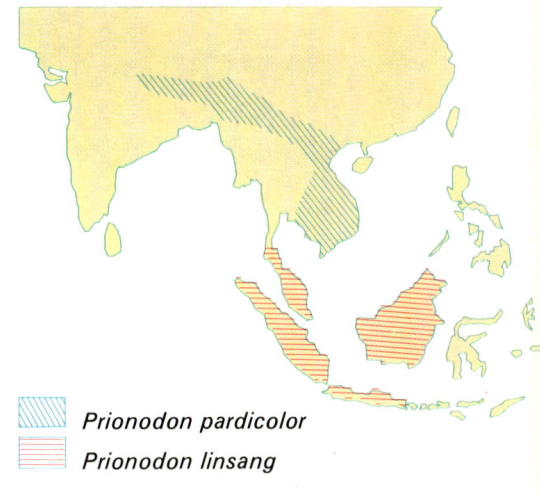

Prionodon pardicolor
Prionodon linsang

Geographical distribution of spotted linsang (*Prionodon pardicolor*) and banded linsang (*Prionodon linsang*).

The linsang, with its elegant, streamlined body and long tail, is one of the most agile of tree-dwelling predators, moving snake-like through the foliage in quest of small rodents and birds.

Binturong
(*Arctictis binturong*)

Leopard cat
(*Felis bengalensis*)

Banded linsang
(*Prionodon linsang*)

Malay civet
(*Viverra tangalunga*)

Indian palm civet
(*Paradoxurus hermaphroditus*)

LINSANGS

Class: Mammalia
Order: Carnivora
Family: Viverridae
Diet: small birds, rodents, frogs, eggs
Gestation: about 2 months
Number of young: 2–3

BANDED LINSANG
(*Prionodon linsang*)

Total length: 29½–31½ inches (75–80 cm)
Length of tail: up to 14 inches (35 cm)
Weight: about 1¾ lb (750 g)

Very long body. Short, thick, compact fur is greyish with 4–6 broad, transverse, brownish-black bands on back. Large eyes, pointed muzzle, large ears. Very long tail with light and dark rings. Small feet.

SPOTTED LINSANG
(*Prionodon pardicolor*)

Total length: 25½–29½ inches (65–75 cm)
Length of tail: 12 inches (30 cm)

Similar to banded linsang but smaller. Regular rows of dark spots on buff coat. Black rings on tail.

The black-striped weasel (*Mustela strigidorsa*), by contrast, is an inhabitant of highland districts above 4,000 feet, scurrying among rocks and across woodland clearings for small birds and rodents. The ratel or honey badger (*Mellivora capensis*), also found in Africa and Arabia, feeds on bees and termites, although its valuable honeyguide ally has no counterpart in India.

Otters gambol in the rivers and mangrove swamps, including one species similar to the African clawless otter, namely the Indian small-clawed otter (*Amblonyx cinerea*).

In addition to the common badger (*Meles meles*) there are related jungle species which are stranger and more colourful. Thus the hog badger (*Arctonyx collaris*) has a white throat and a very long muzzle resembling that of a wild boar, used, it would seem, for a similar purpose–rooting into soft earth to extract the small animals and plants which constitute its diet. The Malay badger (*Mydaus javanensis*) has a rounded body with short, slender legs. Its nose is long and flexible and it feeds mainly on worms, insects, fruit and small mammals. Sometimes it finds its way into farms and causes damage to domestic stock. A nocturnal animal, it sleeps during the day in a burrow about 3 feet below ground. When it senses danger it lifts its tail and ejects a nauseous opaque secretion which is capable of blinding a dog at close quarters and which long ago was used by the sultans of Java as a perfume base. The Palawan or Calamian stink badger (*Suillotaxus marchei*) looks and behaves much like this last species but lives only on islands to the north and east of Borneo.

Finally, brief mention should be made of the ferret-badgers of the genus *Melogale*, consisting of three species. These animals weigh rather less than other mustelids – seldom more than 4–5 lb – and from close quarters somewhat resemble polecats. Although frequently encountered in fields and on cultivated land, they are also residents of the tropical forest, being just as much at home in trees as on the ground. They too possess anal glands which secrete an unpleasantly smelling substance, used, skunk-like, as a weapon of dissuasion; and the white band across the muzzle may be regarded as a warning signal. The badger-weasels are omnivores and the females give birth each year to one, two or three babies. The youngsters are apparently suckled for a considerable period.

Fierce hunting cats of Asia

The majority of the Felidae of the Old World are inhabitants of Asia. In fact, of all the wild cats to be found in Africa and Europe, there are only three that do not have Asiatic counterparts – the serval, the African wild cat and the black-footed cat. On the other hand, there are a number of species which are exclusive to the Asian continent, such as the tiger, the clouded leopard, the snow leopard or ounce, and the leopard cat. Some of them, to narrow it down even further, are to be found only in the Oriental region.

Apart from the leopard, the tiger and the clouded leopard, which may all be defined as medium or large felines, there are at least nine species of smaller types of wild cat which roam the jungles of South-east Asia, both on the mainland and on the islands. Some of them, including the small Philippines cat (*Felis minuta*), are only to be found along a restricted stretch of the mainland or particular island chains; but others, such as the leopard cat, have a much broader range of distribution which covers almost every part of the region.

Like the civets and the linsangs, as well as the majority of small hunters of the undergrowth, these wild cats are nocturnal animals; and because they are as frequently found in trees as on the ground they are notoriously difficult to observe at close quarters. For this reason information about their behaviour in the wild leaves much to be desired.

Characteristic features of all species are the large eyes and narrow, vertical pupils, typical of animals that habitually hunt in half-light and darkness. Their vision, as might be expected, is extremely good but the sense of smell is not particularly well developed. The head is rounded, the muzzle short and the jaws less powerful than in many other carnivores, for in fact they do not consume as much flesh as their larger relatives. Most of these smaller felines are omnivores, and a diet which may include rodents, birds, frogs and reptiles is frequently supplemented by a variety of vegetation. Thus the leopard cat in particular, which is a fairly common predator both in India and the Malay peninsula and archipelago, is especially partial to the leaves of a local tree known as the tjongkok (*Cuculige capitulata*). Another species, the fishing cat (*Felis viverrina*) specialises, as its name suggests, in the capture of fishes. As for the rest, although they spend the greater part of their time up in trees – sleeping

The ungainly binturong, with its long black hair and white whiskers, is yet another nocturnal hunter of the jungle, related to the civets and the linsangs.

BINTURONG
(*Arctictis binturong*)

Class: Mammalia
Order: Carnivora
Family: Viverridae
Total length: 46–73 inches (117–185 cm)
Length of tail: 22–35½ inches (56–90 cm)
Weight: 20–31 lb (9–14 kg)
Diet: small vertebrates, carrion, fruit, leaves, shoots, etc
Gestation: about 3 months
Number of young: 2–3

Strange-looking animal with long, shining black hair, thickest on tail. Greyish head, white vibrissae. Long lynx-like hairs form a kind of brush on back of ears.

The leopard cat, handsomely spotted like other felines of the tropical forest, is an intrepid hunter; but because it feeds on vegetable substances as well as flesh, its jaws and teeth are not as powerful as are those of larger carnivores.

during the day—most of their prey is caught at ground level.

The wild cats of the Oriental region tend to lead solitary lives, except of course during the mating season and when the females are rearing their offspring. Moving silently through the jungle, they have the habit of leaving their claw marks on branches and on tree trunks—presumably as a sign to other animals that these are territorial boundaries.

Probably the most familiar of these felines is the leopard cat (*Felis bengalensis*). Births in this species evidently occur at more or less any time of the year, after a gestation of eight weeks, and a litter may comprise from one to four cubs. The latter open their eyes after ten or fifteen days and reach sexual maturity when aged one-and-a-half years.

The golden cat, also known as Temminck's cat (*Felis temmincki*), is an inhabitant of the Tibetan highlands and the eastern parts of the Himalayas. Pfeffer has described this species as a beautiful tawny animal, about the size of a lynx and with the fierce appearance of a puma. In Burma it is reputed to prey, above all, on sheep, goats, cattle, young water buffaloes and small deer. It often lurks near villages and causes considerable havoc in hen-houses. Litters of either one or two cubs have been discovered in tree hollows.

The marbled cat (*Felis marmorata*) and the bay cat of Borneo (*Felis badia*) are of small build, almost like miniature versions of the clouded leopard. Tradition has it that the former is a fierce hunter of squirrels, birds and frogs, and that in the feline world it is in fact second to none for savagery—a claim that should perhaps be treated with caution. Even less is known about the habits of the flat-headed cat (*Felis planiceps*), apart from the fact that it is one of the smallest of all wild cats of the Oriental region. B. E. Smythies, who was for twenty years commissioner of forests in Sarawak, came across one of these animals in a hen-house but this was the only individual he ever saw on the island.

The rusty-spotted cat (*Felis rubiginosa*) is another small, elegant and apparently very agile member of the family which hunts birds and small mammals. It is found both in Ceylon and in southern India, and these are also the homes of the wide-ranging African wild cat or caffer cat (*Felis libyca*).

The clouded leopard

The clouded leopard (*Neofelis nebulosa*) is a slender animal with short, powerful feet, a very long tail and a most beautifully spotted coat. It has a wide range from Nepal and Sikkim to Malaya, southern China and the offshore islands of Hainan and Taiwan, and across to Sumatra and Borneo. Habitats include jungle, steppe and swampland.

The suppleness and agility of the clouded leopard are proverbial. It hunts silently by night and sleeps during the day, outstretched either on the ground or along a branch (its claws being perfectly adapted for arboreal activities). According to E. M. Selons and E. Banks, who have studied the species closely

and reared a young clouded leopard, the animal is also active by day, hunting prey on the ground to a much greater extent than has been believed. They regard it as a typical predator of the secondary forest. It is fair to mention, however, that they are virtually alone in holding these opinions and that other observations strongly suggest it is a nocturnal animal.

The clouded leopard preys on small and medium-sized mammals and, apparently, birds. But reliable information on this and other aspects of behaviour is so patchy—as it is for almost all the jungle-dwellers—that even its food habits are uncertain. It has been known to launch sporadic attacks on cattle, wild boars, deer and porcupines, but reports from Borneo that it is a man-eater lack confirmation and are highly doubtful.

One young clouded leopard captured in the Sunda Islands, during the long sea journey back to Europe, became very attached to several dogs on the same ship, but its normal aggressive instincts were markedly in evidence when it attacked and killed a chicken destined for the galley.

The female clouded leopard gives birth to two, three or four cubs, usually in the hollow of a tree, after a gestation of ninety days. They are born blind but begin to show signs of activity after a couple of weeks. Within another month they weigh more than 2 lb, reaching adult size and weight (up to 50 lb) at the age of eight months.

Clouded leopards hunt mammals and birds by night. They are found throughout the Oriental region, often at high altitude.

CLOUDED LEOPARD
(*Neofelis nebulosa*)

Class: Mammalia
Order: Carnivora
Family: Felidae
Total length: 47½–75 inches (120–190 cm)
Length of tail: 24–36 inches (60–91 cm)
Height to shoulder: 31½ inches (80 cm)
Weight: 31–50½ lb (14–23 kg)
Diet: flesh
Gestation: 90 days
Number of young: 2–4

Fawn or greyish coat with black marks—oval, round or rectangular—on back, and black spots on shortish legs. Cubs blind at birth, weighing 5–6 ounces, with blurred pattern on fur.

CHAPTER 88

Fine feathers and sharp talons: the birds of the Oriental region

The immense expanse of tropical forest which is bounded to the north by the great mountain chain of the Himalayas, to the south by the Indian Ocean and which then extends eastwards beyond the mainland of Asia to Sumatra, Java, Borneo and the islands of the Philippines archipelago, has some features in common with the rain forests of Africa and South America but is far more varied in character. This is partly because it is subjected to the vagaries of the seasonal monsoons and partly because of natural contrasts of vegetational growth from region to region. In India, Bangla Desh, Burma and Thailand, for example, the areas of true jungle are relatively limited, whereas in Indo-China and the Malay peninsula, as well as on the islands, it stretches without interruption for hundreds of miles. There are many parts of this dark, mysterious world which civilised man has never penetrated—a realm barricaded by towering trees and tangled undergrowth, whose criss-crossing network of streams and rivers can only be plotted from the air; and because of its seclusion it is an ideal habitat for a multitude of animals. As in every forest biome there is life at every level, on the ground, among the lower branches and at progressively greater heights up to the very tree-tops. The fauna includes a number of ravishingly beautiful birds which live on the forest floor.

Wild ancestors of domestic poultry

An unusual bird struts with measured steps through the undergrowth. The delicately streaked reddish-yellow plumage is set off by the glistening green or blue of the scapulars and tail feathers. The two central rectrices are abnormally long, arched and curved so that they practically trail along the ground. The

Facing page : The common ring-necked pheasant is a hybrid, descended from three wild subspecies from Asia. It has been successfully introduced to many countries and is one of the world's most coveted game birds.

1   2

During the courtship ritual the cock summons the hens by crowing loudly, meantime scratching the soil with his feet (1). The birds peck at the ground for food (2) in the same manner as domestic fowl.

The jungle fowl spends most of its time on the ground but flutters up to a tree branch to sleep or to escape an enemy.

face of this bird is equally striking, with a red serrated comb surmounting the head, fleshy wattles hanging loosely down from the chin, similar fleshy growths around the eyes and white discs over the ears. Despite the exotic appearance there is an obvious resemblance to the familiar cockerel of the farmyard, and no wonder. For this is the cock of the jungle fowl (*Gallus gallus*), the wild ancestor of all domestic fowl, with a distribution that extends from India and Ceylon through Indo-China to Java.

This bird is normally found in sparse woodland and forest clearings, often on the fringes of rice paddies and cultivated land; but during the rainy season it frequently seeks shelter in the dense jungle foliage.

Jungle fowl live in mixed flocks comprising up to fifty birds. They are most active either in the early morning or late afternoon and at such times can be seen busily pecking at the ground for seeds, tender shoots and insects. They perch on trees only to sleep and to escape predators, the cocks crowing lustily, like their farmyard counterparts, shortly before dawn, rocking from one foot to the other. As soon as the sun rises the birds flutter down to ground level.

The breeding season is in the spring, reaching a peak in March. Each cock stakes out his territory with strident cries and defends it fiercely against potential rivals. Fights often break out and these may be violent, with each cock ruffling his plumage and lashing out at the enemy with his spurs.

The natural aggressiveness of the males is, as in the case of all animal species that breed prolifically, a guarantee of survival. If all cocks of breeding age were to mate and have progeny the community would become impossibly large and there would not be sufficient food to go round. By destroying its habitat the species would eventually be doomed. So ritual combat serves to sort the strong from the weak. Death is not the usual outcome, except accidentally, but the loser will in all probability be spurned by the hens and have no opportunity of mating. Yet provided he does not sustain a serious injury he may still perform a valuable defensive function for the benefit of the community.

Nature can be trusted to turn the aggressive instincts of the individual to the advantage of the species. But man has cruelly exploited this innate tendency for his own amusement and profit. By cross-breeding particularly belligerent birds he has produced a strain of fighting cock which, as a result of special feeding and rearing, can be trained for no other purpose than to attack others of its kind. After being fitted with sharp metal spurs two cocks are pitted against each other and proceed to fight, sometimes

The jungle fowl is the wild ancestor of all our familiar domestic poultry, some strains of which still bear a resemblance to their forest ancestor. As is the case with domestic species the hen is less gaudy than the cock and the comb is either rudimentary or absent.

to the death, sometimes until one is too badly injured to continue. Cockfighting was a popular sport in India, China and Persia in ancient times and later spread to the West by way of Greece and Rome. It was a favourite pastime in Elizabethan and Stuart England and found its way to America as well. Although now illegal in most countries it still has its devotees.

The courtship display of the jungle cock is similar to that of his domestic counterpart. He entices two or three hens into his territory, performing a kind of dance around the partner of his choice, bending his body towards her as if to show off the beauty of his plumage, then stretching out a wing in her direction and dragging the primary remiges along the ground, producing a characteristic vibrating sound rather like that of a harp being plucked.

The hen, like her farmyard relative, possesses neither comb nor wattles (although in some individuals these may be visible in rudimentary form) and her plumage is much less colourful, predominantly brown. She lays from five to ten eggs in a hollow on the ground, this being roughly lined with grass. She then slips quietly away from the nest and begins a noisy clucking, deliberately trying to attract the attention of predators. It is

The jungle fowl is a pheasant but its tail is distinctive, being arched, with the two long central feathers almost touching the ground.

JUNGLE FOWL
(Gallus gallus)

Class: Aves
Order: Galliformes
Family: Phasianidae
Total length: 25½–27½ inches (65–70 cm)
Length of tail: 11–19½ inches (27·5–50 cm)
Diet: seeds, shoots, fruit, insects
Number of eggs: 5–10
Incubation: 21 days

Plumage basically reddish-yellow; wings and tail dark green or blue. Cock has red comb and wattles, with similar fleshy areas around eyes and white discs surrounding ears. Hen drabber, usually brown; comb small or absent.

interesting to note that this diversionary tactic persists among domestic hens although in their case it can serve no useful purpose. Some domestic strains have modified the habit by clucking before they quit the nest—a procedure which would be suicidal were it to be emulated by wild jungle fowl.

The chicks hatch after twenty-one days' incubation. They are very precocious and quite capable of running about and hiding themselves only a few hours after birth. The hen watches over them with great solicitude, scraping at the earth with her feet and pecking away incessantly for suitable food, clucking softly whenever she wishes to summon her brood. Should danger loom she puts on a brave front, launching violent and indiscriminate attacks with beak and claws against anything that moves in the bushes nearby. Meanwhile the chicks scamper off into the thickets and remain absolutely motionless until the threat has receded. The hen sounds the 'all clear' by emitting a sharp cry—a signal which immediately brings the chicks scurrying back to her side. They grow very rapidly and at the end of the first week their wing feathers are strong enough to support them in flight as far as the branches of nearby trees where from now on they sleep, safe from roaming predators.

Domestication of the jungle fowl stretches back into the mists of time. There was a flourishing export trade from India to China, Mesopotamia and other markets from about 3000 B.C. onwards. In due course tame birds were taken to Egypt, across the Mediterranean to Greece and Rome, and subsequently to the rest of Europe. The conquistadors later introduced them to America.

It is hardly necessary to underline the importance of domestic fowl in modern life, although many of the numerous strains bred for their egg-laying capacity or edible quality have little in common with the jungle stock from which they are descended. Thus a domestic hen is expected to lay an egg almost every day.

Jungle cock
(*Gallus gallus*)

Pheasants on parade

The jungle fowl is actually a pheasant although the cock is distinguishable from other members of the family Phasianidae by his comb and arching tail. But whereas the former has been domesticated and selectively bred for egg and meat production, the true pheasants have been encouraged to roam wild or, if domesticated, have been reared with a view to conserving those qualities of strength and beauty which have since ancient times made them the most coveted of all game birds.

The word 'pheasant' is a corruption of the Latin 'phasianus' and legend has it that the well-known Eurasian ring-necked pheasant (*Phasianus colchicus*) was introduced into Europe by the Argonauts from the River Phasis in the Asia Minor province of Colchis. Most of the species and races of pheasant that are nowadays found in Europe, North America, New Zealand, Hawaii, etc, have been introduced during the last couple of centuries, so that the present populations cannot be regarded as genuinely wild, being largely descended from artificially reared stock.

This method of repopulation, which may satisfy the demands of the hunter who cannot be expected to care too much about the origin of the game he is shooting, is not wholeheartedly approved by naturalists and conservationists who are mainly concerned to preserve wildlife in its original form. In those parts of the world where nature's equilibrium has undergone little if any disturbance for centuries the sudden irruption of thousands of semi-domesticated birds may have disastrous repercussions on the local fauna, which is unexpectedly deprived of its traditional food resources and compelled to share an already restricted habitat with the new arrivals.

All the true pheasants came originally from Asia and those that are now inhabitants of many European countries are hybrids, resulting from cross-breeding among three wild subspecies—the Chinese ring-necked pheasant (*Phasianus colchicus torquatus*), the common black-necked pheasant (*Phasianus colchicus colchicus*) and the Mongolian pheasant (*Phasianus colchicus mongolicus*). There are in fact a number of other subspecies as well, some of them isolated communities, inhabiting a vast area extending from Japan and the shores of the China Sea westwards to the Caucasus.

The majority of Asiatic pheasants are large birds with long

Impeyan pheasant
(*Lophophorus impejanus*)

tails, cocks of the genus *Phasianus* sometimes being distinguished
from others by their wattles. The tragopans (genus *Tragopanus*)
and impeyans or monals (genus *Lophophorus*) differ from the
majority in having shorter tails and are more closely related to
partridges. In most species the area surrounding the eyes is
naked and the top of the head either bare or adorned with a
crest of feathers. The males have spurs and it is they of course
who boast the more glowing colours, the hens being much
drabber. As a group the pheasants of the Oriental region are
remarkable for their outstandingly beautiful plumage.

The mating instincts of these pheasants of South-east Asia
begin to be aroused towards the middle of March. The fleshy
areas around the eyes of the cock expand to cover almost the
whole of the cheeks, and the feathers surrounding the ears,
which at other times of the year are scarcely visible, now stand
up like little horns on either side of the head.

Unlike the majority of Galliformes typical pheasants do not
appear to have strong territorial instincts. Although under
favourable conditions a local population may be quite dense,
with up to fifty pairs of birds per square mile, nobody has seen
them fighting one another in defence of a feeding ground. Each
cock is usually accompanied by two hens who respond obediently
to his vocal summons and will not permit another suitor to
approach. When he comes across a particularly succulent item
of food he softly calls his partners who hurry to share the meal.
Since this type of behaviour is only observed during the breeding
season and is frequently followed by coupling it is widely assumed
to be a feature of the courtship display. The cock normally mates
first with the hen that is nearer, after which he slowly circles his
partner, head lowered, neck feathers ruffled and wings sweeping
the ground, simultaneously uttering curious whistling cries.

After mating, the birds go their own ways, the cock to continue
his solitary life and the hen to prepare for egg-laying. Her nest
consists of a simple depression in the ground, hurriedly lined
with grass and feathers. She lays from eight to twelve eggs and
incubates them for twenty-three days. A few hours after hatching,
the chicks are already running about confidently, ready to take
shelter, if the need should arise, in the undergrowth. They are
capable of flying when two weeks old but remain with their
mother for about a year.

The pheasants belonging to the genus *Lophura* are among the
most decorative members of the subfamily Phasianinae. Distinctive features include the drooping crest of long, soft feathers
(present in both sexes) extending from the head down the neck
to the back, and the amply feathered, flattened tail. Perhaps the
most beautiful representative of the group, as well as the most
secretive, is the black or Kalij pheasant (*Lophura leucomelanus*)
which lives in remote forest areas from Burma to Nepal and
Kashmir, often at heights of over 12,000 feet. The cock would
appear to be monogamous and, contrary to the custom of the
majority of Galliformes, takes a fairly active part in the rearing
of the chicks.

The silver pheasant (*Lophura nycthemera*) which lives in
south-eastern China and eastern Vietnam has been semi-

Facing page: Sexual dimorphism
(difference in outward appearance between
male and female) is of prime importance
for the survival of the species. The hen
pheasant, who has to incubate the eggs for
over three weeks, and the nestlings, which
are initially weak and unprotected, can,
because they are drably coloured, go
unnoticed in the thickets, whereas the
cock, with his bright plumage, draws upon
himself the attentions of predators.
Stationed some distance away from his
small family he is vigilant to all sounds
and movements, flying off if danger looms.

RING-NECKED PHEASANT
(*Phasianus colchicus*)

Class: Aves
Order: Galliformes
Family: Phasianidae
Length: male 30–34½ inches (78–87.5 cm)
female 20½–24½ inches (52.5–62.5 cm)
Diet: seeds, bulbs, shoots, insects, fruit
Number of eggs: 8–12
Incubation: 23 days

Plumage copper, streaked with black. Fairly long,
reddish-yellow tail, delicately marked with black.
White band around neck. Head feathers metallic
blue-green, with two ear-tufts. Bright red
wattles. Female drabber, generally dark brown.

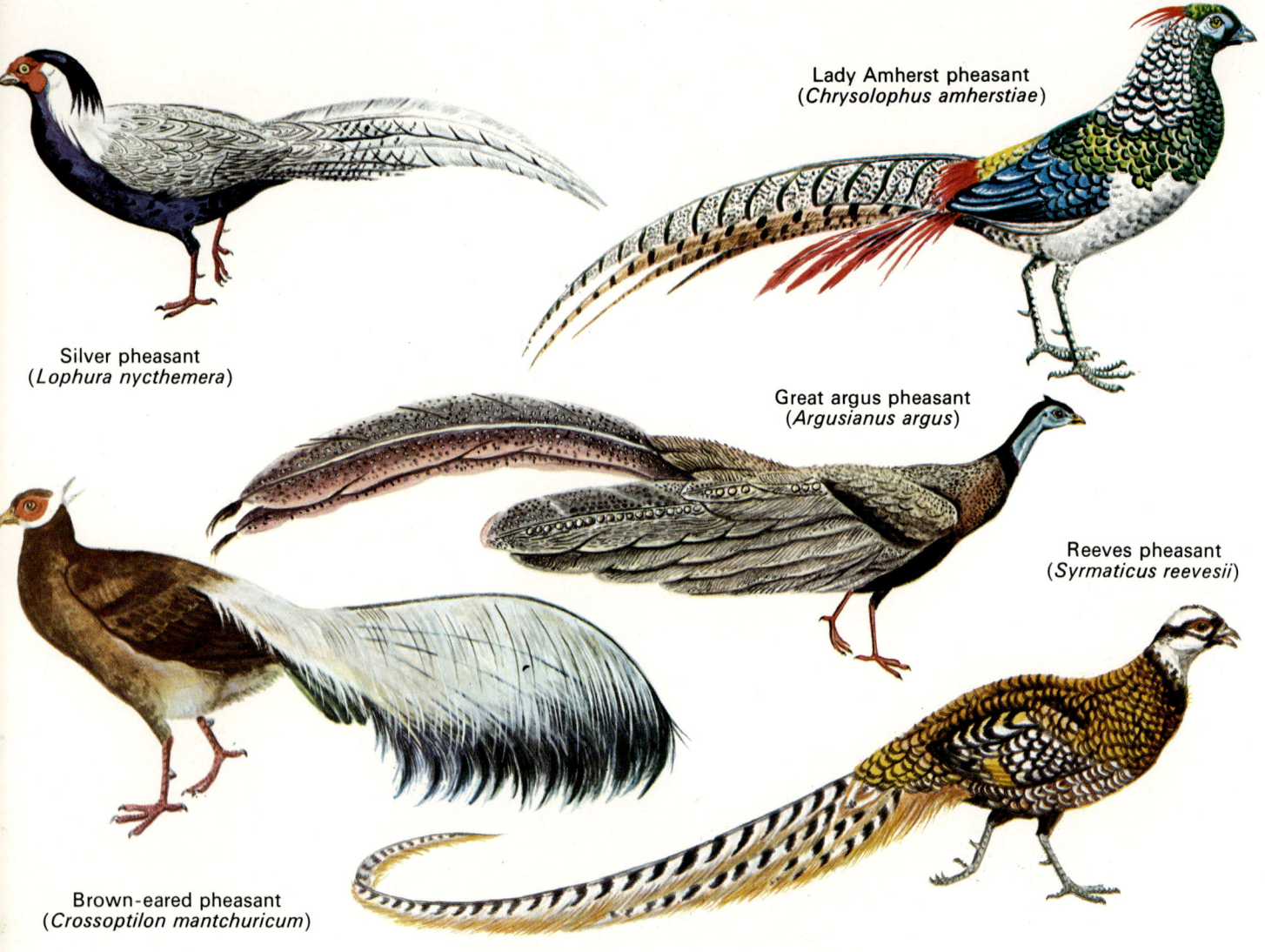

Lady Amherst pheasant
(*Chrysolophus amherstiae*)

Silver pheasant
(*Lophura nycthemera*)

Great argus pheasant
(*Argusianus argus*)

Reeves pheasant
(*Syrmaticus reevesii*)

Brown-eared pheasant
(*Crossoptilon mantchuricum*)

domesticated since antiquity, as is testified by Chinese art dating back more than five thousand years. The cock is extremely belligerent in the breeding season and will engage in ferocious combat with rivals in order to win and then defend his nuptial ground. Once he has got rid of the intruder he attracts the attention of several hens, mating with them in turn. Each fertilised female subsequently lays from six to eight pinkish eggs in a hollow on the ground, incubating them for twenty-five days. Cocks of this species are sexually mature at the age of one year but the hens are unable to breed until they are two years old. In captivity individuals may live for more than twenty years.

Other pheasants of the genus *Lophura,* including the Siamese fireback pheasant (*Lophura diardi*) are inhabitants of lowland forest regions of the Malay peninsula, Sumatra and Borneo. Both cocks and hens bear an ornate, stiff crest of feathers on the head. Although normally diurnal by habit, these birds frequently take advantage of a clear, moonlit night to go out in search of food. They generally live in small groups consisting of one male and four or five females, and several of these family bands may collaborate in looking for food, which comprises grass, shoots, berries, insects and earthworms.

Bulwer's pheasant (*Lophura bulweri*) lives in the rain forests of Borneo and is notable for its spectacular courtship ritual. The

Ring-necked pheasant
(*Phasianus colchicus*)

Swinhoe's pheasant
(*Lophura swinhoii*)

Golden pheasant
(*Chrysolophus pictus*)

Common peafowl
(*Pavo cristatus*)

Siamese fireback pheasant
(*Lophura diardi*)

cock in breeding garb takes on a very strange aspect, for the dark
blue fleshy areas around the eyes swell up and extend backwards
and forwards so that the normal shape of the bird's head is
obscured. All that can be seen are two long blue pouches out of
which shine the brilliant scarlet eyes. Furthermore, when the
cock performs his nuptial dance he struts to and fro, spreading
the thirty-two white feathers of his tail. Outstretched like a
magnificent fan, the tail gleams like a full moon against the dark
background of the jungle foliage. The outside feathers are trailed
along the ground and this produces an odd noise, as if the bird
were dragging its feet through a heap of dead leaves. The purpose
of this dramatic display, which is the immediate prelude to the
sexual act, is to attract the hens and keep rival males at a safe
distance.

The great argus pheasant (*Argusianus argus*) is not as gorgeously
coloured as other species but is notable for its size, the length
of its two central tail feathers and the huge spread of its secondary
wing feathers, these being beautifully adorned with a pattern of
eye-spots (hence the name of the species, after the legendary
hundred-eyed Argus). An inhabitant of the mountain forests of
the Malay peninsula and Sumatra, it is a very shy bird and for
that reason seldom seen. Its approximate location, however, is
given away by its unmistakable melodious call, that of the hen

In the breeding season the red wattles of the cock silver pheasant swell enormously until they cover almost the entire head.

Facing page : The silver pheasant (*above*) is well named. A very rare species, it ventures out only in the half-light and its gleaming silver plumage shows up clearly, signalling its presence to other members of the group. The great argus pheasant (*centre*) is an inhabitant of mountain forests, notable for its spectacular courtship displays. Swinhoe's pheasant (*below*) is an imperilled species, now found only on the island of Taiwan.

being more sustained than that of the cock, who confines himself to short notes that are promptly echoed by rivals in the vicinity. Each male takes up a solitary stance in a forest clearing and his song alone is sufficient for marking the bounds of his nuptial territory, there being no recourse to aggressive action. Disinclined to move about very much, a bird may remain in more or less the same place all the year round, although it will hide when moulting.

The courtship dance of the cock is an unusual spectacle. Having enticed the hen vocally to his domain he then advances on her with measured tread, stamping his feet violently on the ground. In a state of considerable excitement, he walks around her, trailing his long tail feathers and tossing twigs and pebbles in her direction. If this does not frighten her off he swells his plumage, spreads his wings wide and begins to jump up and down, rapidly fluttering his tail. She, by contrast, remains quite calm during this passionate performance and shows little obvious interest when he approaches and mates with her. When it is over she wanders away alone to prepare her nest in a natural hollow sheltered by thickets. The cock plays no part whatsoever in incubating the eggs or rearing the chicks.

The elaborate courtship display of the cock argus pheasant is undoubtedly effective in its main purpose—seducing the hen—but all this noisy bustle and showy exposure of plumage can be a disadvantage if predators are in the neighbourhood. As is the case with the capercaillie, which in the excitement of its courting activities is virtually deaf to the approach of hunters, so the argus pheasant, impervious to everything but his love-making, is completely at the mercy of carnivores.

The cock's vulnerability during the breeding season would not in itself represent a real threat to the future of a species which is highly prolific. Yet the argus pheasant may be in peril for an allied reason. The splendid ornamental feathers which serve as a sexual inducement to the female may themselves be indirectly responsible for the gradual decline in numbers which has been noted by ornithologists. The elongated secondary wing feathers carried by the majority of cocks are a severe handicap to flying and hence escaping from enemies, since it is precisely those males with the largest and most spectacular feathers who attract the hens but who have most difficulty in taking wing when attacked by predators. On the other hand, males who are not so well endowed but who are, as a result, much more efficient at flying, are spurned by the females and given no opportunity to procreate.

The probable consequence of this state of affairs is that the wrong genes will be transmitted from generation to generation, producing an increasing number of birds with magnificent plumage but little ability to protect themselves. So the fine feathers of the argus pheasant may yet prove to be its downfall— an evolutionary trap comparable to the over-developed antlers of that prehistoric member of the Cervidae, the Irish elk. In the opinion of many palaeontologists, this species became extinct simply because the top-heavy antlers transformed the once-agile animal into a standing target for its foes.

Above and facing page : The ornamental train of the peacock is made up of elongated upper tail coverts and is supported by the tail proper, the individual feathers of which are much shorter. Adorned with multicoloured ocelli or eye-spots, the train, when fully outspread, is the peacock's principal means of attracting the hen during the breeding season.

The peafowl is an adaptable bird. Having originated in Asia it has since been introduced into almost every part of the world.

Krishna reborn

The peacock—male of the common peafowl (*Pavo cristatus*)—is an incomparably beautiful bird. In its domesticated form it is a familiar inhabitant of parks and zoos and like other pheasants it has an ancient history. According to mythology the hundred eyes of the giant Argus, killed by Mercury, were transferred by the goddess Juno to the peacock's tail. In its native India the bird is not only valued for the way in which it attacks and kills young cobras but is also worshipped in many districts as the reincarnation of the god Krishna. The date of its domestication is unknown but it may have been a gradual, almost accidental process as a result of wild birds straying from their forest habitat towards villages in search of food. Certainly the species was imported into Mesopotamia by 300 B.C., was later introduced to Egypt, Greece and Rome, and thence found its way to other parts of the Western world.

The peacock is perhaps the most handsome bird alive. The colour of the head, neck and breast is variable, ranging from emerald green to rich, dark blue, with glistening metallic reflections. The small head is surmounted by a flat fan of feathers of similar hue and there are white areas around the eyes. But the most resplendent attribute of the peacock is its magnificent flowing train—green and blue with rows of large, brilliantly coloured ocelli or eye-spots. These dazzling ornamental feathers are not in fact part of the tail proper, but elongated upper tail coverts. The long train is actually supported by the tail itself.

Peafowl, among the largest of the Galliformes, are inhabitants of the densest parts of the jungles of southern Asia, especially of impenetrable areas where hills and valleys are traversed by streams which sometimes turn the lowlands into swamps. The birds are gregarious, gathering in small family groups, and are active at more or less any time of the day, except in regions where they are likely to be disturbed, in which case they only venture out at dawn and sunset.

The breeding cycle of the peafowl depends largely on the pattern of local rainfall, normally coinciding with the onset of the dry season. The polygamous cock, after performing his

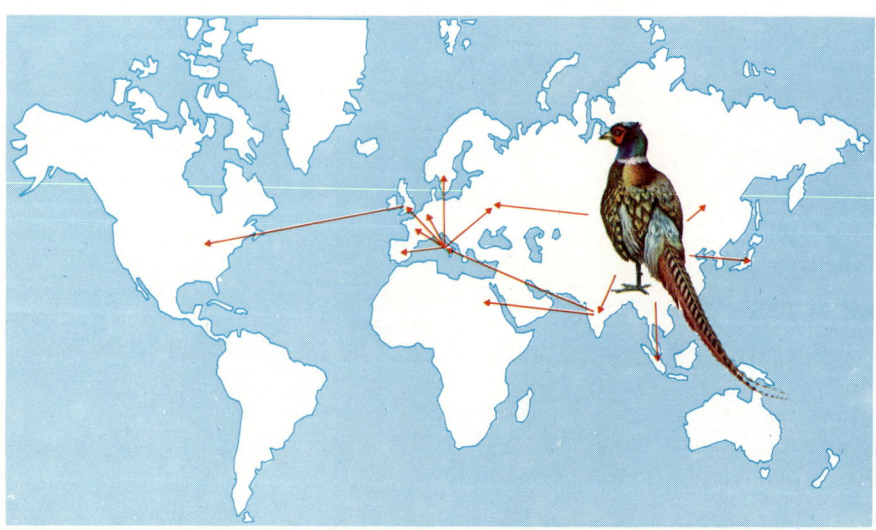

Above and facing page : The gorgeous colours of the common peafowl blend admirably with their jungle surroundings and in the wild the birds spend much of their time perched on branches, ready to emit piercing cries should a predator appear. The birds which strut across park lawns and zoo clearings are semi-domesticated.

COMMON PEAFOWL
(*Pavo cristatus*)

Class: Aves
Order: Galliformes
Family: Phasianidae
Length: very variable, depending on train, but
usually 5—6½ feet (1·5—2 m)
Diet: seeds, shoots, fruit, ants and small
reptiles
Number of eggs: 3—5
Incubation: 28 days

Cock splendidly coloured, with green or dark blue breast, belly, head and neck; wings brown and white at base, turning russet at tips. The elongated upper tail coverts are adorned with green, blue and orange eye-spots. The stiff crest terminates in dark blue feathers. Hen comparatively drab.

---

spectacular train-spreading display, mates in turn with from two to five hens. Unlike other pheasants, the peacock does not face the hen when courting her but turns round and round with his back towards her, so that she is compelled to run about frantically to keep up with him. Schenkel is of the opinion that she pursues her mate as though he held a morsel of food in his beak, the outspread train being the signal that stimulates her to do this. Among other pheasants a genuine food offering is part and parcel of the courtship ritual, so that if Schenkel is correct it would seem to be a residue of an ancient pattern of behaviour which has lost its original meaning.

Eventually the hen stretches out on the ground and the cock, bringing the display to a conclusion by closing the train, lowers his tail and couples with her. Later she conceals herself in the thickets, sometimes selecting a hollow tree trunk or the abandoned nest of a bird of prey, and lays from three to five eggs. After an incubation of four weeks the chicks are born, covered with light brown down, and within a few hours are pecking at the ground for food. They are closely protected by the mother and often take refuge under her tail. The young males soon begin to spread their tiny tails but the ornamental feathers of the train only reach full development after two years.

The decorative parrakeets

Despite their vivid colours the pheasants of the tropical rain forests, including the peafowl, do not normally attract undue attention because of their timid, retiring habits. This can hardly be said of the equally brilliantly coloured parrakeets which seem to go to special lengths to make their presence known, screeching and chattering as they perch fearlessly, almost insolently, on the lower branches of the trees and flutter across the jungle clearings.

Parrakeets are members of the Psittacidae, a family with a worldwide distribution that embraces Australasia, tropical Asia, Africa, South America and Oceania. There are obvious resemblances between the different representatives of this large family, which includes huge macaws and cockatoos to tiny lovebirds and budgerigars. The parrakeets might be described as small, long-tailed parrots, and they are rather curious birds. Although their plumage is far more attractive they seem to have a number of features in common with the nocturnal birds of prey – a sharply hooked bill, sturdy feet and a voluminous brain case indicative of high intelligence. Nevertheless this is probably just coincidence, an example of convergent evolution; for the behaviour patterns of the two groups are completely contrasted. Thus the short hooked bill of the owl serves to rip apart its animal prey whereas that of the parrakeet is an ideal instrument for breaking hard seeds or peeling fruit. Another special peculiarity of the beak of the parrakeet is that the upper mandible is mobile and that the lower mandible can be moved freely backwards and forwards, all of which is invaluable for feeding. Moreover, the bill can be used for grasping branches as an additional support to the feet.

The parrakeet has four toes on each foot, two facing the front

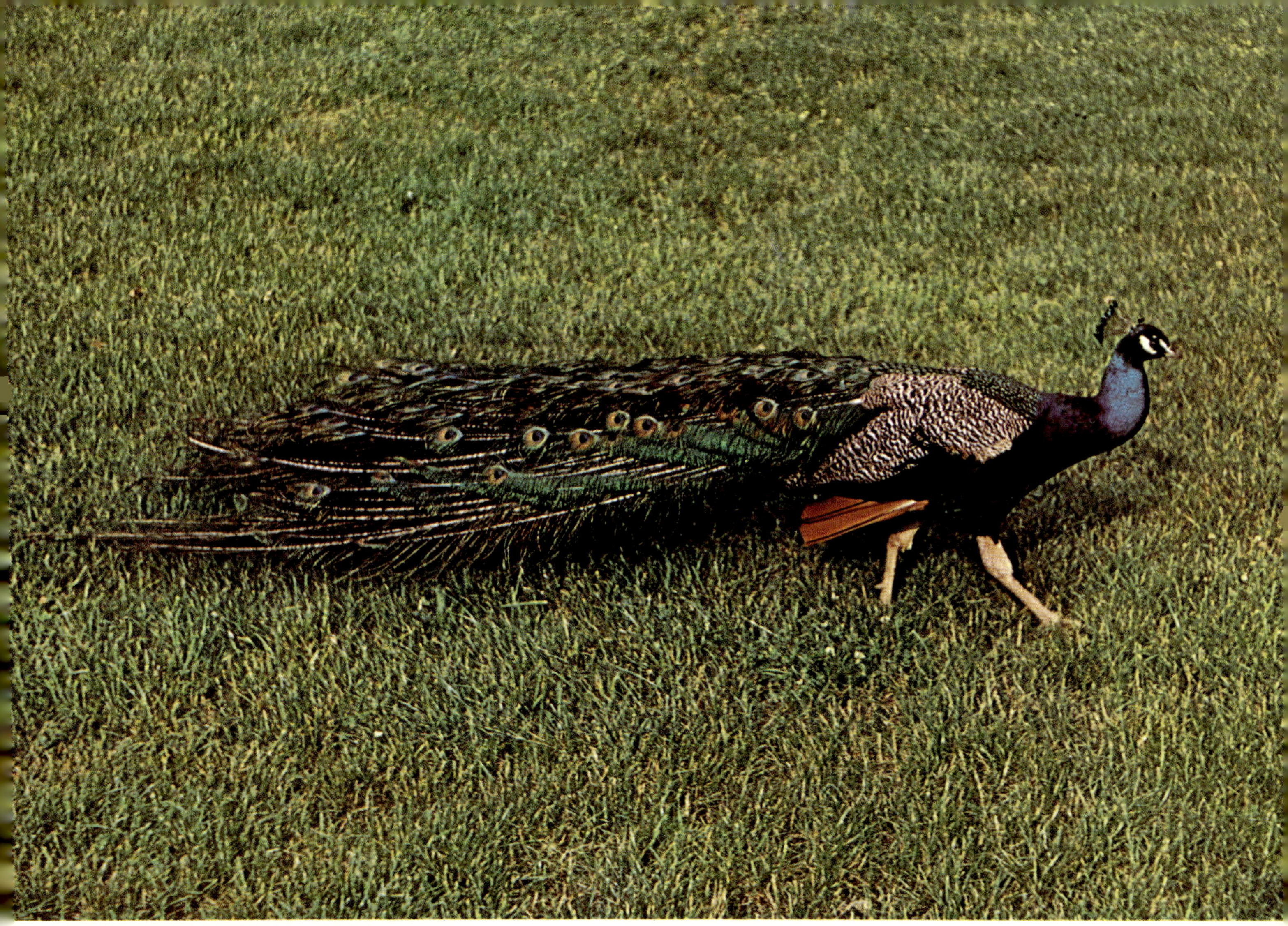

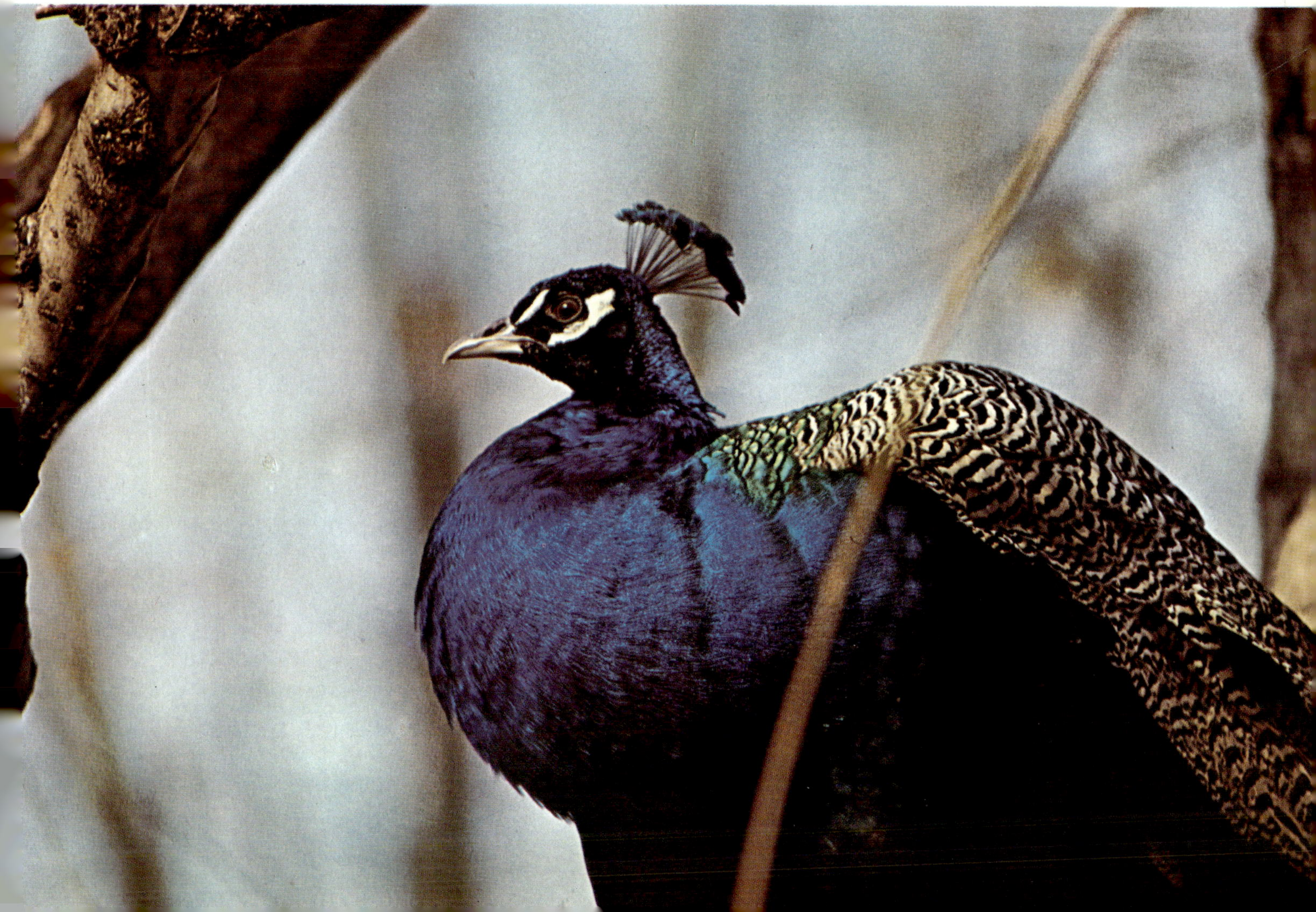

Blue-crowned hanging parrakeet
(*Loriculus galgulus*)

and two to the rear, whereas the owl has three, one of which is opposable to the others. The parrakeet uses its claws, not for catching live prey but for perching securely on a branch and for carrying food to the bill.

As a rule there is little difference in outward appearance between the male and female parrakeet and the birds usually pair for life. The female takes sole responsibility for incubating the eggs, these generally being laid in a tree cavity or rock cleft. The baby parrakeets are born naked and defenceless, and are fed by the parents on fruit and predigested seeds. The nestlings develop very slowly. When they leave the nest the face and head are still bare of feathers and these do not in fact sprout properly until the birds are adult.

Fossil remains discovered in many parts of the world indicate that such birds flourished in the Oligocene epoch and that some species were native to Europe. Millions of years later they were reintroduced to Europe by soldiers of Alexander the Great, returning from Asia. Along with other exotic birds such as flamingos and peafowl, they were much prized in ancient Rome, though not necessarily in the manner of the emperor Helio-

gabalus who tossed all three species as food to his lions.

Among the parrakeets of the Oriental region there are a dozen species of the genus *Psittacula*. These are to be found throughout Asia and although tropical inhabitants of the jungle they may sometimes be seen on cultivated land as well as in parks and gardens. The characteristic features of this group are the graduated tail, which is longer than the wing, a high, prominently curved bill and an exceptionally short lower mandible.

The Cingalese Alexandrine parrakeet (*Psittacula eupatria*) is not one of the typical forest dwellers of this group, preferring rocky terrain. The female may lay her two or three eggs either in a rock fissure or even a ruined building, but more frequently in a tree hollow. At one time there were six distinct subspecies but one of them, the Seychelles parrakeet (*Psittacula eupatria wardi*), has become extinct.

The ring-necked parrakeet (*Psittacula krameri*) is regarded as the most beautiful member of the group. The plumage is brilliant green, except for the blue nape and tail, and the broad black band which encircles the throat like a collar. It is one of the most abundant of parrakeet species. Sometimes, particularly in

Above and facing page : The parrakeets are among the noisiest of the jungle birds. Strangest of them are the hanging parrakeets which sleep upside-down like bats. The species shown on this page and on the left of the facing page is the Cingalese Alexandrine parrakeet; in the centre is the ring-necked parrakeet. Both species nest in tree cavities.

Cave swiftlet
(*Collocalia fucifaga*)

the breeding season, these birds form huge, noisy flocks, causing extensive damage to millet and sorghum crops.

More curious are the tiny green hanging parrakeets of the genus *Loriculus* whose tails are so short that they disappear completely under the wing tips when the birds are perched on a branch. Like other parrakeets they are active and agile in trees but are also to be seen hopping about quite effortlessly on the ground. They feed principally on fruit, pollen and, to a lesser extent, seeds and tender shoots. But what gives the group its name is the strange habit these parrakeets have of sleeping head-downwards, suspended by their feet from a branch, like bats.

Another special feature of the behaviour of the hanging parrakeets concerns their nest-building activities. During the breeding season they carefully select a cavity in a tree trunk as a nesting site and then proceed to line it with pieces of bark, twigs, leaves and grass. What is unique is the manner in which these miscellaneous building materials are transported to the nest. The parrakeet collects the bits and pieces in its bill and then tucks them securely between the feathers of the back, shoulders, breast and throat.

The commonest bird in this strange group is the blue-headed hanging parrakeet (*Loriculus galgulus*) which is about the size of a robin. The plumage is magnificent—green on the back, bright red on the breast and tail, blue on the forehead.

The source of bird's nest soup

Not all the birds of the South-east Asian jungles are remarkable for their glowing colours. One of them is positively drab in comparison with the multicoloured parrakeets and would hardly attract attention were it not for the fact that it constructs what might be termed the most expensive nest in the world—this being the principal ingredient, in fact, of that Oriental gastronomic delicacy known as bird's nest soup.

This rather nondescript grey or brown bird which by its shape and manner of flight might be mistaken for a swallow or martin, is actually a member of the swift family—the cave swiftlet (*Collocalia fucifaga*). It is one of several related insectivorous species found throughout the Indo-Australian region, most of them nesting in rock clefts and caves. During the breeding season the salivary glands of the bird swell, exuding large quantities of transparent saliva which hardens immediately it comes in contact with the air. This fluid, sometimes mixed with pieces of bark, lichen or vegetable fibre, but in some cases without any other additive, is used for building a cup-shaped nest which adheres to the rock surface. The edible quality of these nests has been recognised in the East for many centuries, particularly by the Chinese, who use it to prepare a tasty soup.

The trade in cave swiftlets' nests is colossal. To collect the raw material specialised gatherers prop rickety bamboo ladders against the cave walls and knock the nests down with long poles. Each nest weighs only an ounce or thereabouts but they are found in enormous numbers for the cave swiftlets live in colonies containing many thousands of birds. They breed two or three times

a year and build a new nest on each occasion, replacing it immediately if destroyed. The replacement nests, however, generally contain impurities and the highest prices are paid for nests that consist entirely of pure hardened saliva. The coastal limestone caves of Indo-China are particularly rich hunting grounds for best-quality nests.

Another insect-eating bird of the jungle is the frogmouth (*Batrachostomus javensis*), but unlike other members of the family Caprimulgidae, which includes the nightjars, it catches its prey not on the wing but on the ground or in a tree. During the daytime it perches on a branch, so stiff and motionless that its presence can hardly be detected. Twice a year, in March and September, it builts a nest on low, lateral branches. One egg is laid and is incubated by both the male and female. So tiny is the nest that there is a serious risk of it being dislodged by wind or rain, and it is probably for this reason that the parents do not abandon the egg for an instant.

Although insects are the main constituent of the frogmouth's diet these are often supplemented with fruit, in which respect too it differs from related species.

The nest of the cave swiftlet is made of hardened saliva, sometimes reinforced, as here, by feathers and vegetable fibres. It constitutes the main ingredient of the delicacy known as bird's nest soup.

CAVE SWIFTLET
(*Collocalia fucifaga*)

Class: Aves
Order: Apodiformes
Family: Apodidae
Total length: up to $5\frac{1}{4}$ inches (13 cm)
Wing-length: $4\frac{1}{2}$–$4\frac{3}{4}$ inches (11·3–12·1 cm)
Diet: insects
Number of eggs: 1–2
Incubation: 28 days

Colour basically grey or brown, slightly lighter on underparts. Dark eyes, bill and feet black.

Raptors of the Asiatic forests

For millions of inhabitants of South-east Asia the annual miracle of the monsoon is a life-restoring event. Torrential rains, of a force and intensity that we in the West seldom experience and can hardly imagine, sweep in from the ocean, transforming arid tracts of semi-desert into swampland. Streams swell into rivers, rivers into raging cataracts. Even the thickly knit canopy of trees cannot effectively protect the jungle inhabitants from the lashing fury of the elements. Where only the previous day the forest echoed to a varied chorus of animal noises now there is just the monotonous drumming of rain on leaves, punctuated occasionally by the strident, terrified screech of a peafowl. For hunter and hunted alike the search for food is for the time being a secondary consideration, their main concern being to remain under shelter until the rain abates. Only when the skies clear can the normal life of the jungle be resumed.

The forest birds are soon fluttering once more through the dripping foliage; and the raptors too are again on the wing, soaring high above the trees and mountainsides for their prey. There are several species here that will be familiar to the student of European birdlife. A medium-sized bird swooping on a rodent in a jungle clearing turns out to be a black kite; and both the peregrine falcon and Bonelli's eagle may be sighted circling their eyries high above the rock faces that jut up beyond the solid mass of jungle greenery. The buzzard tribe is represented here by the upland buzzard (*Buteo hemilasus*); and the ornithologist will have little difficulty in recognising other birds of prey that clearly have counterparts in the Palearctic region, even though closer examination reveals small differences in local races. But apart from minor variations in appearance the behaviour of these raptors is the same as that of their European relatives.

The fishing eagles

Nevertheless there are a number of other birds of prey living in and around the rain forests of South-east Asia which are not to be found anywhere else in the world. Particularly distinctive is a small raptor whose habitats are forest-fringed streams and rivers. From its vantage point high in a tree this bird will swoop down, dive into the water and emerge seconds later with a fish clutched in its talons. Known as the lesser fishing eagle (*Ichthyophaga nana*), this specialised hunter has a wide distribution from Kashmir and the Himalayan massif in the north down through Burma, Thailand and northern Indo-China to Malaysia and across to Sumatra, Borneo and the Celebes. Within this area there are two subspecies, similar in appearance and habit, *Ichthyophaga nana nana* and *Ichthyophaga nana plumbea*, the latter slightly the larger.

The lesser fishing eagle is a sedentary bird whose hunting grounds extend for perhaps a couple of miles along a river bank. From the branch where it spends the greater part of its time the bird keenly scans the surface of the water, alert for the slightest movements of the fishes which constitute its only food.

Facing page : Stiff erectile feathers surround the head of the Philippine monkey-eating eagle, enlarging the face like the mane of a lion. This powerful raptor, which hunts mammals such as flying lemurs in addition to monkeys, is an endangered species, the last sizeable population being confined to the island of Mindanao.

LESSER FISHING EAGLE
(*Ichthyophaga nana*)

Class: Aves
Order: Falconiformes
Family: Accipitridae
Total length: 19½ inches (50 cm)
Wing-length: 16 inches (40 cm)
Diet: fish
Number of eggs: 2–3
Incubation: 32–35 days

General colour brownish-ochre, slightly bluish on head and neck. Belly and rump white. Breast brownish-green.

Black baza
(*Aviceda leuphotes*)

Jerdon's baza
(*Aviceda jerdoni*)

Grey-headed fishing eagle
(*Ichthyophaga ichthyaetus*)

Osprey
(*Pandion haliaetus*)

Shikra
(*Accipiter badius*)

Philippine monkey-eating eagle
(*Pithecophaga jefferyi*)

Should it be unable to detect any activity below it will fly off to another tree and settle down to a further vigil. Sometimes it abandons its perch to cruise slowly around over the water. It only needs a ripple or shadow near the surface for the eagle to come diving down. The claws strike home and the raptor heads back for land with its meal.

After the monsoon rains the gushing streams and rivers criss-cross the forest in every direction; but in the long dry season their courses are marked only by strings of pools and puddles where fishes in their thousands mill together, gasping for life. Now the task of the fishing eagle is enormously simplified. It sits and waits as usual but hardly a moment passes without a fish leaping out of the shallow water. Food is there for the taking and the bird expends the minimum of energy in the process. In particularly dry districts where there are no watercourses even in the monsoon season, the raptor survives in the swampy ground fringing the jungle.

Male and female live together all the year round and may sometimes be seen chasing each other over the water or through the trees. As a rule, however, they spend hours on end perched on a branch, for there is no purpose in wasting energy.

The breeding season is between November and March, vary-ing according to latitude. The birds build a large nest in the higher branches of a tree bordering the river and line it with leaves and grass. On this bed of soft, moist greenery the female lays two or three white eggs, measuring about $2\frac{1}{2} \times 2$ inches, which are incubated by both birds for 32–35 days. The eaglets are born naked and defenceless, relying entirely on their parents who provide them with copious amounts of food. The same eyrie may be used by the adults for several consecutive years and since they add fresh vegetation every time the nest eventually assumes considerable dimensions.

The lesser fishing eagle plays an important role in the lives of some jungle tribes who believe that it is capable of predicting the future. If the bird moves its head in one direction the auguries for the coming season's crops are favourable; if in the other they are bad.

Closely related to the lesser fishing eagle is the grey-headed fishing eagle (*Ichthyophaga ichthyaetus*). Not only is the latter somewhat larger and heavier but its behaviour differs to the extent that it prefers to establish itself near lakes and large river estuaries rather than in forests.

The secretive bazas

We have to plunge much deeper into the jungle, far from the rivers and streams, to discover the habitats of Jerdon's baza (*Aviceda jerdoni*), a beautiful bird of prey which, although not especially rare, is so secretive that a great deal of luck and patience are required to catch a glimpse of it. When this does happen it is likely to be quite by chance and probably during the early evening. Its flight pattern is very distinctive, an alternation of flapping and gliding, but it is not a particularly aggressive or powerful hunter, feeding on insects, amphibians, reptiles and, very occasionally, other birds.

January to March is the breeding season as both male and female perform spectacular aerobatic displays high above the trees, to the accompaniment of shrill cries. Later they mate in a tree close to their nest which is securely placed in the fork of two main branches. Both birds use their claws to carry large quantities of twigs to the site and the finished nest is quite roomy. Between February and April—the exact time varies from one region to another, and in exceptional cases may be in September—the female lays two or three greenish eggs measuring $1\frac{3}{4} \times 1\frac{1}{2}$ inches, incubating them alone for about 35 days. The nestlings are fed on various vertebrates caught by the father and the mother remains with her brood as long as they need protection and care. Being well fed they grow rapidly and soon after their wing feathers appear they are fluttering from branch to branch. Within three months of hatching they are able to fend for themselves.

In contrast to most raptors which make use of the same nest year after year, Jerdon's bazas do not normally reconstruct their old eyrie but build a new one.

Similar in appearance and habit is the black baza (*Aviceda leuphotes*); but whereas Jerdon's baza is black and brown, the underparts white with chestnut bars, this species is a handsome grey, almost black on back and wings, and the orange breast is striped with broad black bars. Both species have erectile crests on the head.

The bazas of the genus *Aviceda* all possess a characteristic type of double-toothed bill. Another distinctive feature is the stiff crest of head feathers.

Jungle sparrowhawks

Among the numerous small and medium-sized birds of the Asiatic rain forests several species of sparrowhawk make their particular mark. One of the most famous is the shikra (*Accipiter badius*), much prized by falconers for its astonishing persistence and skill in manoeuvring its way through the trees. This bird is found from Kashmir into central India and is renowned for its bold, aggressive temperament. In pursuit of a small bird, for example, it will be quite undeterred by the presence of people nearby and will even chase its prey through an open window. Its normal hunting technique is to flutter from tree to tree, finally settling on a high branch and then swooping like lightning on its victim. Should the quest prove unrewarding it may hover low above the ground, flapping its wings violently, which sometimes has the effect of driving a frightened animal into the open.

The courtship displays of these birds are characterised by much excited and noisy fluttering through the foliage. The male then perches on a branch and summons his partner with shrill cries, after which they mate.

The crested goshawk (*Accipiter trivirgatus*), another jungle resident, is especially handsome, with strikingly contrasted plumage. The cheeks are pale grey, the forehead and crown of head black, topped by a splendid black crest. The dark back is flecked with white and the whitish underparts are streaked with dark to russet bars, lengthwise on the breast, transverse on the belly. The female is usually browner than the male and her underparts are not so finely marked. This is a very small raptor, which weighs at most 12 ounces, yet this does not prevent it being an extraordinarily brave and belligerent bird. Both male and female are capable of killing birds almost as large as themselves, but in general they prefer to catch smaller birds, supplementing these with reptiles and tiny mammals.

Little is known about the biology and behaviour of this species, for it is extremely shy. Both birds evidently share the building of the nest, this being assembled from twigs and small branches and lined with fresh vegetation. Here the female lays two or three blue-white eggs which she incubates while her partner occupies himself with hunting. The nest is used in successive seasons.

The Besra sparrowhawk (*Accipiter virgatus*) is another small jungle raptor, whose nestlings look so much like immature crested goshawks that they might easily be mistaken for miniature versions of the latter species. The adult plumage is dark brown on the back, streaked with white on the nape of the neck. The underparts are white and there are large black marks on the throat and upper part of the breast. The belly is streaked with transverse ochreous bars but these show great individual variation.

This little bird conceals itself in the densest vegetation, whether it be on the wooded slopes of the Himalayas or in the mangrove swamps of the coastal regions. With its extremely powerful wings it is a swift flier, much used in India for falcony. Although obviously incapable of going after large prey it is a specialist in the capture of other birds, including swifts. In the

---

CRESTED GOSHAWK
(*Accipiter trivirgatus*)

Class: Aves
Order: Falconiformes
Family: Accipitridae
Total length: 16 inches (40 cm)
Weight: maximum 12 ounces (340 g)
Diet: birds, reptiles, small mammals
Number of eggs: 2–3

Dark upper parts flecked with white. Crest of black feathers. White throat; breast and belly with dark brown to russet bars. Lower tail coverts white.

wild it will seldom pursue prey much larger than a sparrow and may vary its diet with insects and lizards. Its versatility is shown by the fact that it will hunt on the ground, up in the trees or on the wing.

The nest of this species is generally situated high in a tree and is very often the abandoned property of a crow or pigeon. The female lays from two to four eggs, white or pale blue with dark brown speckles. The breeding season is very variable, depending on distribution. Although generally in the spring it may occur at any time between January and September.

The forests of the Oriental region, as has been pointed out, assume a variety of guises and this lack of uniformity has resulted in diverse forms of animal life especially adapted to the conditions of a particular habitat. This helps to explain why these three species alone are subdivided, according to their range through South-east Asia, into twenty separate subspecies or races.

Two other notable birds of prey belonging to the same family are to be found in the forests of the Celebes and adjoining islands. One is the Celebes crested goshawk (*Accipiter griseiceps*), which is similar in general appearance and colour of plumage to an immature common goshawk; the other is the spot-tailed accipiter (*Accipiter trinotatus*), with a wine-coloured breast and a pure white belly. Because of their secretive habits not much is known about the behaviour of these birds.

The shikra, a fierce bird of prey, waits patiently for its victims, sometimes perched on a branch or post. It looks more like a small goshawk than a sparrowhawk and is much prized by falconers in the Orient for its fearless, aggressive temperament.

The Indian black eagle is the pirate of the South-east Asian jungle. To obtain its staple diet of eggs and nestlings it swoops down and removes the entire nest while the parent birds relax their guard.

The nest robber

The goshawk, with its silent, speedy flight, has a claim to be considered the most remarkable hunting bird of the deciduous forests of the Palearctic region; but in the jungles of South-east Asia there is a bird which can boast just as much skill and cunning in comparable conditions but whose habits are even more extraordinary. The Indian black eagle (*Ictinaetus malayensis*) is in the literal sense the pirate of the jungle. A bird that preys on eggs and nestlings is in no way unusual but a bird that filches the entire nest, as this one does, to get at the contents, is clearly a very exceptional individual.

One might well receive the impression that this bird, which flaps so slowly and noiselessly over the treetops, is ill-equipped to carry out a 'commando' operation of this nature, for surely success must depend on the element of surprise – a sudden, swift and precise attack followed by equally rapid withdrawal. This is in fact what happens. Having pinpointed its target, the eagle holds back until it sees the adult birds abandoning their brood for a few moments. It then swoops down, grabs the nest and contents, and flies off to safety. Eggs and nestlings are promptly consumed and the empty nest tossed away.

Principal victims of the Indian black eagle are doves and passerines, and when nests are not freely available it will catch adult birds, lizards, small snakes and rodents. The stomachs of three raptors were found to contain a bat, two swifts and three rats respectively, evidence of a broadly based diet.

As a rule this astute nest robber sites its own eyrie in the fork of a main branch of a large tree. Building materials include branches which are secured with stems and fibres. December is the breeding season and the female lays usually one egg, rarely two. While these are being incubated and later when the eaglets are being reared, the adults defend the nest with great determination.

Hawk eagles

When the sun is high in the sky and the cool freshness of dawn has given way to searing midday heat, almost all the animal inhabitants of the Oriental tropical forest – even those that are best adapted to withstand extremely high temperatures – are asleep, finding relief in the shade of rocks and trees. These are not propitious hours for hunting and in this knowledge the smaller animals of the forest may rest reasonably secure until the shadows lengthen and the sun sinks low over the horizon.

Yet there are exceptions to almost every rule. A strange mewing

---

CRESTED HAWK EAGLE
(*Spizaetus cirrhatus*)

Class: Aves
Order: Falconiformes
Family: Accipitridae
Total length: 27½ inches (70 cm)
Wing-length: 16–18 inches (40–46 cm)
Number of eggs: one
Incubation: 35–37 days

Adults
Colour very variable. Upper parts usually reddish-brown; lower parts of wings and belly white, streaked with red. Erectile crest. Some individuals almost black.

Young
Pale plumage; breast and belly whitish with light brown bars; head and wing coverts white.

The crested hawk eagle is a versatile hunter. Widely used in falconry, it is equally effective over open ground and among shrubs and thickets.

sound from on high is a warning to some of these lethargic animals, for it indicates that one powerful raptor at least is abroad and scouring the forest below for prey. Its silhouette is reminiscent both of a goshawk and Bonelli's eagle and it is known as the crested hawk eagle (*Spizaetus cirrhatus*), favourite of falconers throughout the Orient from Persia to Japan. Identifiable by its small head and erectile crest, this magnificent bird has all the attributes of a hunter, the flashing eye, the short, curved bill, long legs with powerful talons, broad wings and sweeping tail. This is obviously a raptor that combines strength and speed, perfectly adapted to special forest conditions. Whether hunting over open ground and swooping on its prey like an eagle or zigzagging through the undergrowth like a goshawk, this compact bird is an efficient killer of birds and mammals much larger and heavier than itself.

The hawk eagles of the genus *Spizaetus* have an enormous distribution range which embraces almost the entire Asiatic jungle region from India to Japan. The species already described is essentially a bird of the southern lowlands, particularly those stretches of open savannah that are punctuated by clumps of woodland. The mountain hawk eagle (*Spizaetus nipalensis*), as its name implies, is an inhabitant of the highlands and colder forests of the north.

Despite its extraordinary hunting capacity the crested hawk eagle generally concentrates its attacks on small and medium-sized birds and mammals, a preference evidently dictated by the fact that such a prey can more conveniently be carried back to the eyrie, without the need to linger at the site of the kill.

The raptor is sedentary by habit and each male and female pairs for life, roaming the same hunting grounds at all seasons, year after year. At the beginning of the breeding period (and in the most southerly parts of the distribution range this is in December) both birds soar up on rising warm air currents until they are thousands of feet above the trees. Even at this height the characteristic sharp mewing cries are clearly audible from the ground. Now and then they flap their wings to gain altitude but for the most part they indulge in long graceful glides, ceaselessly patrolling their territory, vigilant for the possible intrusion of an unwelcome rival.

This is the time when the nest of the previous year has to be restored to serviceable condition or a new one built. It is usually situated in a tall tree in the forest, often close to a river or stream. The eyrie much resembles that of a goshawk but is somewhat larger. Both birds carry small branches and twigs to the site, these serving as the framework. Later it is lined with grass and green leaves, usually by the more capable female. This soft layer of vegetation is replaced at regular intervals so that the nest is always kept fresh and clean.

When this task has been completed the female lays a single egg (whitish or pale yellow) and incubates it for about five weeks, leaving it only for the briefest of moments when she flies off to a nearby branch to eat whatever prey has been provided by her mate. But the eggs are not abandoned for she is immediately replaced by the male. After the baby hatches he continues to hunt and bring back food to the eyrie while the female devotes all her attention to her offspring. The latter is initially covered with down and its rate of growth is fairly slow. Its adult plumage appears only by stages, after a series of moults.

The monkey-eating eagle

Each of the world's jungles is the home of one enormous and powerful bird of prey specialising, though not to the exclusion of all other prey, in the capture of monkeys. South America has the great harpy eagle (*Harpia harpyja*), Africa the crowned hawk eagle (*Stephanoaetus coronatus*), and Asia the Philippine monkey-eating eagle (*Pithecophaga jefferyi*). All these fierce-looking raptors stand 3 feet or more high and have certain physical features in

The piercing gaze, broad wings, long legs and powerful talons of the crested hawk eagle are the marks of one of the most proficient raptors of the Oriental region.

common – a very long tail, similar to that of a goshawk and ideal for following a twisting, turning route through closely packed trees; short, broad wings that enable them to pursue victims in the undergrowth; huge talons that are far more powerful than those of most other birds of prey; and a viciously hooked bill for tearing flesh.

They are all impressive birds and the similarity between them, despite their geographical remoteness from one another, is no mere accident. All three have evolved along parallel lines in response to the needs imposed upon them by their surroundings. The short, compact wings are essential attributes of a bird which has to zigzag through dense forest, avoiding collisions with tree trunks and stout branches, and which furthermore must be able to accelerate rapidly over short distances. Similarly, the long tail is invaluable for steering, making it possible for the bird to effect abrupt changes of direction. Speed and manoeuvrability are the real secrets of success in hunting under such conditions. The powerful feet and talons are merely the weapons which guarantee that the kill will be swift, accurate and immediate. There can be no half measures in this respect. An animal that is

The rounded wings and long tail of the crested hawk eagle are clearly visible in flight. The bird is sedentary by habit and patrols its hunting grounds from the air every day.

only stunned or slightly injured has the opportunity of escaping into the depths of the forest, and from the raptor's point of view all the preparatory work for the attack will have been so much wasted energy. Nature has therefore ensured that these three great forest birds seldom miss their mark.

The harpy eagles are prudent birds which do not draw attention to themselves, as do so many other raptors, by soaring on thermal currents to great altitude. In fact they will seldom be seen above the treetop level at all, gliding noiselessly like huge dark shadows through the half-light of the jungle or simply perching on a dead branch, not moving a muscle for hours on end. Yet they are far from lethargic and the nickname which African tribes have given to their local species – 'leopard of the air' – is singularly appropriate. For within seconds of a potential victim coming into their orbit, whether it be a tree-dwelling mammal, a reptile or (more rarely) a bird, they will be in action, swooping unerringly on their prey.

The species with which we are concerned here is the least known of the three – the Philippine monkey-eating eagle – and unfortunately it is on the verge of extinction. Like its relatives it is an imposing bird, for in addition to the strongly hooked beak and glittering eyes the head is adorned with a tuft of long feathers which, when fully erected, resembles the mane of a lion. Nowadays confined to dense mountain forests on a handful of islands in the Philippines archipelago (up to heights of 4,000 feet), the sudden appearance of this fierce predator, like that of a tiger or leopard, unleashes a general state of panic in the jungle. Monkeys start chattering noisily, leaping from branch to branch in frenzied warning to their companions, desperate to escape to safety; and smaller birds twitter in alarm, fluttering in agitated fashion through the foliage. Nevertheless not all the forest inhabitants flee in terror. If a hornbill should happen to sight the eagle it may dive unhesitatingly to the attack, striking at the raptor's head repeatedly with its heavy, oversized beak.

The Filipino naturalist Rodolfo B. González of Silliman University confirms that monkeys are not the only, nor indeed the principal, victims of this eagle. In the zone studied by the scientist in 1963 and 1964 a variety of tree-dwelling mammals were attacked, the majority being flying lemurs (*Cynocephalus volans*). González omitted to make it clear, however, whether these animals were caught while actually in flight or when they were taking refuge in the trees. Because of their protective coloration in the latter circumstances it is reasonable to assume that the eagle would only attack when it could see them clearly, namely when they spread their patagium preparatory to leaping into space; and since, once launched, the lemurs are incapable of varying their flight path to any great extent, it is even more likely that the eagle would choose precisely this moment to swoop.

The naturalist reported that out of 48 identifiable victims, 90 per cent were flying lemurs and 6 per cent macaques. The fact that the latter are strong, aggressive and intelligent is further proof of the eagle's remarkable capacity.

PHILIPPINE MONKEY-EATING EAGLE
(*Pithecophaga jefferyi*)

Class: Aves
Order: Falconiformes
Family: Accipitridae
Total length: 33½–39½ inches (85–100 cm)
Length of tail: 16½–18 inches (42–45 cm)
Wing-length: 22½–24 inches (57–61 cm)
Diet: flesh
Number of eggs: 2, but usually only one eaglet survives
Incubation: about 60 days

Back and wings dark brown with ochre or white marks at tips. Underparts and sides of head white; feet striped light brown. Tail dark chestnut with irregular light brown bars, paler underneath. Long, striped erectile feathers surround head like a mane; stiff crest. Bluish iris. Enormous curved bill tinged with blue, like cere. Feet yellow with powerful black claws.

The raptor constructs its eyrie on the broad lateral branches of a large tree, often on a wooded hill or mountainside. The female lays two eggs around mid-November and incubates them for approximately 60 days. The eaglet is capable of leaving the nest when only three months old but as a rule will not stray far away. Its upbringing is essentially the same as that of other young birds of prey. The male, which has the responsibility of feeding his partner while she is incubating, continues to hunt on behalf of the family when the chicks hatch but the female now assumes the duty of tearing up the food and generally caring for the brood. Some months after asserting its independence the immature eaglet, already closely resembling its parents, begins to accompany them on daily hunting expeditions, returning each night to sleep close to the place where it was born.

This bird, once an inhabitant of four of the Philippine islands, has recently vanished from two—Samar and Leyte. It had been thought to have disappeared too from Luzon, but in 1963 and 1964 two birds were killed and several others were later sighted there, leading to the conclusion that there is still a small population left on this island. Most of the survivors, however, are clustered on Mindanao, although even in this last remaining sanctuary of the species numbers have dwindled alarmingly in the past twenty years. One reason is that the local population keep up a running battle with the birds, knowing that in various shapes and forms they will fetch high prices from wealthy trophy hunters. The increasing availability and use of firearms, dating from the Second World War, has resulted in a terrible wave of destruction throughout the Philippines, and the monkey-eating eagle is just one of many victims. The naturalist Lee Talbot has pointed out that to exhibit a stuffed specimen of this bird in one's living room is a sign of real prowess. A local journal recently carried a story of a game warden who boasted how he had shot an enormous monkey-eating eagle in a protected area for which he was responsible. Far from being punished the man was extolled as a local hero for his feat. If this can win applause what hope is there for the few remaining birds in the wild?

The IUCN and the World Wildlife Fund, keenly aware of the peril facing this magnificent species, has launched an information and propaganda campaign in the islands, hoping to appeal to public conscience and to save the birds. Furthermore zoologists all over the world have undertaken not to use any more live monkey-eating eagles in laboratory experiments. But hunting and poaching are not the only causes of the species' decline. As in so many other places, the gradual destruction of the forest habitat to make way for agriculture is an additional factor. By and large, therefore, the outlook is somewhat gloomy. In 1964 Talbot estimated that there were about one hundred eagles still left in the Philippines. By 1969, according to information supplied by Tom Harrisson to the World Wildlife Fund, the numbers of monkey-eating eagles had dropped considerably in the intervening five years. More recent information, provided by Charles A. Lindbergh in 1971 and based on the findings of Álvarez, paints an even blacker picture, suggesting that there may now be no more than forty of the birds left on Mindanao.

A raid through the undergrowth by the formidable Philippine monkey-eating eagle provokes general confusion and panic among most of the jungle inhabitants. Trophy hunters have brought the species to the verge of extinction.

CHAPTER 89

Pygmies and giants of the jungle waterways

There are few parts of the world where water plays such a vital part in shaping the character of the landscape as it does in the Oriental region. The enormous quantities of rain that pour down every year on the jungles, savannahs, mountains and plains of South-east Asia form a complex network of streams and rivers that eventually make their way to the sea. Some of these are comparatively short and their names mean nothing to people in the West; but others, such as the Indus, the Ganges, the Brahmaputra and the Mekong are recognised as being among the most impressive and imposing waterways in the world. From time immemorial they have served as navigable highways to the interior of the Asian continent and along their banks remarkable civilisations have sprung up, great religions have been born and—to man's discredit—frightful wars have been fought. Yet large or small, famous or obscure, the rivers of tropical Asia have influenced the lives of millions more profoundly than their counterparts, for example, in Europe.

Each year, during the monsoon season, torrents of water crash down on the burned, parched lands of the suffering continent and flood thousands of square miles of countryside, frequently bringing further devastation and misery in their wake. Ton upon ton of soil is dislodged by an overflowing river and swept along in its relentless passage towards the sea. Eventually it is either deposited as a fertilising agent on cultivated land or it is simply swallowed up in the vast expanse of ocean. In Burma, the Irawaddy contains some 20 ounces of silt to the cubic yard, which is roughly ten times as much as exists in the Seine or the Rhine. The rivers of Java spill a greater quantity of alluvium into the ocean in a single day than the Marne does in 200 years! On this island the Solo (Bengawan), one-third longer than the

Facing page : The Siamese fighting fish, with its flowing fins and brilliant colours, is renowned for its aggressive nature and has been selectively bred for the freshwater aquarium. Here the male begins his courtship 'dance' around the female.

Rhine, carries along sixty times as much sediment, and because of this the river delta becomes 300 yards or so wider every year; and the comparable figure for the Mekong delta is about 250 yards. The pattern is repeated throughout South-east Asia and it is not surprising to learn that the most extensive delta in the world is that formed by the confluence of two mighty rivers, the Ganges and the Brahmaputra.

South-east Asia is also crisscrossed by innumerable smaller streams whose courses depend in large measure on the monsoon rains. When the wind sweeps away the clouds and the sun burns down mercilessly for months on end, these streams are reduced to mere trickles or more frequently dried up completely, so that their former passage is marked only by undulating white streaks across the arid plains. The only permanent watercourses are those that have their source in the Himalayas, as a result of the extremely slow thawing of the snows that shroud the slopes. The levels of the Indus, the Ganges and the Brahmaputra, all of which rise in the heart of that massive mountain range, drop considerably in the long dry season but never dry out completely.

After flowing through the narrow pass that separates the plains of the Punjab from those of Sind in Pakistan, the Indus widens out into an immense alluvial plain which extends from the Kirthar Mountains in the west, on the borders of the Baluchistan plateau, to the edge of the Thar Desert in the east. Like all the waterways of the Oriental region this river is unpredictable and has evidently changed its course many times over the centuries, as is shown by the traces of old canals and the sites of many abandoned villages in Sind province. Even the delta of the Indus has shifted position. At one time the river had its outlet in a huge inland sea, the Rann of Kutch. The latter was gradually filled up with silt until it was transformed into a vast stretch of marshy terrain; and doubtless this swamp will vanish completely beneath the sand carried westwards by the blustering desert winds. The modern delta of the Indus, with its outlet in the Gulf of Oman, becomes larger every year and is a paradise for innumerable species of water birds. Along the banks of the estuary one may also be fortunate enough to catch a glimpse of a basking salt-water crocodile or a gavial, largest of the world's crocodilians; but although both species were at one time abundant these huge reptiles are nowadays becoming rare for, as has happened with the African crocodiles, they are ruthlessly hunted for their skin.

The Ganges, like the Indus, also flows through an immense plain, the highest point of which is 1,000 feet above sea level. The Ganges delta is immense, for it is at this point that it is joined by an even longer river, the Brahmaputra. The latter flows through a much narrower valley. After a long sweep towards the east, the river coils its way via rapids and cascades through the mountain barrier formed by the Himalayas, changes direction and finally emerges in the fertile valley of Assam. Here, in a region 500 miles long and 50 miles wide, the river forms a number of tributaries which come together and then separate again to make up an enormous tract of swampland, thickly covered with vegetation, inhabited by a wonderful variety of animals.

Each year torrents of monsoon rain flood huge tracts of land in South-east Asia and quantities of alluvium deposited by streams and rivers stimulate the growth of staple crops. This is an aerial view of the plateau of Laos which is irrigated by the Mekong River.

Here, in the marshes of Assam, Indian elephants tranquilly feed in the tall grass, single-horned Indian rhinoceroses wallow in the mud, and majestic water buffaloes either graze peacefully or simply lie motionless, up to their necks in mud. Groups of barasingh (swamp deer) ruminate in the shade, and wild boars dig for tasty roots, though ever on the alert for a prowling tiger or leopard. All these species are comparatively rare and it was to check their steady decline in numbers that the Indian government decided in 1908 to establish the Kaziranga Wildlife Sanctuary. Stretching for 166 square miles along the left bank of the Brahmaputra, the reserve is open to visitors who can admire its wildlife safely from the back of an elephant.

Beyond the valley of Assam the Brahmaputra joins the Ganges to form the world's largest delta, extending over an area of more than 13,000 square miles and scattered with innumerable muddy-shored islands, the homes of many animals, large and small. At one end of the scale are the extraordinary fishes known as mudskippers, which feed on the mud at low tide; at the other the famous and terrible man-eating tigers of the Sundarbans.

Like that of the Indus, the delta of the Ganges and Brahmaputra has altered position in the course of time, having shifted progressively eastwards, transforming central Bengal into a land of dead rivers and huge swamps, some of which have been converted into rice paddies.

For many centuries the forest of the Sundarbans, extending inland for 60–80 miles but now comprising only a narrow coastal belt, has provided occupation in the form of hunting, fishing and woodcutting for the people of the region. This forest is one of the last remaining haunts of the Bengal tiger, a fierce carnivore which, being an excellent swimmer, has no difficulty in making

The Kaziranga Wildlife Sanctuary on the banks of the Brahmaputra River in Assam harbours many rare animal species, including the great Indian rhinoceros. The elephant provides the only means of transport for visitors to the 166 square-mile reserve which is largely swampland.

its way from islet to islet, usually in quest of its favourite prey, the wild boar. Not infrequently the tigers kill a fisherman or woodcutter, but contrary to traditional belief they do not, it seems, eat human flesh and should properly be called man-killers rather than man-eaters.

The once-splendid forest of the Sundarbans is now only a shadow of its former self; and its gradual disappearance has had disastrous repercussions, not only on its wildlife, but also on the climate and tidal pattern of the Gulf of Bengal. Conservationists have been especially concerned about the fate of the Bengal tiger, a number of deer species and the many rare and beautiful orchids that once grew in such profusion. In 1960, after an exploratory expedition by the British ornithologist Guy Mountfort, and following the advice of the World Wildlife Fund, the then Pakistani government declared it a protected area and established the 100 square-mile Sundarbans Game Sanctuary.

The water buffalo

Probably the most typical animal of the river-studded plains which are completely submerged during the season of the monsoon rains, is the Indian buffalo or water buffalo (*Bubalus bubalis*). But although there are some thousands of the animals in South-east Asia, as well as in other continents where they have been introduced, these are domesticated beasts, differing in some respects from the genuine wild water buffaloes which are nowadays protected in reserves.

The water buffalo is a magnificent animal with smooth black hair and powerful, backward-curving horns which, unlike those of the African buffalo, are flat-fronted. It may stand up to 6 feet at the shoulder and weigh between 1,100 and 2,200 lb. Its habitat is swampy grassland with plenty of muddy pools for wallowing, this being the only way the animal can escape the incessant torments of swarms of insects.

In days gone by large herds of water buffaloes roamed the immense plains through which the Ganges and Brahmaputra coursed, and their range extended southward along the east coast of the Indian peninsula. But as land was given over to cultivation the habitat of the animals shrank and their numbers were further decimated by cattle plague, spread by domesticated buffaloes. According to estimates made in 1966 there are today about 2,000 buffaloes left. About one-quarter of these live in Assam, of which 700 are in the Kaziranga reserve. Smaller groups are scattered through other national parks.

Water buffaloes are as a rule found in herds of from ten to twenty individuals. When in rut the bulls collect a harem of cows and display great aggressiveness. The animals mate at more or less any time of year but the majority of calves are born between October and December.

Water buffaloes have few enemies. From time to time a tiger, spurred to rashness by the pangs of acute hunger, may attack a solitary buffalo but risks life and limb in challenging such a powerful animal. There is also the additional hazard of being charged and trampled by the whole herd.

Above and facing page : Water buffaloes are gregarious animals, living in small herds in marshy areas or close to rivers. The mud-bath ritual is essential for ridding the skin of parasites and keeping insects at bay. Numbers of genuinely wild buffaloes are nowadays reduced for the species has been used for domestication for centuries, performing the same role in Asia as do domestic cattle in Europe and America.

African buffalo

Indian or water buffalo
(wild form)

There is a notable difference in the horn structure of the Indian and African species of buffalo. Those of the water buffalo are not as heavy, curve backwards and are flat at the front.

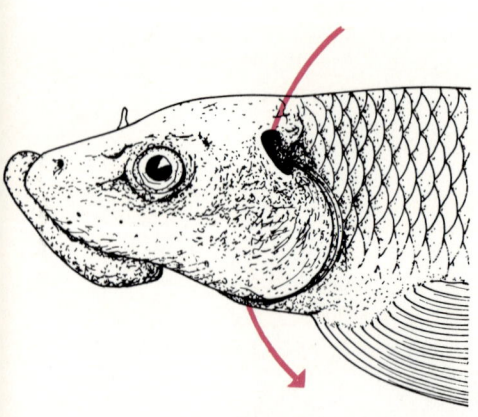

The two species of sucker loach of the genus *Gyrinocheilus* have enlarged lips which enable the fishes to attach themselves to the pebbly beds of swiftly flowing mountain rivers. The problem of respiration is solved ingeniously by the juxtaposition of the two apertures, one at the top and the other on the underside of the head, used respectively for inhaling and exhaling Networks of tiny membranes open and close rapidly like valves. Thus the mouth is not used at all for breathing as it is in the majority of fishes.

The rare tamarau

The largest native animal of the Philippines is the tamarau (*Bubalus mindorensis*) which lives only on the island of Mindoro and is for that reason known by the alternative name of Mindoro buffalo. Standing 3 feet high, it has thick brown hair and large horns. There are only about one hundred of these animals left as a result of human persecution, and these have been driven into the last remote corners of the island, for although reserves have been established for their protection they are purely nominal and afford little guarantee of survival. Intensive hunting has also led the animals to adopt nocturnal habits. Not that such expeditions are necessarily crowned with success, judging by the exploits of one hunting party which loosed off 167 rifle shots and managed to kill only one tamarau!

The name of the Mindoro buffalo appears, predictably, in the list of threatened mammals in the IUCN's Red Data Book. Probably the only way to save it is to organise a detailed study of the species and its needs in the wild, followed by appropriate action by the Philippines government.

Playground of freshwater fishes

The continental waters of the Oriental region teem with a marvellous array of freshwater fishes, notable for their number and variety. They have adjusted in different ways to the demands of their particular environment, whether these are permanent watercourses or areas of flooded land, giving rise, in some cases to remarkable physical adaptations and behaviour patterns. Little wonder, therefore, that the freshwater fishes of South-east Asia are an extraordinarily rewarding field of study for the ichthyologist and ecologist.

Asia (or, more precisely, China and India) is the original home of the huge family of Cyprinidae which subsequently colonised Europe, Africa and America; and it is in eastern Asia that the greatest number of cyprinid species are still to be found, as well as some of the most unusual representatives of the group. Thus the largest cyprinids in the world live in India, including *Barbus tor*, one of the Indian mahseers, the rohu (*Labeo rohita*) and the catla (*Catla catla*), which may grow to more than 5 feet long and weigh over 100 lb. In contrast to these giants there is the tiny sleeper (*Pandaka pygmea*) from the Philippines, a member of the goby family, which at ½ inch long qualifies as the smallest vertebrate in the world.

The ichthyologist Sunda Lal Hora, who specialised in Asiatic fauna, has pointed out that there are interesting modifications in the shapes of the mouth of *Barbus tor*, some projecting farther out and being more lobate than others. In quiet waters the shape of the mouth tends to be normal, but it is markedly larger in fishes that live in swiftly flowing rivers and streams. The reason for this phenomenon is that the fish uses its mouth as a sucker for attaching itself to the pebbles and stones on the river bed in order not to be swept away by the current. The faster the flow the more efficient the sucker apparatus must be.

The mountain rivers of Thailand and Borneo are the homes of another strange cyprinid, the sucker loach. There are two species belonging to the genus *Gyrinocheilus* and like *Barbus tor* they too affix themselves to the pebbly river bed by suction; but in their case the mouth is situated low in the head and it is the lips which are enlarged and pulpy.

What is even more interesting is the way in which these species have solved the problem of breathing. Fishes normally use their mouth to swallow water containing oxygen and expel it through the gills; but those fishes with sucker apparatus cannot do this without being swept away by the current. *Barbus tor* has no alternative but to remain in a state of apnoea or suspended breathing, while attached to the bottom, prising itself loose occasionally for brief moments in order to take a few rapid breaths. The sucker loach, on the other hand, resorts to a much more ingenious method, not using its mouth at all for respiration. Breathing is effected by means of two openings, one for inhaling and one for exhaling. The former is situated at the top of the head and is directly linked to the front part of the lung cavity, the latter is on the lower side of the head. Both are furnished with membranes which serve as valves and these open and close 240 times every minute, giving some idea of the rate of breathing. Thus the mouth can be used exclusively for suction and feeding.

In discussing the freshwater fishes of the Oriental region special mention must be made of one species that has achieved popularity as a pet all over the world—the goldfish (*Carassius auratus*). In the wild it lives in swamps but in its domesticated form, selective breeding has resulted in a number of ornamental variants with unusual features, such as prominent eyes, large fins, etc. In fact the distinctive gold colour is found only in these aquarium specimens.

The Siamese fighting fish

One of the special delights of studying the animal kingdom is that there are always new surprises in store. Let us suppose for a moment that we were coming fresh to the subject, having read nothing and been taught nothing, and were asked to name the most aggressive fish in the world. Even if we were stuck for a precise answer, logic would compel us to choose one of the larger species, perhaps one of the sharks. We would be wrong.

Tamarau
(*Bubalus mindorensis*)

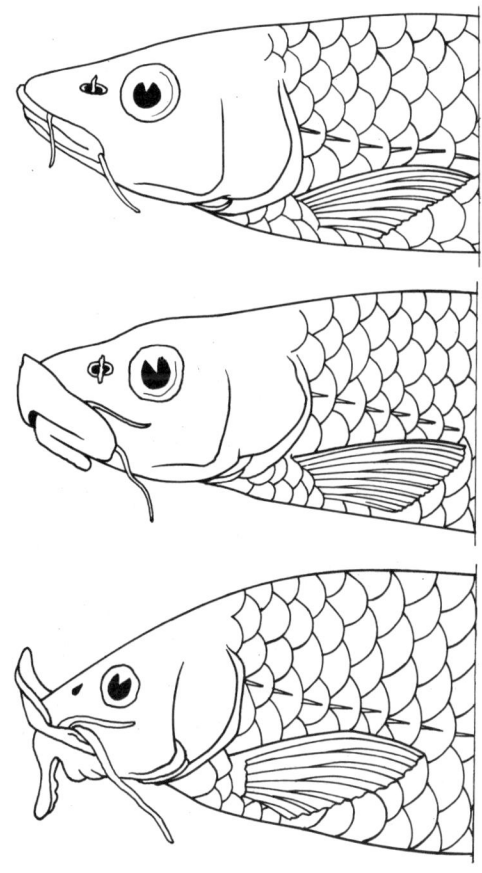

The shape of the mouth in the cyprinid *Barbus tor* has been found to undergo modifications according to the habitat of the individual fish. Thus in tranquil waters the mouth shape is normal but in fast-flowing currents, where the mouth is used as a sucker and affixed to stones and boulders on the river bed, the lips become fleshy and enlarged.

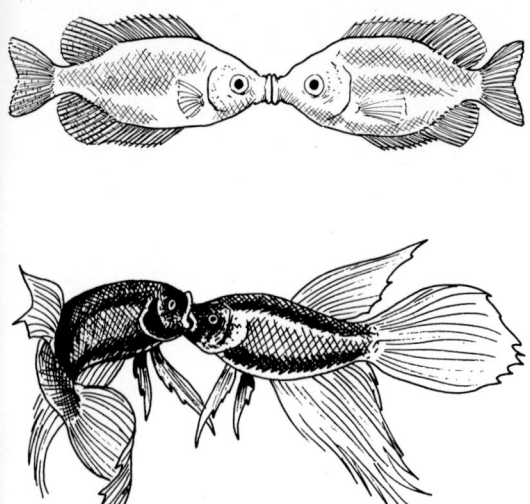

Two of the best known species of freshwater fishes in the Oriental region are the kissing gourami (*Helostoma temmincki*) and the Siamese fighting fish (*Betta splendens*), the former growing to 12 inches, the latter only to about 2½ inches. Although related to members of the Anabantid (Labyrinth) family, their behaviour is very different, the former not building a bubble nest like the latter when spawning. The name of kissing gourami derives from the fishes' habit (*above*) of bringing their lips in contact. The significance of this is debatable and may be a form of ritualised fighting. Intraspecific combat among Siamese fighting fishes (*below*) is much more serious for rival males will battle ferociously, snapping at each other's jaws, fins and flanks, often resulting in grave injury or death.

Facing page : The goldfish, related to the carp, is one of the commonest Cyprinids of South-east Asia. Centuries of selective breeding have produced many ornamental forms, one of which is shown in the upper picture, popular in aquaria all over the world.

The largest member of this family, the whale shark, is in fact a peaceful eater of plankton and quite inoffensive; and although some sharks are certainly dangerous if provoked, their reputation for ferocity has been greatly exaggerated. The truth is that for sheer savagery and belligerence none of the larger species can begin to match the 2½-inch-long *Betta splendens*, one of seven related species living in streams, rivers, lakes and canals through South-east Asia from Thailand to Borneo, more familiar under the name of Siamese fighting fish.

These little fishes grow to maturity very rapidly for their life is short, seldom exceeding two years. In spite of their well merited reputation for fierceness they do not play an important predatory role in the wild. Partly because of their modest size, they feed principally on minute aquatic animals such as water fleas, mosquito larvae, worms and the flesh of dead fishes.

Although all varieties of Siamese fighting fish are aggressive, combats between rival males in the wild seldom last more than about fifteen minutes. But the aquarium varieties, products of many decades of selective breeding, are far more bellicose and may battle for more than an hour (aquarists have reported fights lasting as long as six hours). In some Asiatic countries fights are elaborately staged and followed with as much enthusiasm as the equally cruel sport of cock-fighting.

The duel begins as two brightly coloured males (red, blue or green) size each other up, large flowing fins outspread, gill covers open, evidently with a view to mutual intimidation. Then they leap to the attack, showing an extraordinary turn of speed, each endeavouring to tear the other's fins with its teeth. After the battle is over the previously handsome fins may be reduced to ragged stumps. The antagonists also attempt to bite each other's flanks (sometimes pulling out scales) or clash head-on, which results in deep gashes around the jaws. Fights between females of the species never occur.

For the less bloodthirsty observer, the courtship rites of Siamese fighting fishes are equally spectacular and fascinating. Before approaching the female, the male prepares a large floating nest of transparent bubbles, expelled from the mouth and each enveloped in a mucous secretion which makes the nest quite strong. The actual spawning is preceded by a kind of dance in which the male, whose colours are exceptionally vivid at this stage, swims around the female, fins outspread. He compels her to perform a somersault until she is floating belly upwards and then embraces her, providing the stimulus for her to extrude from three to six eggs. The male then fertilises them as they float to the surface, collects them in his mouth and spits them into the bubble nest. Repeated embraces result in the production of several hundreds of eggs, each batch being fertilised, gathered in the mouth (the female now assisting) and spat into the nest.

The male now takes over parental responsibility. When the female has finished spawning he chases her unceremoniously from his nuptial territory and incubates the eggs alone for twenty-four to thirty-six hours. When the tiny fry hatch his interest gradually diminishes, and when they are capable of swimming freely he takes himself off and pays no further attention to them.

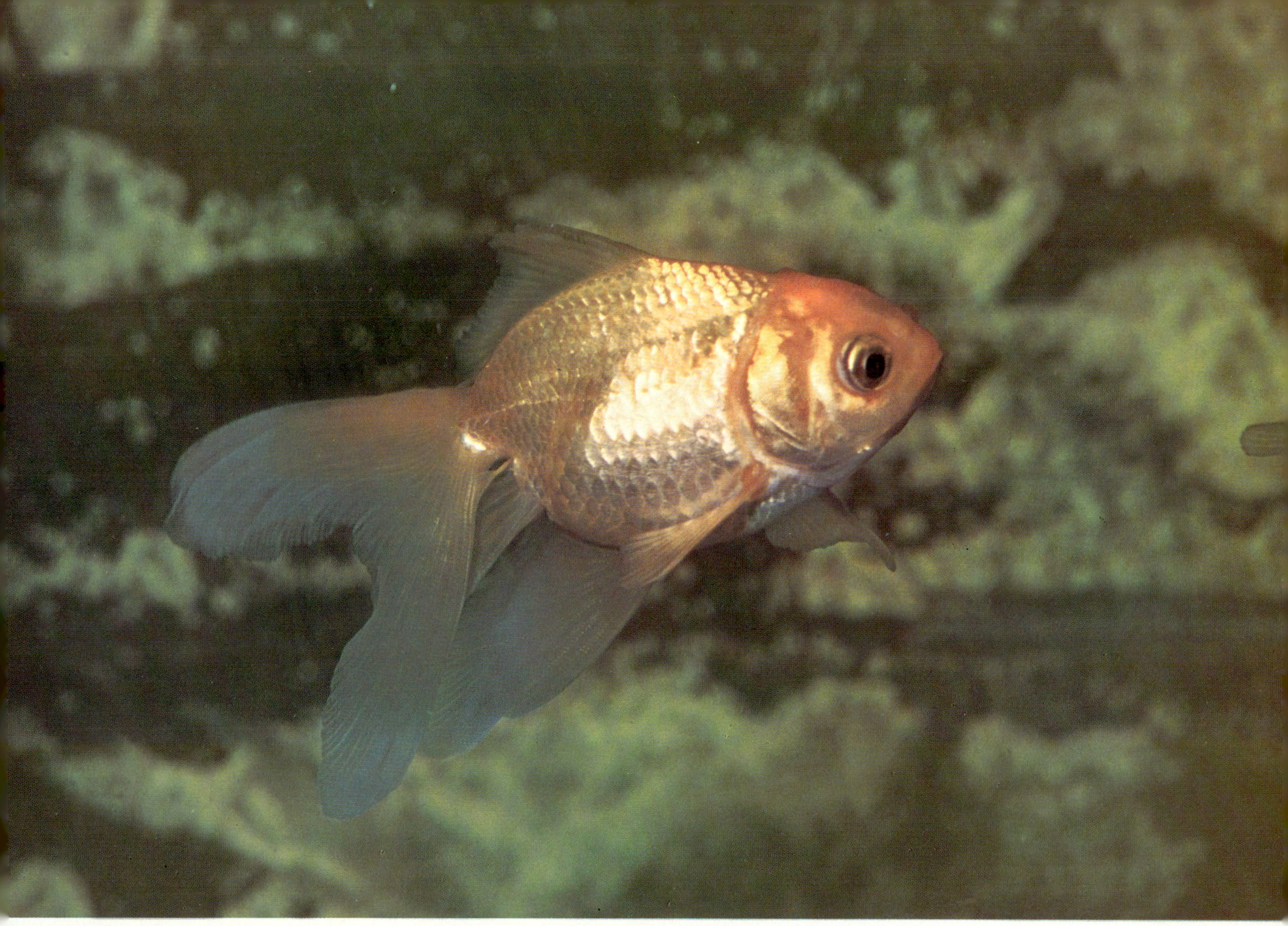

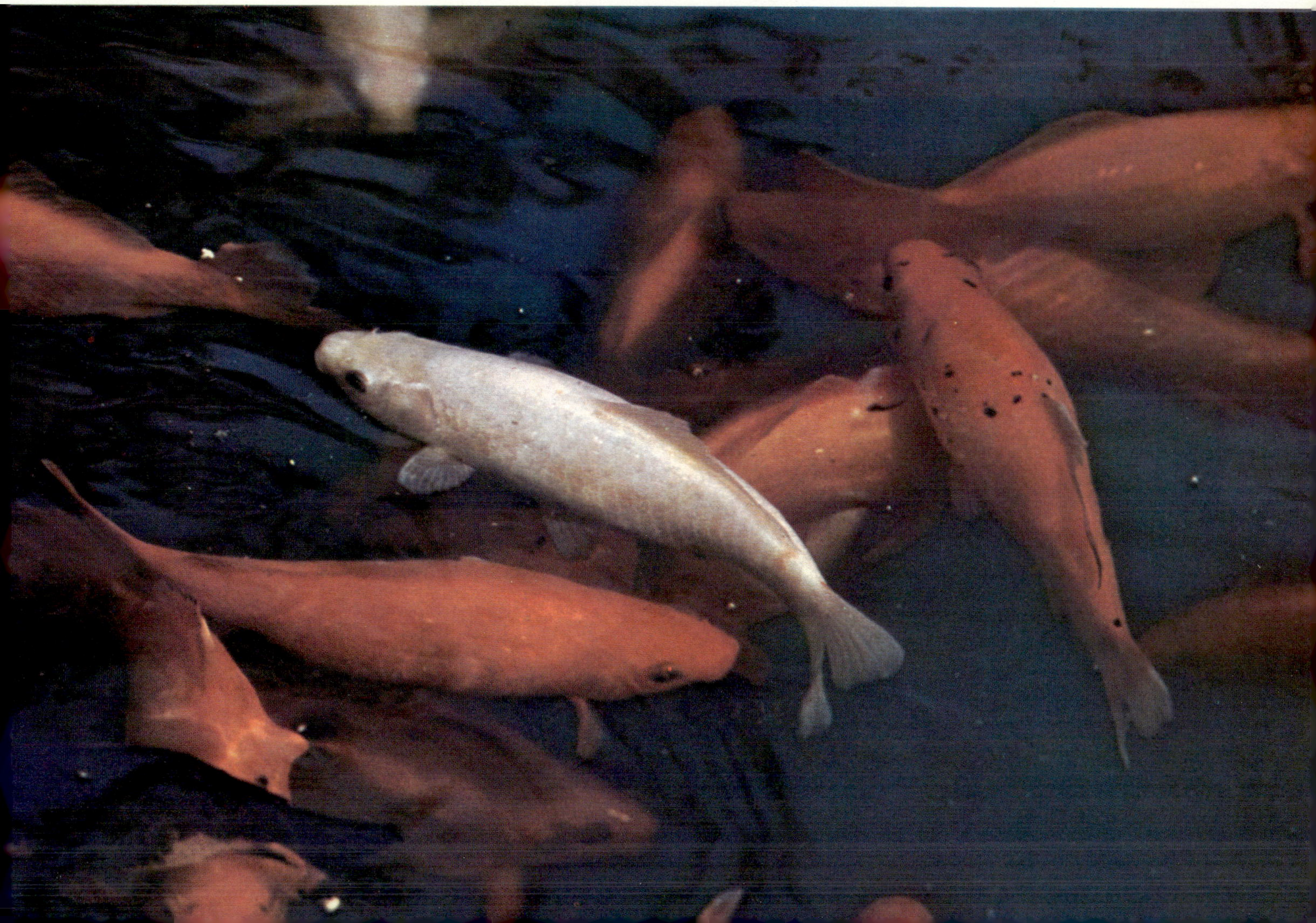

Combats between rival male Siamese fighting fishes are often deliberately staged in the Orient as a spectator sport and the strains that have been artificially bred for such a purpose are particularly aggressive and spectacular in appearance. Fights may continue uninterruptedly for many hours.

Siamese fighting fishes belong to the family Anabantidae, otherwise known as labyrinth fishes. The latter name is derived from the accessory breathing structure on either side of the gill chamber, the labyrinth, which will be described in more detail when we come to the most remarkable fish in the group, the so-called climbing perch. This vascular organ enables the Siamese fighting fish to dispense with normal gill breathing and to live, like other members of the family, in dirty ponds and ditches with little oxygen—habitats in which other species could not survive. The bubble-nest building is also typical of many of the Anabantidae. It is a remarkable method of dealing with eggs and presumably an adaptation to the low concentration of oxygen.

Closely related to the Siamese fighting fish is the kissing gourami (*Helostoma temmincki*). It is, however, much larger, growing to about one foot. There are two colour varieties, the common one being pink, the rarer one green. The common name of the species is derived from the curious 'kissing' habit which consists of two individuals bringing their mouths into contact. It is possible, though not generally agreed, that this may be a form of ritual dissuasion, for the males do not fight each other. Disputes over territory are therefore settled not by violent combat, as in the case of the Siamese fighting fish, but by mutual appraisal and measurement of mouth size. The heavier individual will probably have the larger mouth and thus will automatically get the decision over his smaller rival.

Catfishes and loaches

As already mentioned, a diversity of freshwater habitats has given rise to a number of strange and fascinating adaptations on the part of the fishes of the Oriental region. They include the ingenious mechanisms which enable certain species to survive in swiftly flowing river currents and others to lead a partially amphibious existence.

It is reasonable to wonder why so many of these fishes have chosen to live in rapidly flowing streams and rivers—habitats which on the face of it seem unusually inhospitable. The answer is simple. Animals that have discovered the means of surviving in such surroundings have little cause to fear enemies. It is, after all, difficult enough to live, let alone pursue prey, under these challenging conditions.

None of the special adaptations to life in this environment occurred overnight, but were the outcome of centuries of evolution. We have already seen how some of these freshwater species gradually acquired the external physical apparatus enabling them to attach themselves to the pebbles, stones and rocks on the river bottom, thus avoiding being swept away by the force of the current. In addition to the aforementioned *Barbus tor* and sucker loach, various representatives of the freshwater catfishes have shown themselves to be similarly adaptable. The sucker-mouthed catfishes of South America have no counterparts in the Oriental region, but the Asiatic species of the family Bagridae, known as naked catfishes (notably those belonging to the genus *Glyptosternum*) affix themselves to the bottom with the aid of their pectoral and ventral fins, the surfaces of which act as suckers. A similar phenomenon is present among the loaches of the family Homalopteridae. Here too the pectoral and ventral fins are clamped against the rock face, forming a semi-vacuum which improves the adhesive capacity. The flattened body shape of these catfishes and loaches is additionally valuable in resisting the horizontal pressure of the current.

The suction apparatus among the Homalopteridae is seen at its most effective in the more highly evolved representatives of the family. Thus in the species of the genus *Sinogastromyzon* the pectoral fins partially overlap the ventral fins and the latter are joined at the rear to form a single broad adhesive surface.

Before the female Siamese fighting fish spawns the male constructs a floating nest made of bubbles, each of which is surrounded by a mucous secretion. The male then fertilises the eggs, spits them into the nest and incubates them alone, chasing away the female.

SIAMESE FIGHTING FISH
(*Betta splendens*)

Class: Osteichthyes
Order: Perciformes
Family: Anabantidae

In the wild the fish measures 2–2½ inches long. Colour brownish-yellow or bluish, with indistinct darker markings on flanks. During the breeding season colours are darker and scales brighter—metallic green. Dorsal fin is the same colour with a red mark at the tip. Large anal fin has blue border and caudal fin is rounded. Female smaller and less brilliant. There are a number of domesticated aquarium strains, of various shapes and striking colours—red, blue, green, etc.

Among the larger freshwater fishes of South-east Asia capable of leading an amphibious existence is the snakehead, a powerful predator, often measuring several feet in length. A special accessory breathing apparatus enables the fish to survive for hours, even days, out of water.

CLIMBING PERCH
(*Anabas testudineus*)

Class: Osteichthyes
Order: Perciformes
Family: Anabantidae

Measures up to 9½–10 inches in wild. Body greyish-green or silvery grey. Fins brown. Dark marks behind each gill cover and at base of caudal fin.

The remarkable climbing perch

Even stranger than the species which live in fast-flowing rivers are those freshwater fishes that lead a partially amphibious existence, leaping out of water onto dry land for various purposes, perhaps to bask in the sun, escape a predator or catch an insect. Such fishes are perfectly adapted to life in regions where dry land may be transformed in a matter of hours into a swamp or where the shallow water of a pond or puddle may evaporate in the searing heat of the tropical sun. This ability to transfer abruptly from one element to another and to survive equally easily in both is of tremendous biological interest. This is how certain groups of fishes began to evolve in geological times, gradually adapting to life both in and out of water, culminating in the ancestors of our present-day amphibians.

Perhaps the most extraordinary Anabantid is the climbing perch (*Anabas testudineus*), fish with a wide geographical distribution, including Africa south of the Sahara, and the whole of South-east Asia from India eastwards to the Philippines and the Malay archipelago. Rather drab in appearance, it grows up to 10 inches in the wild and is notable both for its amphibious habits and its ability to withstand sharp, sudden temperature changes.

The climbing perch might more accurately be described as the travelling perch, for what it does is to leap out of the water from time to time and propel itself overland by means of its pectoral fins to another pond or ditch. During the dry season it buries itself in the mud in a similar manner to the African lungfishes. Normal metabolism is slowed down and the fish does not require any food until the rains return. Reports of its climbing abilities, however, seem greatly exaggerated. Although it can apparently wriggle its way up a gentle slope there is probably no truth in assertions that it can clamber vertically up a tree. If, as so many local stories attest, the dead body of a climbing perch is occasionally found at the top of a tree, the simple explanation is that it has most probably been deposited there by a bird of prey and left to putrefy.

The climbing perch, as previously mentioned, is one of the labyrinth fishes and herein resides the secret of its ability to survive in foul water and on dry land. The highly specialised breathing mechanism consists of two sections. The lower and smaller partition contains the gills, and the larger structure, called the labyrinth, is made up of concentric plates covered by a fold of skin with numerous blood vessels near the surface. Although there is no structural resemblance the labyrinth plays the same role in respiration as the lung, for it is here that the gaseous exchange vital to any breathing process takes place. Air is gulped in at the surface via two valve-like openings on either side of the throat and passes directly into the labyrinth, later to be expelled in bubble form through the gills. The fish depends entirely on this method of respiration for survival in poorly oxygenated water and may die of asphyxiation if it remains too long below the surface.

Another interesting fact about this curious species is that the adults do not devour their fry as is the case with most other fishes. After spawning, the female allows the eggs to float with the current and pays them no further heed. The fry hatch about twenty-four hours later.

Other fishes with amphibious habits include the Ophicephalidae or snakeheads, found both in Africa and Asia. The members of the genera *Ophicephalus* and *Channa* are predators with powerful, protruding jaws. None is under 6 inches and the largest are more than 3 feet long. They too have an accessory breathing organ, similar to the labyrinth of the climbing perch but much simpler in structure, comprising a single cavity, richly furnished with blood vessels, and linked with the gill chamber. This enables them to breathe air and to survive for hours, sometimes days, out of water. Snakeheads are eaten fresh or dried in many parts of Asia. In the Philippines, for example, market traders carry baskets with live fishes that are cut up on the spot.

Gavials and other crocodilians

In a previous chapter we described the way in which the bill structure of ducks and geese varies in accordance with the type of food eaten by a particular species. Thus most of the dabbling and diving ducks have flat bills with lamellae for straining plankton; and geese possess bills with serrated edges which are suitable for plucking grass. But the sawbills, which live on a diet of fish, have a delicate, long bill with rows of small but sharp teeth that close, pincer-like, around the slippery prey. Much the same type of diversification is in evidence among the

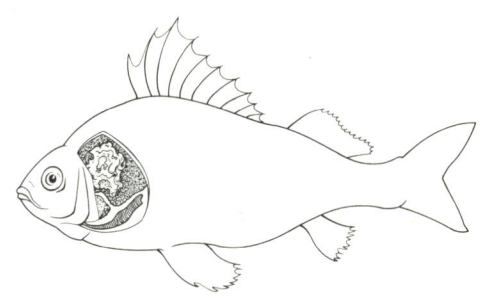

The climbing perch is the most remarkable of the Anabantidae or labyrinth fishes and is capable of crawling considerable distances on dry land, supported by its pectoral fins. A two-chambered breathing mechanism allows it to inhale and exhale air. The smaller, lower partition contains the gills; the upper section, or labyrinth, is a vascular organ with numerous tiny blood vessels which functions in the same manner as a lung.

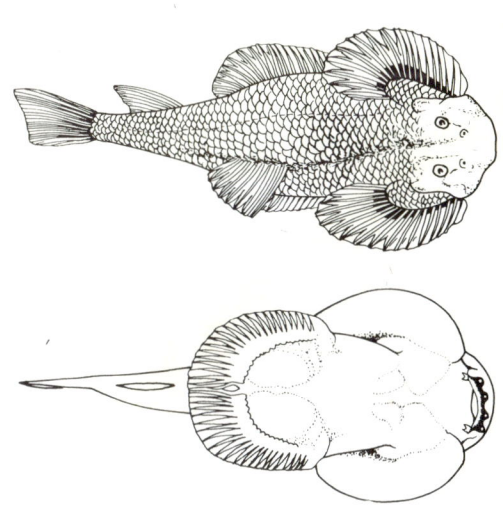

The loaches of the family Homalopteridae, many of which live in swiftly-flowing mountain streams and rivers can attach themselves to stones and rocks on the bottom by creating a partial vacuum between their pectoral and pelvic fins. This suction effect is seen at its most effective in the species belonging to the genus *Sinogastromyzon* (depicted here), whose ventral fins are joined below the body, making up one continuous adhesive surface.

TRUE GAVIAL
(*Gavialis gangeticus*)

Class: Reptilia
Order: Crocodilia
Family: Gavialidae
Total length: 13–16½ feet (4–5 m), sometimes
  up to 18 feet (5·50 m) or more
Diet: fishes, occasionally birds and small
  mammals
Number of eggs: up to 40

Adults
Large dark green plates on back and tail, smaller
plates on neck and sides. Elongated, slender
snout with rows of long, thin teeth, all alike and
widely spaced. Eyes positioned high in head;
nostrils at tip of upper jaw, forming a swollen
projection in male. Limbs rather larger than in
other crocodilians; digits, particularly those of
hind feet, partially webbed.

Young
Measure 12–14 inches at birth with a disproportionately long snout. Colour greyish with five
dark bands on body, nine on tail.

crocodilians. Most crocodiles and alligators consume large prey
and are characterised by a massive head and enormously powerful
jaws. Two crocodilians, however, feed almost exclusively on
fishes, supplemented from time to time by birds and mammals
as well as scraps of carrion. They are the true gavial (*Gavialis
gangeticus*) sole member of the family Gavialidae–and the false
gavial (*Tomistoma schlegeli*), one of the Crocodylidae or true
crocodiles. Both have an extremely long, slender snout (the
length being about four times the width) and enormous teeth,
both in the upper and lower jaws, all of equal size, well spaced
out and pointing slightly to the rear. The true gavial has more
than 22 teeth in either jaw, whereas this is the maximum for
true crocodiles and alligators. The specialised dental structure
is ideal for grasping and retaining slippery fishes while the
length and mobility of the snout give the reptile a broad range
of operation. The flexibility of the jaws and the speed with
which the snout can be moved from side to side offer the gavial
a freedom of movement and action which other crocodilians
conspicuously lack (as an analogy, Maurice Burton likens it to
moving a stick, as against a plank or a log, through water).

Although it is the largest freshwater crocodilian (males may
measure up to 18 feet or more) the true gavial poses no danger
to humans, as is testified by the fact that in India villagers
frequently wash and bathe in rivers inhabited by the reptile,
which is in fact regarded as a sacred animal. It is true that the
stomachs of gavials have been found to contain watches, rings,
bracelets and other pieces of jewellery but this is because

drowning is an all too common occurrence in these rivers. It cannot be taken as evidence of man-eating habits.

The true gavial spends more time under water than other crocodilians, allowing only the high-positioned eyes and nostrils to appear above the surface. If alarmed it will sink deeper, leaving only the nostrils exposed, and in case of imminent danger it will plunge completely below the water.

The male, which is usually larger than the female, may be distinguished by the projection at the tip of the snout; in the centre of this swollen area are the nostrils. Some authors assert that this protuberance is similar in shape to a type of Indian clay vase known as a *ghara* and that this gave rise to the name 'gharial', which was later transcribed as 'gavial'.

In the breeding season the female lays about forty eggs, each measuring $3\frac{1}{2}$ x $2\frac{3}{4}$ inches, on the river bank, then covers them with sand. The process is spread over a couple of days but once the eggs are laid and buried the mother abandons them. When hatched, the baby gavials are 12–14 inches long and the length of the snout is out of all proportion to the size of the rest of the body. They are greyish-brown with fourteen irregular dark stripes, five on the body and nine on the tail.

The false gavial, which was discovered in Borneo in 1838, looks much like the true gavial and has similar habits and food tastes, but is slightly smaller. It is normally found in rivers and freshwater lakes in the southern part of the Malay peninsula and is fairly abundant in Sumatra.

Because of their imposing size and unusual shape, as well as

Above and facing page : The true gavial is one of the largest crocodilians, notable for its extremely long, slender snout and the rows of widely spaced teeth which are designed for catching fish. Although the gavial looks fierce, the teeth are not very strong and the reptile is not considered to be a danger to humans.

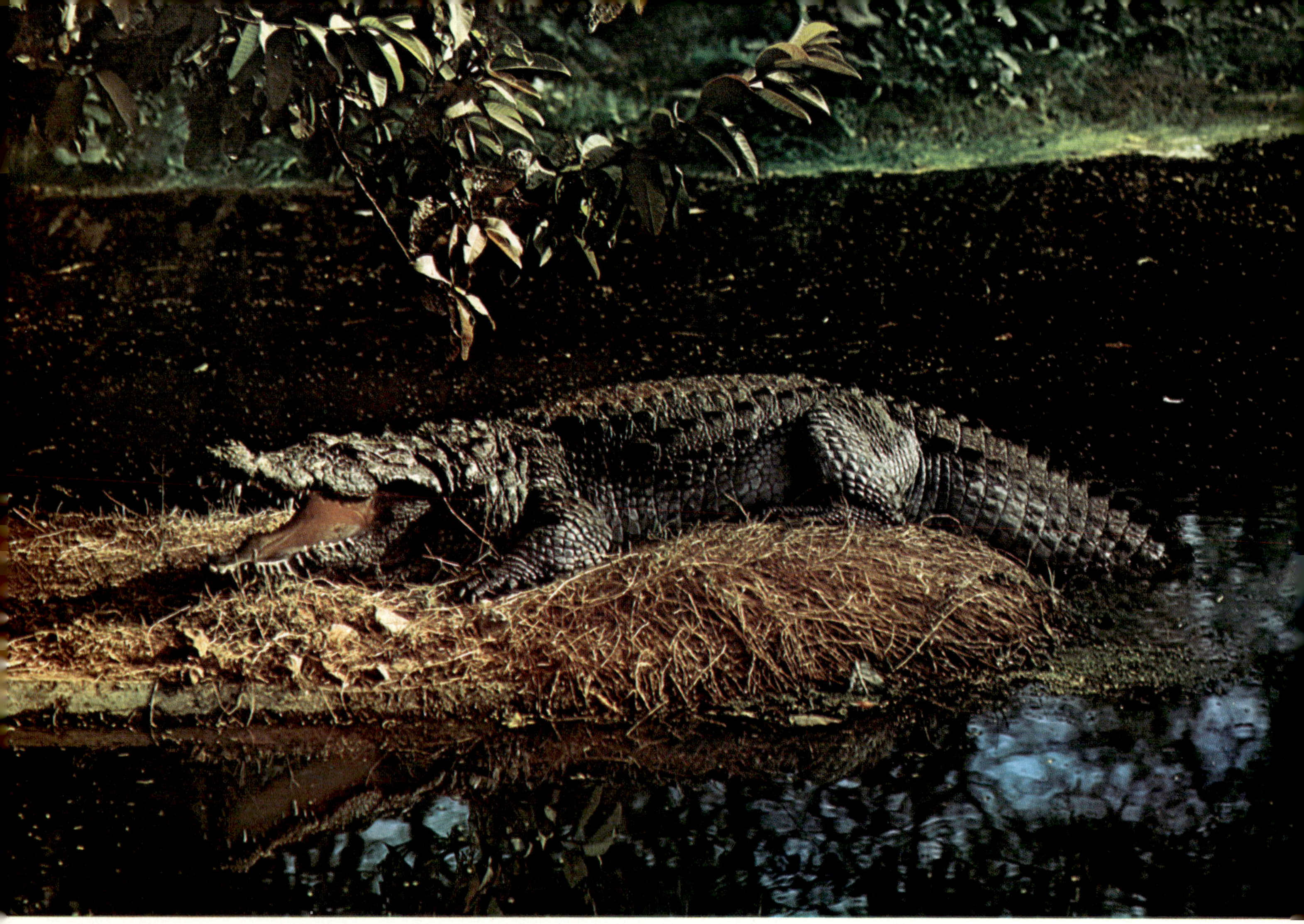

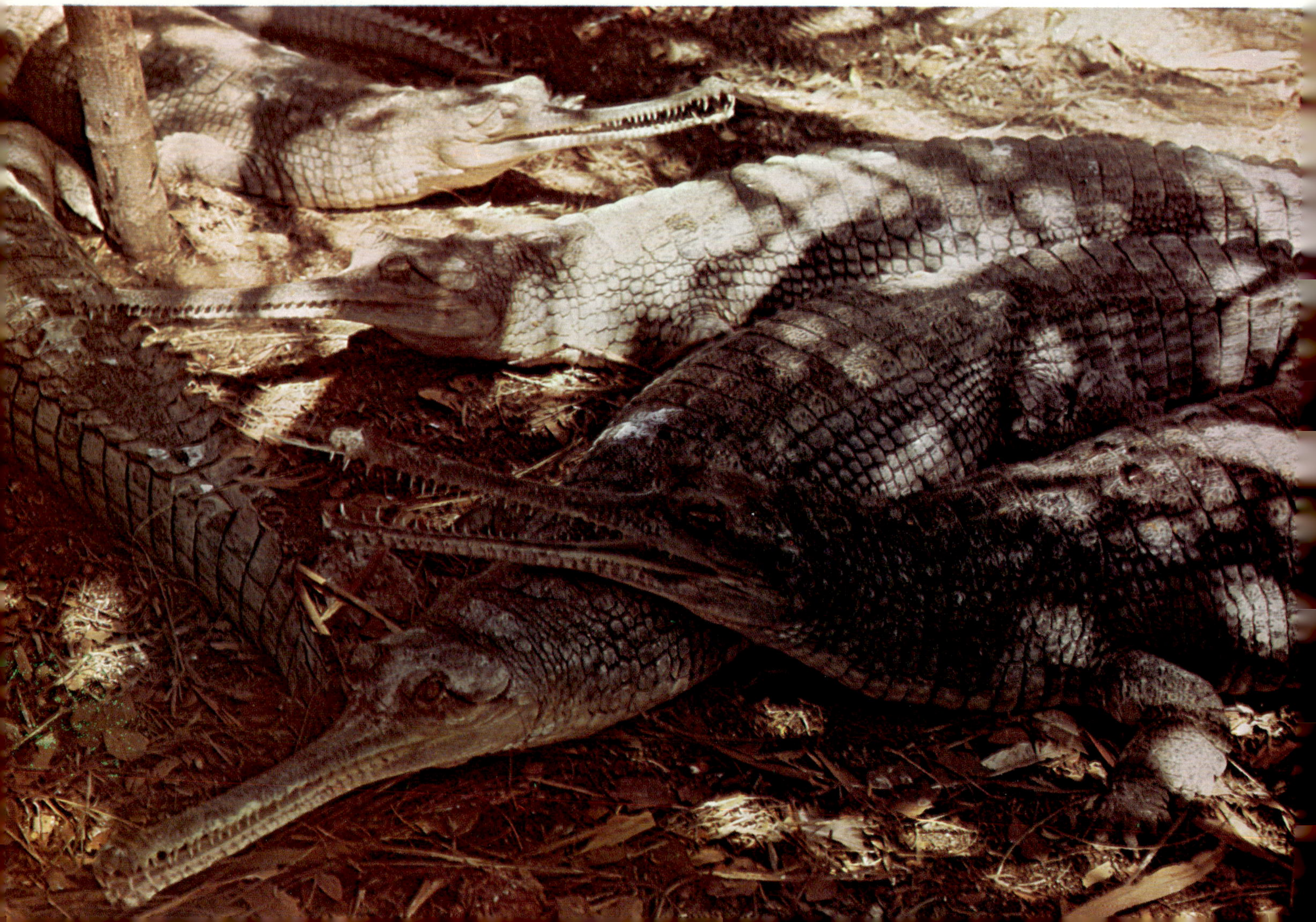

the dangers which they represent for man and his domestic stock, crocodilians have been known to science for quite some time. Thus as long ago as 1825 fifteen of the twenty-five extant species had already been described by zoologists. Since the beginning of the present century only three new species have been discovered, all by the same naturalist, Dr Karl P. Schmidt. These are the short-snouted crocodile, the New Guinea crocodile and the Mindoro or Philippines crocodile.

Yet merely to be aware of the existence of a crocodile is not necessarily to know everything about its biology and behaviour; and this is especially true of one of the typical Asiatic species, the very rare Siamese crocodile (*Crocodylus siamensis*), which is an inhabitant of the rivers and flooded plains of South Vietnam and Thailand, of northern Malaysia and of Java. Measuring up to 13 feet long, it is not a danger to man, and local village people bathe fearlessly alongside it.

The Philippines crocodile (*Crocodylus mindorensis*), which is found only on the islands of Luzon, Mindanao and Sulu, where it is abundant, was first discovered in 1935. Robert Mertens has stated that in Mindanao alone, where it inhabits the rivers and swamps of the interior, one American hunter despatched more than 12,000 skins to the local tannery industry in the space of only six months.

An inhabitant of the rivers, lakes and swamps of India and Ceylon, worshipped and kept in semi-captivity in some districts, the marsh crocodile (*Crocodylus palustris*) plays an ecological role similar to that of the Nile crocodile in Africa. Measuring more than 10 feet long, it feeds essentially on fishes but often supplements such a diet with other animals, especially those to be found in small pools and ponds when fishes are scarce. It seems that this crocodile is gradually establishing itself in forest regions. Roaming far afield in the rainy season when immense areas are flooded, it is often trapped in woodland ponds and streams when the waters recede.

The marsh crocodile is a formidable predator. One hunter who shot several of the reptiles discovered in their stomachs the remains of humans, scraps of skin from leopards, muntjacs, deer, four-horned antelopes and domestic cattle, and a variety of feathers from aquatic and riverine birds.

At the conclusion of the dry season the female lays fifteen to twenty eggs in the sand. The young hatch after fifty days' incubation and measure 10 inches at birth.

Eastern China is the home of the Chinese alligator (*Alligator sinensis*), the only representative in the Old World of the characteristically American family of Alligatoridae (alligators and caymans). Furthermore, together with the American alligator, it is the only crocodilian not to be found in the tropics. In China, where it is known variously as *to* or *ugo*, it was once believed to possess magical powers and, together with the dragon, symbolised the ancient Empire. The phrase 'as old as the *to*' is still used today despite the fact that the reptile does not often live for more than fifty years. Today the species is almost extinct. Individuals measure about 5 feet but judging by fossil remains the reptile once grew to greater dimensions.

Facing page : The marsh crocodile (*above*) is one of the Crocodylidae and is closely related to the Nile crocodile of Africa. During the rainy season it roams far and wide throughout India and Ceylon, extending its range to many forest regions. The huge jaws and powerful teeth are the hallmarks of a predator which, although feeding principally on fishes, is known to hunt birds and mammals as well. The true gavials (*below*) comprise the only species of the family Gavialidae. Their food consists overwhelmingly of fishes which they catch most efficiently with side-to-side movements of their long, mobile snout.

MUGGER OR MARSH CROCODILE
(*Crocodylus palustris*)

Class: Reptilia
Order: Crocodilia
Family: Crocodylidae
Total length: 10–11½ feet (3–3·50 m), up to 14 feet (4·25 m)
Diet: basically fishes but also other vertebrates
Number of eggs: 15–20
Incubation: 50 days

Adults
Large plates on back, smaller plates on sides and feet. Comparatively small head; snout larger than that of other crocodiles; powerful teeth of varying size in both jaws. Nostrils and eyes positioned high in head. Short legs.

Young
Measure about 10 inches at birth.

Facing page : The Gangetic dolphin, a freshwater species, is completely blind. Its long snout and rows of conical teeth, resembling those of the gavial, are similarly adapted to a fish diet—an illustration of convergent evolution.

IRRAWADDY DOLPHIN
(*Orcaella brevirostris*)

Class: Mammalia
Order: Cetacea
Family: Delphinidae
Total length: 7 feet (2·15 m)
Diet: fishes
Number of young: one

Colour greyish or bluish; underparts lighter. Compact body; short snout; each half of jaw contains 12–19 teeth. Dorsal fin very small, situated well to the rear.

GANGETIC DOLPHIN
(*Platanista gangetica*)

Class: Mammalia
Order: Cetacea
Family: Platanistidae
Total length: up to about 8 feet (2·5 m)
Diet: fishes and water animals
Number of young: one

Colour greyish or nearly black. Very long snout with 28–29 pointed teeth to each half of jaw. Atrophied eyes. Dorsal fin little developed and placed well back. Young measure 18 inches and weigh about 15 lb.

CHINESE RIVER DOLPHIN
(*Lipotes vexillifer*)

Class: Mammalia
Order: Cetacea
Family: Platanistidae
Total length: 7–8 feet (2–2·5 m)
Weight: 350 lb (160 kg)
Diet: fishes and other water animals
Number of young: one

Similar to Gangetic dolphin. Upper parts blue-grey and belly almost white. Long snout, curving slightly downwards. 33–36 teeth to each half of jaw. Atrophied eyes.

River dolphins of South-east Asia

The Oriental region, home of many strange fishes that seem to deny their true identities by proving themselves capable of surviving on dry land as well as in water, is also one of the few places in the world where one can find freshwater representatives of an order of mammals which is normally associated with marine life—the Cetacea.

The freshwater dolphins of South-east Asia look something like their ocean counterparts but most of them have longer beaks and rows of pointed teeth. The Irrawaddy dolphin (*Orcaella brevirostris*) in fact belongs to the same family as the marine species—the Delphinidae. Its distribution extends along many coastal areas of South-east Asia (it has been sighted, for example, in the Gulf of Bengal, on the east coast of the Malay peninsula and in the Malacca Straits) and is also found in the Irrawaddy River almost 900 miles from the sea.

Even more unusual is the susu or Gangetic dolphin (*Platanista gangetica*), which lives only in the fresh waters of the Ganges, the Indus and the Brahmaputra. This is one of the true river dolphins of the family Platanistidae, a strange, inoffensive animal measuring about 8 feet long (although a 13-foot female was once discovered), which is completely blind. After a gestation of 8–9 months the female gives birth between April and July to a baby which measures about 18 inches long and weighs 15 lb.

Closely related to the Gangetic dolphin is the Chinese river dolphin (*Lipotes vexillifer*), found only in the Tung Ting lake of Hunan province, some 600 miles up the Yangtse River. It is 7–8 feet in length and weighs about 350 lb. Groups comprising from three to twelve dolphins are frequently seen together.

These two species look very similar and also resemble other river dolphins from South America belonging to the same family. They tend to be smaller than marine dolphins and the characteristically elongated beak and rows of conical teeth put one in mind of the snout and jaws of the gavial. This is not just coincidental but yet another example of convergent evolution, stemming from the fact that both the mammals and the reptile feed on fishes and that the so-called homodont structure of their teeth serves to grasp slippery prey and not to chew or cut food. Furthermore, the cetaceans use their long beak to stir up the mud of the river bottom in search of fishes concealed there.

Certain zoologists have advanced the theory that the river dolphins are all that remain of a primitive phase of cetacean development, the majority of that order having passed through the intermediate freshwater stage and gone on to conquer the ocean. Although it is undeniable that the anatomy of the Platanistidae is somewhat primitive, indicating a descent from very ancient stock, the truth is probably exactly the reverse. The river dolphins are more likely to be the last representatives of an ancient evolutionary branch which proved incapable of holding its own against more highly evolved cetaceans and other predators, the only survivors being those species that succeeded in returning to inland waters. Whatever the truth, the present-day populations of these mammals are very small.

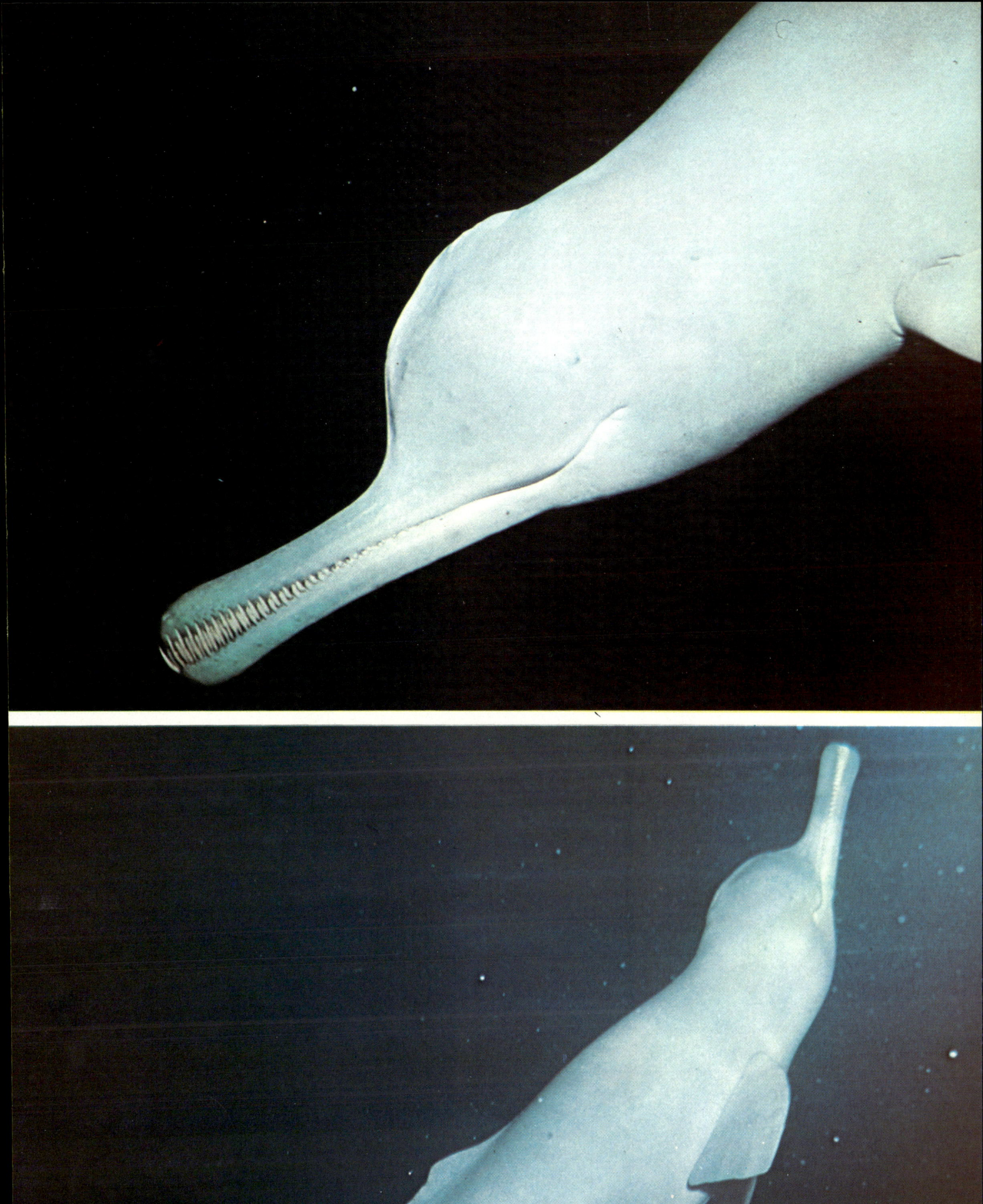

CHAPTER 90

The strange amphibians of the mangrove swamps

Rain is the most important climatic factor in the Oriental region, shaping the pattern of the landscape and controlling the life rhythms of plants, animals and people. The transformation wrought by the monsoon rains is so complete that it is sometimes impossible, in certain areas, to distinguish dry land, now flooded, from permanent tracts of water. Hills that normally break the monotony of the flat, arid countryside are unrecognisable, jutting up as muddy islands in a newly created expanse of swamp.

Of all the watery habitats of the Oriental region the swamp is without doubt the most typical, for in total area it is exceeded only by the jungle. In South-east Asia swamps and marshes cover a much greater area than they do in any of the world's other tropical regions, and there are several reasons for this. In the first place, torrential rains, wherever they occur, have an erosive effect not only on the more or less bare slopes of hill and mountain but also on field and forest, causing massive, sometimes catastrophic, landslips. Such upheavals are not confined to those countries subjected to heavy, continuous rainfall (features of an equatorial climate) but are equally characteristic of those lands influenced by the monsoons, where there is a cyclic alternation of wet and dry seasons. Extreme drought merely hastens the erosive process, cracking open the hard, parched earth and breaking up the surface layers so that the soil can offer little resistance to the subsequent impact of the raging monsoon torrents.

Another factor in the formation of swamps is that almost all the rivers of South-east Asia are swift and turbulent, stirring up a vast quantity of mud that is never allowed to settle and which, when the floods occur, is swept along for hundreds of miles, forming a crust of sediment when the waters retreat. In

Facing page : The mudskipper is a strange-looking fish inhabiting mud flats and mangrove swamps and one of the many small animals of South-east Asia which has managed to adapt to life both in and out of water.

A tangled network of aerial roots characterises certain species of mangrove, providing a firm support for the tree as they thrust deep into the mud surface of the swamp, as well as concealment for a multitude of small animals.

many places the silt is deposited in hollows which gradually fill up until they form a huge swampy zone of fresh water.

The third formative element — and also the commonest — stems from the fact that there are comparatively few stretches of coastal plain in the Oriental region. On the shores of the Sunda Islands, where the sea is calm and shallow, banks of mud are thrown up on both sides of the river estuaries. Thus the coast is literally on the move. In the deltas of the Tijmanuk and Solo, for example, it advances some hundred yards every year.

It is a combination of all these factors which has resulted in the formation of immense swamps on the island of Sumatra. Here — especially along almost the whole length of the east coast — there are over 50,000 square miles of marshes and sand banks, approximately one-third of the island's total area!

The advancing army of trees

The vast expanses of swampland in South-east Asia would be merely geological curiosities, devoid of interest for the biologist, were it not for the fact that even in these muddy wildernesses life exists in many fascinating forms. The vegetational substratum, foundation of every ecosystem no matter how simple or complex, consists characteristically of some thirty species of strange shrubs and trees known collectively as mangroves; and we cannot comprehend the real nature of these swamps and their inhabitants unless we examine the form and unusual development of the mangrove in some detail. For the story of the mangrove swamp illustrates how a species that is apparently approaching an evolutionary dead-end can adapt to a new environment and miraculously survive.

When they first began to take root in these areas of brackish slime the mangroves were trapped in a narrow belt of unpromising terrain to which they had to adapt or die. There was no going back, for behind them was a flourishing, well balanced plant community of great antiquity – the tropical rain forest – comprising species ideally suited to local conditions. Nor, to all appearances, was there any way forward, for ahead lay the seemingly impenetrable barrier of the ocean. Yet expansion was essential for survival and nature solved the dilemma in unexpected fashion. For if the land route to the rear seemed blocked the way forward, by comparison, must have appeared non-existent; and yet it was in the latter direction that the move came – an army of trees literally invading the ocean, transforming the coastlines and creating a new biome.

Capable, in varying degrees, of building up a tolerance to salt water, the mangroves eventually started growing along the lines of newly formed mud banks, providing a stabilising influence and preventing the slime being washed away by the tides. By thrusting their network of strongly growing roots into the mud the plants also checked the flow of river-borne sediment into the ocean, banking it up to increasingly high levels. In due course this led to the formation, ever farther inland, of freshwater swamps, while all the time the belt of trees advanced slowly and implacably towards the sea, extending the areas covered by land.

The numerous mangrove species are broadly divided into three groups, each recognisable by a different type of root structure, although in all cases the roots are partially above ground. The species that belong to the genus *Sonneratia* are often found semi-submerged for they are not adversely affected by salt water. The roots of these species snake out horizontally and do not delve deeply underground, producing numbers of vertical shoots above the surface which form a kind of stockade around each tree. The mangroves of the genus *Rhizophora* are supported by a sturdy entanglement of aerial roots that arch outwards and plunge back into the mud so that the trunk has the appearance of being supported by props. Finally there are the trees of the genus *Bruguiera,* the underwater parts of whose roots intertwine to form tight knots. These mangrove species tend to grow farther inland and are only occasionally reached by the highest tides. In spring, this happens twice a month.

The function of the aerial roots of the mangrove has not been satisfactorily explained. The suggestion that they have something to do with respiration seems improbable, given the fact that they are submerged by water for much of the time. But they certainly help to anchor the tree more firmly and to build up a solid muddy foundation for new growth. Wherever mud envelops the roots the tree produces new seeds which subsequently throw out aerial roots for further expansion.

The extraordinary adaptive capacity of the trees to their seashore surroundings is clearly demonstrated by a truly remarkable

Mangrove swamps already cover thousands of square miles of terrain in the Oriental region and steadily expand their area year by year. Unaffected by salt water, the mangroves take firm root in the mud banks formed by river sediment, forming dense belts of coastal greenery.

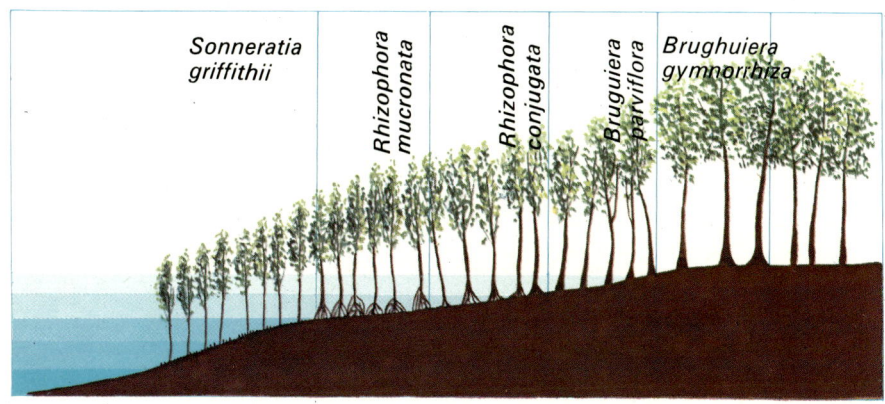

Sonneratia griffithii — *Rhizophora mucronata* — *Rhizophora conjugata* — *Bruguiera parviflora* — *Brughuiera gymnorrhiza*

mechanism of reproduction. If this process were to follow the pattern of the majority of plants, the seeds, scattered at random and requiring some time before beginning to germinate, would have no chance of penetrating the mud, washed as it usually is by high tides twice a day. This problem has been ingeniously resolved, for instead of being tossed to the mercy of the tides the seeds are not shed until already germinating, and each individual shoot is only separated from the plant when fully formed. When it breaks and falls, the smooth, spindle-shaped shoot embeds itself in the mud and within a matter of hours has thrown out a network of primary roots. What is equally amazing is the fact that the shoot is normally dropped at low tide, but even if this happens at high tide the seed is not invariably swept away and lost because of its tolerance to salt. In fact the ebb and flow of the tides may be beneficial inasmuch as this often helps to disperse the seeds over a wide area.

Distribution of mangrove species

The various species of mangrove are distinguished from one another not only by appearance (notably the external root structure) but also by their varying degrees of tolerance to salt. Because of this diversity and the fact that the substratum in which they are rooted is not everywhere identical, the coastal distribution of these trees assumes a fairly well defined pattern, being arranged broadly in series of parallel lines. At the water's edge and practically at sea level, so that their roots are almost continuously submerged, are the serried ranks of mangroves belonging to the genus *Sonneratia*, these trees being the most highly resistant to salt. Next come the species of the genus *Rhizophora*, only slightly less tolerant of salt, whose roots are periodically immersed; and finally there are the mangroves of the genus *Bruguiera*, far less adaptable to saltwater conditions, which grow some distance from the shore on what might be termed 'new land' and whose roots are washed by the spring tides only once every two weeks. The genus includes mangroves that grow exclusively in fresh water and these practically merge with the advance guard of forest trees.

Belts of mangroves line the shores of many islands and stretches of mainland throughout South-east Asia, sometimes consisting of a single row of trees, sometimes extending miles inland.

The different species of mangrove show varying degrees of tolerance to salt and tend to develop in parallel rows to the coastline. Closest to the shore, with roots continuously submerged, are the trees of the genus *Sonneratia*. Next come the trees of the genus *Rhizophora* whose arching aerial roots are exposed at low tide; and finally, because least tolerant of salt, the species of the genus *Bruguiera*, growing higher up and farther from the sea, affected by the high tide only twice a month.

Facing page : The mangrove's remarkable adaptation to its swampy habitat is evident in the manner in which it reproduces. The seed does not separate from the parent plant until it has begun to germinate, at which stage it is smooth and fusiform (*above left*). When it breaks off it becomes embedded in the soft mud (*above right*) and soon develops primary roots. The small plants stand firm against the ebb and flow of the tides and eventually grow into larger trees with widely branching roots. In the case of the mangroves of the genus *Sonneratia* the new shoots spring up like a stockade (*below*).

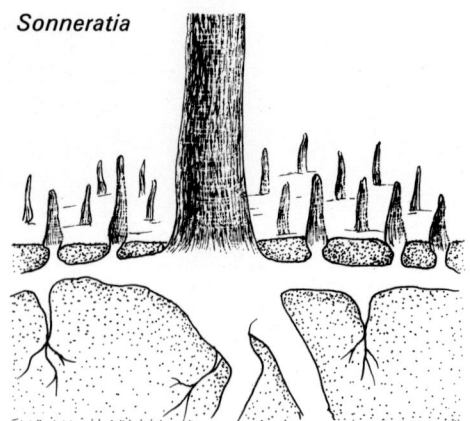
Sonneratia

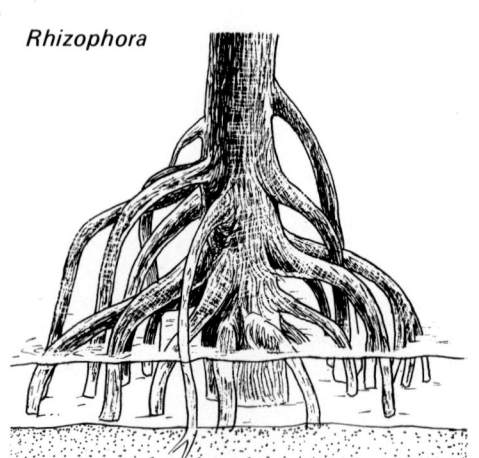
Rhizophora

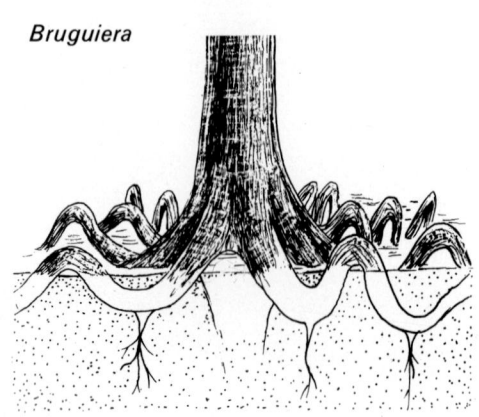
Bruguiera

The three genera of trees that make up a typical mangrove swamp are easily distinguishable by their different aerial root structures.

Facing page : Crabs are the most characteristic inhabitants of the mangrove swamp. The fiddler crab is chiefly remarkable for the disproportionate size, in the male, of one pincer which has both sexual and territorial significance.

Life and death among the mangroves

The naturalist who ventures into the realm of the mangrove swamp for the first time soon becomes aware that there are regular cycles of activity to which all the plant and animal inhabitants of this strange world of water, mud and trees are subjected. Although at first glance everything seems still and inert, there is an underlying rhythmic pulse that governs the lives of all the living organisms, terrestrial and aquatic, and which even determines the shape and movement of mud and sand. It compels the creatures of land and water to dovetail their activities so that the former seek their rest while the latter bestir themselves for food, and vice-versa. Elsewhere this natural life rhythm is dictated by the alternation of light and darkness; but here in the mangrove swamps the absolute controlling factors are the tides. There is a fine knife-edge between life and death, and survival depends directly on the capacity of each individual to adapt to ever-changing conditions at the very moment that the sea advances or retreats.

The vital influence of the tides can be appreciated only when one bears in mind that there are flat coastal regions extending for hundreds of square miles which are flooded by the sea waters twice a day. The flatter the terrain the greater the effect of this regular ebb and flow.

It is because of this constantly changing scene, this rapid alternation of dry and wet conditions, that the animal inhabitants of the mangrove swamps display a diversity of anatomical and physiological adaptations which enable them to adjust to an amphibious life. Indeed there is no other choice. This is a world of unspecialised individuals. Animals that are adapted exclusively either to a terrestrial or an aquatic mode of life have little chance of survival here. The leopard, for example, so speedy, agile and versatile a hunter on dry land, would be helpless on the soft, slippery surface of the mangrove swamp and it is only too likely that it would be caught unawares by the rush of the incoming tide and end up as a meal for a crocodile. As for the dolphin, that marine mammal which is so perfectly adjusted to its ocean environment as to be more than a match for almost any kind of fish, no matter how large and ferocious, it could hardly be expected to live long in a mangrove swamp, stranded by the suddenly retreating tide and exposed to the deadly rays of the tropical sun. Yet another marine mammal, the dugong, prospers in these coastal swamps for although not as venturesome or mobile as the dolphin it revels in shallow waters.

The naturalist will therefore not be unduly surprised to find unusual creatures here – a fish that can scuttle across the mud flats faster than a man can run on dry land, monkeys and felines that are expert swimmers and take to the water as soon as the ground vanishes under their feet. Such animals have the advantage over their more specialised relatives for an ordinary predator could easily be drowned if caught by the receding tide, while a water animal, even if capable of living for some time in the fresh air, would surely die if it accidentally found its way to a place beyond the reach of the incoming tide, without any

At low tide the crabs of the mangrove swamp scuttle along the mud banks in search of small animals living in or on the mud.

hope of rescue. But the inhabitants of the mangrove swamps depend on this very ebb and flow.

Thus we can readily understand why there are so many different kinds of fishing animals in this biome. Some of them have always subsisted on a fish diet but others originally practised alternative hunting techniques and were compelled to adjust their methods and food tastes in accordance with the dictates of their new surroundings. Although fishing is a complicated activity, the modification of behaviour was not necessarily as arduous as might be imagined. Once the first steps had been taken, repeated practice led to increasing efficiency so that after a time it became difficult to distinguish between the original inhabitants and the later arrivals. In the first place the prey was often there for the taking when fishes carried inland by the tide found themselves temporarily stranded and helpless, so that even the more clumsy predators had no need to go hungry. Furthermore, to supplement their fish diet, there were innumerable crabs and snails which were no problem to catch at low tide.

Crabs by the million

Every biome has its pattern of plant and animal life, conveniently presented in the form of an ecological pyramid made up of producers (plants) and consumers (phytophages, predators and super-predators). The same categories apply in the mangrove swamp but with one peculiarity. The second stage of the pyramid, that of the primary consumers, is represented here almost exclusively by a single group—the crabs. These are present in vast numbers and great variety.

Why so numerous? Imagine a typical rocky coastline fringed by tall, vertical cliffs. The sea bed slopes steeply and the surface of rock washed by the tides is a narrow belt on which the changing water levels are visibly recorded. Where the slope is gentler there may be an expanse of sand or shingle and a corresponding increase in seashore activity. But at the other extreme, on these tropical shores bordered by mangroves, the sea bed is almost flat and the influence of the tides extends over many square miles, affording ideal conditions for a multitude of crustaceans, including enormous communities of crabs.

Facing page : Like many other crabs, those of the genus *Dotilla* have strong territorial instincts and engage in fierce combat with one another. When the tide comes in they bury themselves in muddy holes so as not to be swept away.

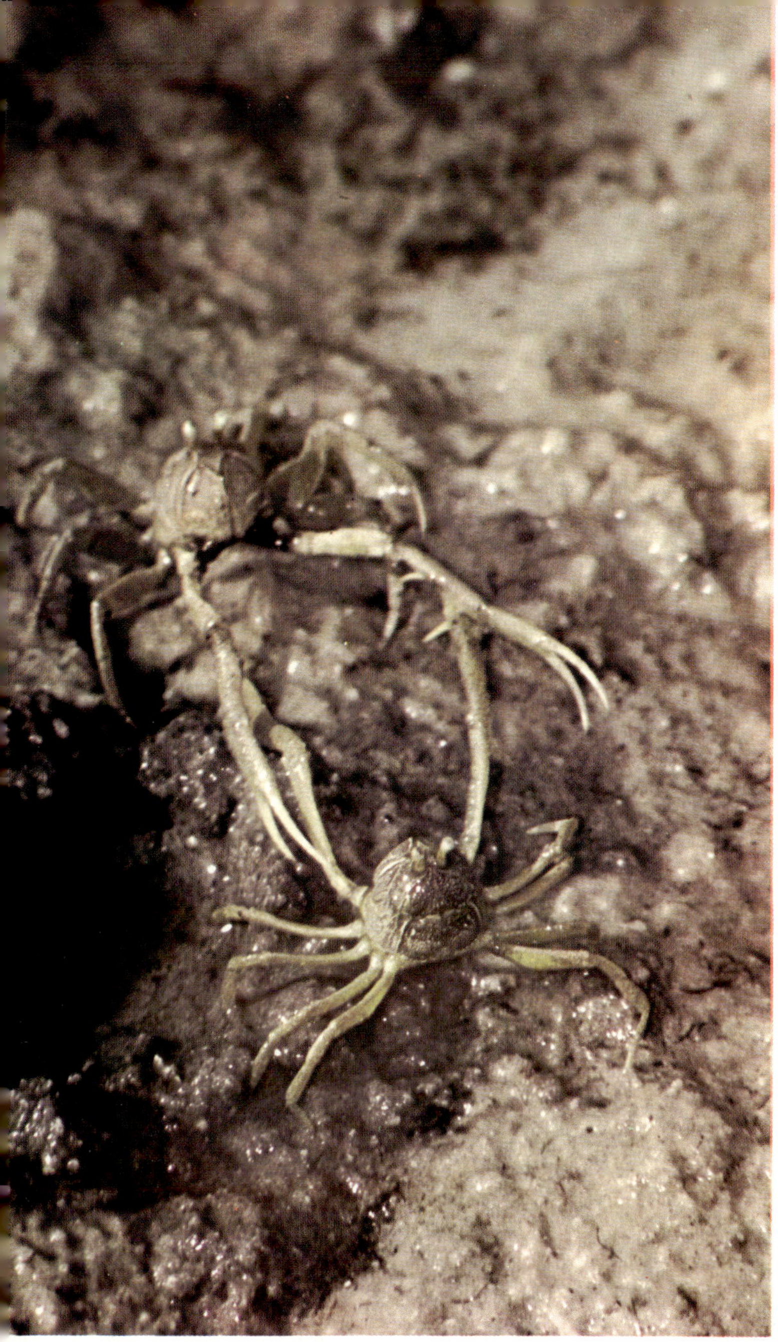
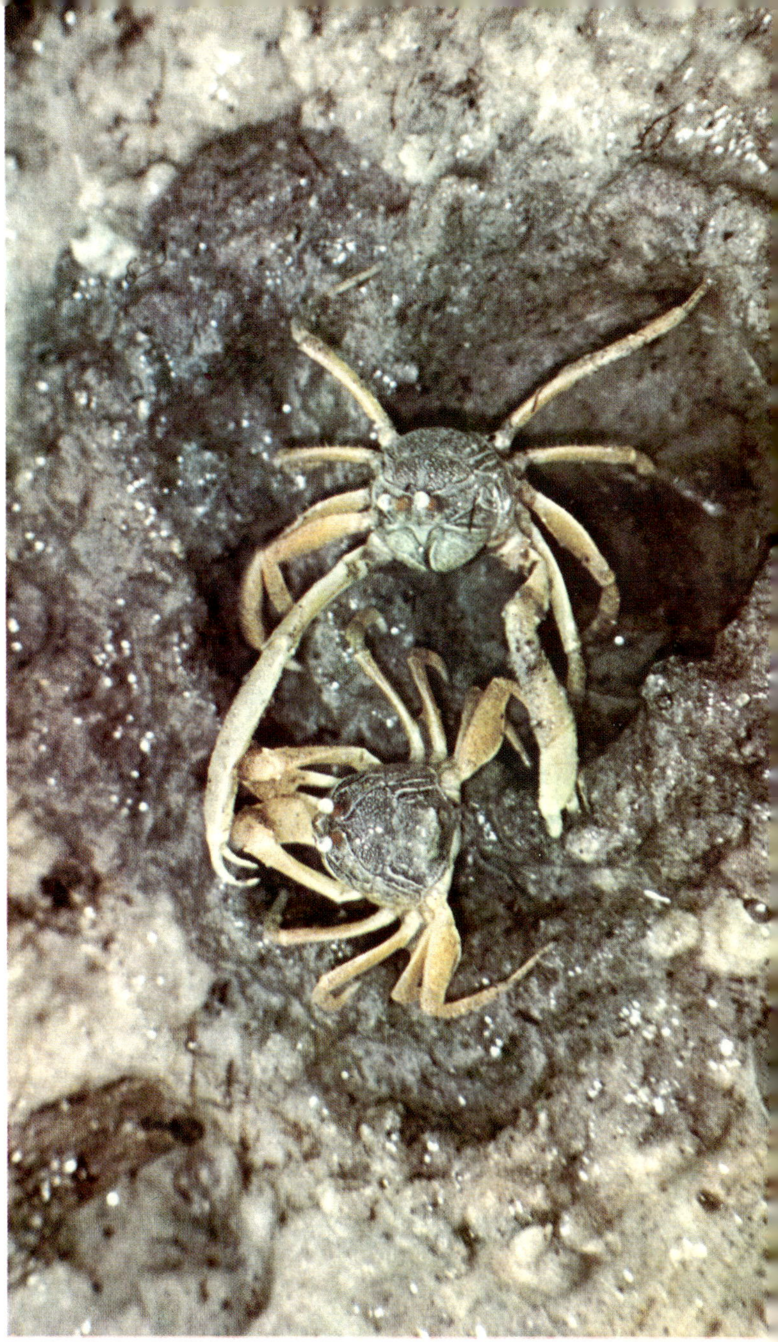
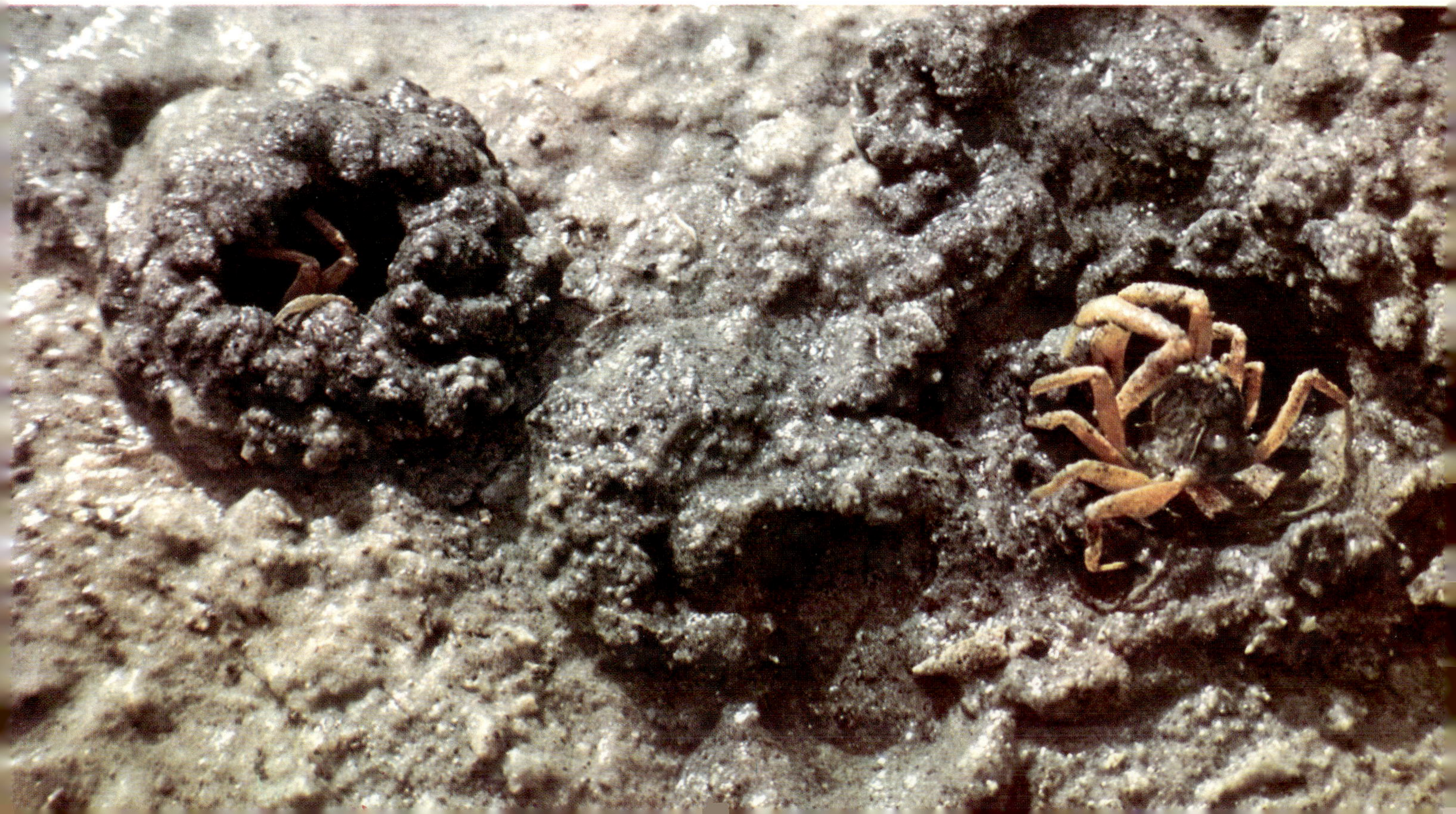

It is also not difficult to understand why there are so many species of crabs. One must remember that the area occupied by mangrove swamp is immense and, as we have seen, increases month by month, year by year; and because the conditions of the environment and the necessities of survival are so variable the animal inhabitants of this biome have evolved in a number of astonishing ways, according to the ecological niche they have come to occupy. Thus the crabs of the mangrove swamps have acquired truly remarkable adaptive features as they have spread out in every direction.

Even to list the many genera and species of crustaceans inhabiting the mangrove swamps of South-east Asia would require a separate volume, and only a handful can be mentioned here. Among the aquatic crabs which are equipped with paddle-shaped legs are *Matuta lunaris* and *Neptunus pelagicus*. Far more numerous, however, are those crabs that lead an amphibious existence, represented by the genera *Sarmatium*, *Dotilla*—including the soldier crab (*Dotilla mictyroides*)—*Sesarma*, *Metopograpsus* and *Ocypode*—notably the ghost crab (*Ocypode ceratophthalma*)—all of which evince strange shapes and unusual colours. Special mention must also be made of the hermit crabs of the genus *Coenobita* which, because their abdomens are naked and vulnerable, seek refuge in empty mollusc shells and then carry them around wherever they go. Certain species of hermit crab have adapted to life on dry land and instead of using a mollusc shell as a portable suit of armour make use of a hollow piece

of bamboo. Other crabs too have virtually abandoned a marine existence. Two of the most remarkable terrestrial species are the land crab (*Cardisoma carnifex*) and the large robber or coconut crab (*Birgus latro*). The latter is capable of climbing up palm trunks. It is said to detach coconuts but more probably it feeds only on damaged coconuts that have already fallen to the ground and have been split open in the process. It is interesting to note, nevertheless, that this crab is still dependent on the sea for perpetuating its kind. In the breeding season the female lays her eggs in the water where they and the larvae hatching from them drift with the tides. The young may perhaps fetch up on a distant island and thus the species can be established in a new habitat which is inaccessible to the adults.

Above and facing page : The mud flats so characteristic of the mangrove swamps of the Oriental region swarm with crabs of many shapes and colours. Illustrated here are (*below, left*) the ghost crab (*Ocypode ceratophthalma*) and (*above, from left to right*) an amphibious crab of the genus *Sesarma*, a hermit crab of the genus *Coenobita* and a crab of the genus *Macrophthalmus*. (The bright green of *Sesarma* may sometimes be lost, as in this individual.)

Colour advertisements

The diversity of shapes and variety of colours to be found in the animal kingdom have long been the source of wonder and delight to zoologists everywhere; but it is only comparatively recently that scientists have realised that these are no mere whims of nature and that they serve particular purposes, vital to the safety of the individual and the survival of the species. It was easy to appreciate the value of *not* possessing bright colours, for a plain coat or drab plumage was obviously an advantage to any animal trying to hide from an enemy. The significance of vivid coloration, however, was less clear since on the face of it this must surely prove a disservice to the animal concerned, attracting unwelcome predators.

Thus, to return to the world of crustaceans, the drab colour of some crabs effectively conceals their presence from enemies; how then to explain the bright green hues of the genus *Sesarma*, the violet tints of *Metopograpsus oceanicus*, the glittering reflections of the claws and eye stalks of the crabs of the genus *Sarmatium*? These colours, far from concealing, positively advertise an individual, and this is precisely what they are intended to do. For the crabs that scuttle to and fro across the muddy surface of the mangrove swamp need to be recognised, especially by others of their kind, be it male or female, the male because he is a potential rival for territory, the female because she is a possible sexual partner.

The crab community has become so dense that it is only by the acquisition of distinctive colours and the adoption of characteristic body movements that a crab of a particular species is able to recognise a congener. Such visual signals are unmistakable. Contact is made immediately, without the unnecessary waste of energy that might otherwise be involved in defending territory against a male of another species or uselessly attempting to court a strange female. Thus the rainbow hues of these crabs have exactly the same territorial and sexual significance as those of the fishes of tropical coral reefs and the brilliant birds of the rain forests.

Strange rites of the fiddler crab

The effect of strange forms and vivid coloration is strikingly illustrated in the case of the various species of fiddler crab belonging to the genus *Uca* which are among the most characteristic mangrove swamp inhabitants. The fiddler crab is so named because one of the legs ending in chelae or pincers is disproportionately enlarged compared with the other normal one (so that it looks something like a violin). It is in fact almost as big as the rest of the crab's body.

Fiddler crabs live in large colonies that may extend for miles along the shore. Seen from a distance these communities look like clusters of tiny flashing lights. For although the body of each crab blends with the surrounding mud, the enlarged claw is brightly coloured, glistening in the sunlight as it is waved slowly and rhythmically or rapidly and jerkily back and forth. Yet when the fiddler crab chooses to pass unobserved or wishes to hide from a predator the claw loses its brilliant hues and reverts to the normal drab shade of the rest of the body.

Within the colony each fiddler crab possesses its own piece of territory, at the centre of which is a hole. The crab retreats into this hole when threatened by an enemy–a sea bird, a heron or a kite. It feeds at low tide on the tiny animals in the mud. When the tide rises the crab scoops up a ball of mud, heaves it onto its back and retires into the hole, sealing up the entrance with the lump of mud to keep out the water.

Like so many shore animals the fiddler crab leads an amphibious life, spending the greater part of its time out of water. This of

The male fiddler crab waves its elongated claw either to proclaim its territorial rights or to attract a female, the action varying slightly among different species. The Malaysian fiddler crab, seen here, moves the claw rapidly up and down when the female is in the vicinity while his territorial signal tends to be more regular and rhythmical.

course requires a special breathing apparatus. The cavities beneath the carapace, which contain the gills, are here enlarged and the gaseous exchanges with the air are effected within the cavity walls with their vascular membranes. Thus the cavity walls with their vascular membranes. Thus the cavities serve as lungs for breathing fresh air. The mechanism is perfected even further in those species, such as the aforementioned robber crab, which live entirely out of water (except when breeding).

What use is the huge claw of the fiddler crab? It is certainly of no value in feeding and far from being an effective weapon would appear to be an encumbrance to its owner. If used at all in fighting a rival it can only be to intimidate, never to inflict injury. Noting that the female fiddler crab possesses two normal-size chelae zoologists concluded, correctly, that the male's large claw must have both territorial and sexual connotations. The male in fact waves its brightly coloured appendage continuously whenever a rival is in the vicinity, the movement varying slightly among different species and according to the message it is intended to convey. Thus in the case of the Malaysian fiddler crab (*Uca annulipes*) the claw movement is usually slow and rhythmical, but when a female approaches the up and down movements are more agitated and irregular. In other species the male vibrates the pair of claws used for locomotion or performs a kind of 'dance'. Furthermore, whereas some fiddler crabs employ a number of distinct signals in different circumstances others appear to confine themselves to only one type of claw movement (although there may be barely perceptible variations) whatever the situation.

A detailed survey on this subject which was carried out on

When the tide rises the fiddler crab digs a hole and seals the entrance with a lump of mud so that the interior remains dry. The enormous pincer which gives the species its name is of no value either as a tool or a weapon but serves merely as a means of interspecific communication.

MALAYSIAN FIDDLER CRAB
(*Uca annulipes*)

Class: Crustacea
Order: Decapoda
Suborder: Reptantia

General colour of male pale blue; pink legs, orange feet. One of the two chelae of the male is much enlarged and coloured bright orange.

a dozen species of fiddler crabs in Panama by the American zoologist Jocelyn Crane has confirmed this dual function. The conclusion was that the principal significance of the claw movement was to mark territorial boundaries but that the same action was employed to attract and stimulate the female. Crane was satisfied too that the colour of the appendage was also important, though to a lesser degree.

Not all females provoke identical reactions in the males. In some cases the presence of a single female may arouse extreme agitation in the ranks of several males, in others there is no responsive movement whatsoever. Nobody knows why two males should react in contrasting ways to what appear to be similar situations. It may have something to do with the individual female – her degree of desirability as a sexual partner – or there may be some feature of her appearance and behaviour, recognised only by an intended mate, which indicates whether or not she is mature and sexually responsive.

Apart from communicating with one another by visual signals, fiddler crabs are also capable of producing chirring sounds, similar to those made by crickets and grasshoppers. This stridulation is commonly practised by a number of the segmented invertebrates known as arthropods and in the case of the fiddler crab it is produced by rubbing the joint of one claw against the edge of the carapace. Although we do not know how another crab can hear such sounds, it is clear that communication can be effected in this manner, for crabs have been observed trying to force their way into holes already occupied and promptly beating a retreat as soon as the incumbent begins stridulating.

Intraspecific fighting among fiddler crabs for possession of territory is quite common but since the enlarged claw is useless as a weapon (it is not even strong enough to nip a person's finger), combat is ritualised. The weaker individual simply withdraws after an exchange of warning gestures.

According to Jocelyn Crane the breeding season of all these species coincides with a spring tide.

Fiddler crabs are found in many parts of the world. Outside Asia they live on the seacoasts of Africa and America and are very common in the Mediterranean region, especially around southern Spain where the meat of the animal's large claw is considered a rare delicacy. Even after the claw is removed – provided the fisherman leaves the animal alive on the beach – it will eventually grow again.

Fish out of water: the mudskipper

Of all the animal inhabitants of the mangrove swamps of tropical Asia the most unusual, though far from being the most beautiful, is a strange-looking fish with a torpedo-shaped body, an enormous head and large periscopic eyes. The fish, one of the goby family, is called the mudskipper (*Periophthalmus*) and as likely as not will be seen hopping across the mud flats or basking in the shallows, for it has the distinction of being amphibious. In the water it swims like any other fish; on land it 'walks' by supporting itself on elongated pectoral fins.

Mudskippers are sometimes seen climbing or clinging to tree trunks. The manner in which this is achieved is shown in the lower picture of two individuals of *Periophthalmus chrysospilos*, where the ventral fins join at the base to form a broad suction surface.

There are many species of mudskipper in various habitats and only a few are characteristic residents of mangrove swamps. Taking both African and Asiatic species into account, these may be classified under three headings, the members of each group being differentiated by behaviour (some are more terrestrial by habit than others) and by diet. Thus the species belonging to the genus *Scartelaos* are comparatively small and slender, living for the most part on low mud flats that are permanently washed by shallow sea water but where mangroves have not yet made an appearance. The mudskippers of the genus *Boleophthalmus* are to be found on mud and sand banks bordering mangrove swamps but since they feed exclusively on diatoms and other algae they are wholly dependent on an ocean habitat. The fishes of the genus *Periophthalmus*, however, are inhabitants only of mangrove swamps. Larger than the representatives of the other two genera (from 5–12 inches long) they spend much of their time on land, feeding on insects, worms, crabs and smaller species of mudskipper.

Taking this last group as our model, it is evident at a superficial glance that the mudskipper is organically well adapted to life both in and out of water. The supporting pectoral fins resemble a pair of arms; and the ventral fins are of two types. In some species they are separated but in others, such as *Periophthalmus chrysospilos*, they extend forwards and are joined at the base below the pectoral fins so that they form a kind of suction pad by means of which the fish can cling firmly to smooth or vertical surfaces.

The mudskipper 'walks' by propelling its body with

Some species of mudskipper are better adapted than others to life on land. Those of the genus *Boleophth almus* do not go far from water and are somewhat larger than the related species of the genus *Periophthalmus* which are characteristic inhabitants of the mangrove swamp and spend much more time out on the mud. Both types are shown here side by side.

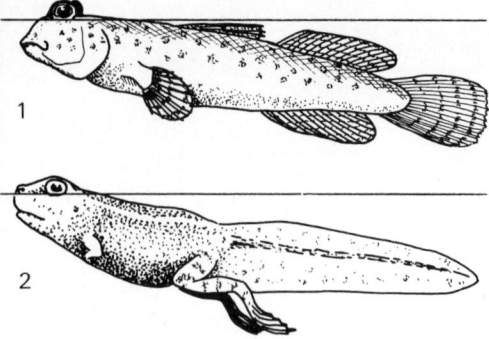

This type of mudskipper is perfectly adapted to an amphibious way of life. While submerged (1) its large eyes remain above the surface so as to command a view of all that is happening on land yet able to remain unobserved. In this it resembles the tadpole (2) which is a genuine amphibian.

Facing page : The mudskippers of the genus *Boleophthalmus*, like other species, possess strong territorial instincts but combats between rivals are ritualised. The spreading of the fins and the display of the blue-black throat probably serve to intimidate each other. Phases of such a confrontation are seen in the upper and centre picture. Those with more terrestrial habits (*below*) have a wider food range, including small crabs.

simultaneous backward and forward motions of the pectoral and ventral fins, the principal support being provided by the former. When the fish is travelling at its normal moderate pace the tail drags along the ground, but when it decides to increase speed it moves the muscular tail vigorously to and fro, with the effect that the body is propelled forwards like a spring and the mudskipper progresses in a series of powerful hops. If, for example, the fish is in a hurry to get back to the water, possibly to escape a predator, the speed at which it can move is quite astounding—faster indeed than a running man. Once in the water it is safe and swims about just below the surface with only its huge eyes protruding like a pair of periscopes. Certain species, however, which are better adapted to spending long periods on dry land, will seek refuge in such situations up a tree.

The eye structure suggests that the mudskipper's vision differs in considerable degree from that of normal fishes. In fact it sees very much better on land than in water, the refractive index of the lens being similar to that of land-dwelling vertebrates. Because the eyeballs are so mobile (they can be raised, moved independently and retracted) the mudskipper enjoys a wide range of monocular and binocular vision and can gauge both shape and distance with accuracy. Furthermore, the retina is divided into two sections, the upper part being well supplied with rods (sensitive to dim light) and the lower part with numerous cones for the perception of fine detail and colour.

Gradual adjustment to life on dry land has brought about a number of ingenious adaptations on the part of these animals. Whilst still on the subject of vision it is worth mentioning an indispensable mechanism which allows the eyes to be continuously moistened and cleaned. In animals that have them the eyelids of course perform these functions but since they are absent in fishes the ancestral forms of mudskipper resolved the matter in a different manner. The eyeballs, as noted above, are retractile; in other words they can be withdrawn into suborbital sockets where they are washed and all impurities removed.

Breathing is naturally the most important problem to be solved by any fish that leads an amphibious existence and we have already seen how this works among the Anabantidae. The way in which it is managed in the case of the mudskipper is something which has long perplexed zoologists and has even now not been fully explained. There is, for example, nothing here resembling either a lung or a labyrinth. Until quite recently it was believed that the fish absorbed oxygen from the water through its tail, partly because it was often seen hanging its tail in water while the rest of its body was on dry land. But this failed to explain how the fish could breathe at those times (and they were frequent) when it was basking on the sand or mud, tail and all. The theory is nowadays discountenanced not only because of this obvious objection but also because breathing through the skin, in this as in other fish species, is feeble and cannot possibly satisfy the oxygen demands of the whole organism. It is now known that when out of water a mudskipper's gill cavities are filled with water. Some breathing also takes place through the lining of the mouth and throat. So the mudskipper resolves its respiration

210

Geographical distribution of the archerfish.

ARCHERFISH
(*Toxotes jaculator*)

Class: Osteichthyes
Order: Perciformes
Family: Toxotidae
Total length: up to 12 inches (30 cm) or more

Silvery body marked with black bands, with fluorescent yellow blotches. Large eyes. Dorsal and anal fins placed well to rear; small caudal fin.

MUDSKIPPER
(*Periophthalmus*)

Class: Osteichthyes
Order: Perciformes
Family: Gobiidae
Total length: 5–12 inches (12–30 cm)

Colour grey-blue or brown, usually well marked with blue but marks varying in size according to species. Flattened, spindle-shaped body with well-developed, spiny dorsal fins, arm-like pectoral fins and ventral fins linked at base to form sucker. Large head and protruding, mobile, retractile eyes.

problem in a surprisingly simple way. One might say that it brings its element with it, for just as a deep sea diver will carry canisters of compressed air so that he can live under water, the mudskipper carries its water supply onto dry land. The water is stored in a large chamber (containing the gills) situated on either side of the head and hermetically sealed therein by the operculum or gill cover. When the fish is on land, also, air is taken in through the mouth and this helps to oxygenate the water in the gill cavities as well. When it captures and swallows prey water is expelled violently from the gill openings and then the fish has to return to the water to replenish its supply of liquid.

Mudskippers fight frequently among themselves and display considerable aggressiveness in the process. They appear to possess strong territorial instincts and stake out their frontiers with a variety of visual landmarks. The representatives of the genus *Periophthalmus* display the colours of their dorsal and caudal fins, opening and closing them in regular rhythm. Those of the genus *Scartelaos*, with their more slender bodies, throw themselves up on their tails from time to time, momentarily standing on the tips, not, as has been thought, for the purpose of catching insects but simply to signal their presence to rivals.

The two sexes can only outwardly be distinguished from each other in the breeding season. Thus in the species *Periophthalmus chrysospilos* the throat and chin of the male assume vivid, almost golden hues, at this time of year, which apparently serve to attract the female. When the latter signifies her sexual interest the two fishes enter the hole which has previously been excavated by the male with his mouth, and it is here that fertilisation takes place. According to Pfeffer, among certain species of mudskipper it is the female who digs the small pit in the mud, mixing the mud and lining it with a mucous secretion. In this way she constructs a miniature pool with walls a couple of inches high, and this is linked to a nearby pond or puddle by means of a narrow canal. It is in this tiny aquarium that the female lays her eggs and in the shelter of these comparatively warm waters that the fry spend the first days of their life, gradually adapting to an amphibious existence.

It is interesting to note that during this period the mother mounts close guard over her offspring. This is exceptional behaviour for an adult fish. Most species either devour their progeny or pay no special attention to them; and on the rare occasions that parental care is displayed it is generally the father who assumes such duties.

The sharp-shooting archerfish

Another fascinating inhabitant of the shallow waters of the South-east Asian mangrove swamps, although also a familiar resident of coral reefs, streams and rivers, is the archerfish (*Toxotes jaculator*). This fish, which grows up to about 12 inches in the wild, belongs, together with several related species, to a family of its own—Toxotidae. Its principal claim to fame (and this explains its common name) is the manner in

which it captures winged insects by directing a thin, powerful jet of water from its mouth upwards towards its prey, thus literally shooting it down.

The curious behaviour of the archerfish was discovered quite by chance and is an example of the way in which the outcome of scientific investigation may depend on pure luck. It was back in 1765 that the Dutch director of a hospital in Batavia, capital of Java, sent a report of the allegedly peculiar habits of this fish to the Natural History Museum in The Hague. He accompanied his report with a specimen of the fish in question, but unfortunately made the error of despatching, instead of an archerfish, another coral reef species. It was hardly surprising that the Dutch zoologists at home should fail to confirm the sender's findings. There was of course no indication that the fish under examination possessed any of the astounding capacities claimed for it and an additional puzzling feature was that, having correctly identified the 'impostor', the scientists could find no evidence of insect life in that particular coral reef habitat. As a result of this misunderstanding the archerfish was declared a figment of the imagination and no further evidence was submitted until 1902 when a Russian scientist managed to capture some live specimens and observed their behaviour under aquarium conditions. It was he who confirmed that the archerfish is indeed a most unusual individual – the only fish that is able to kill its victim from a distance, without any direct contact.

Despite its hunting prowess, however, the archerfish does not depend exclusively or even principally on insects for its food,

The mudskippers of the genus *Scartelaos*, distinguishable by their very slender body and long, pointed dorsal fin, do not rear up out of the water in order to catch insects, as was long believed, but to stake out territory. Their fights are highly ritualised.

for the basic constituent of its diet is plankton. Apparently it concentrates on insects only when unable to find sufficient supplies of its staple food. On such occasions its aim is precise and deadly. At a range of up to 6 feet (and it can operate to twice this distance) it lines up its victim, which may be resting on a leaf or a stem at the water's edge, contracts its gill covers and shoots out a fine stream through the narrow tube formed by the tongue against the roof of the mouth. The jet is usually powerful enough to stun the prey and send it tumbling into the water. Incidentally, young archerfishes practise this hunting technique at a very early age but miss the target as often as they hit it. Only at a later stage does the aim become more sure and the strength of the jet stronger—an interesting example of the way in which practice and experience, on a trial-and-error basis, refine and perfect inborn abilities.

When it spots its victim the archerfish takes up its hunting stance, body completely submerged and mouth just touching the water surface. The eyes are actually below the water at the moment the jet is propelled from the mouth, and this is a phenomenon which has puzzled scientists. A ray of light that passes obliquely from one medium to another—in this case from air to water—changes speed and direction so that the image formed on the retina does not give a true indication of the position of the object under scrutiny. Allowance must be made for the angle of refraction between image and object. How can a fish possibly do this? It seems unlikely that the central nervous system of such a simple organism can store and furnish the necessary information, like a miniature computer, in order to bring about an automatic correction in aim. How then does the archerfish manage to direct its 'fire' with such accuracy?

The solution is in fact quite simple, conforming to the laws of physics. When a ray of light strikes perpendicularly at the surface separating the two media it undergoes no change of direction, and the closer it is to the perpendicular the slighter the deviation. Those who have studied the habits of the archer-fish have noticed that, as if instinctively aware of this law of optics, the predator normally takes up a position almost directly below the victim. Any minor deviation in aim is thus automatically adjusted by the spread and velocity of the expelled jet.

Although the true explanation of the archerfish's prowess is

The archerfish, which shoots down its insect prey with powerful jets of water expelled from the mouth, has long been a puzzle to zoologists. The main difficulty was to explain how a fish could overcome the problem of light refraction since when viewed obliquely from under the water (B) the image of the prey (I′) is distorted owing to the change of speed and direction as the light ray passes from one medium to another. What apparently happens is that the fish stations itself (A) almost immediately beneath its insect prey (I) so that there is the minimum amount of light deviation. Any slight error in aim is compensated by the power and spread of the jet.

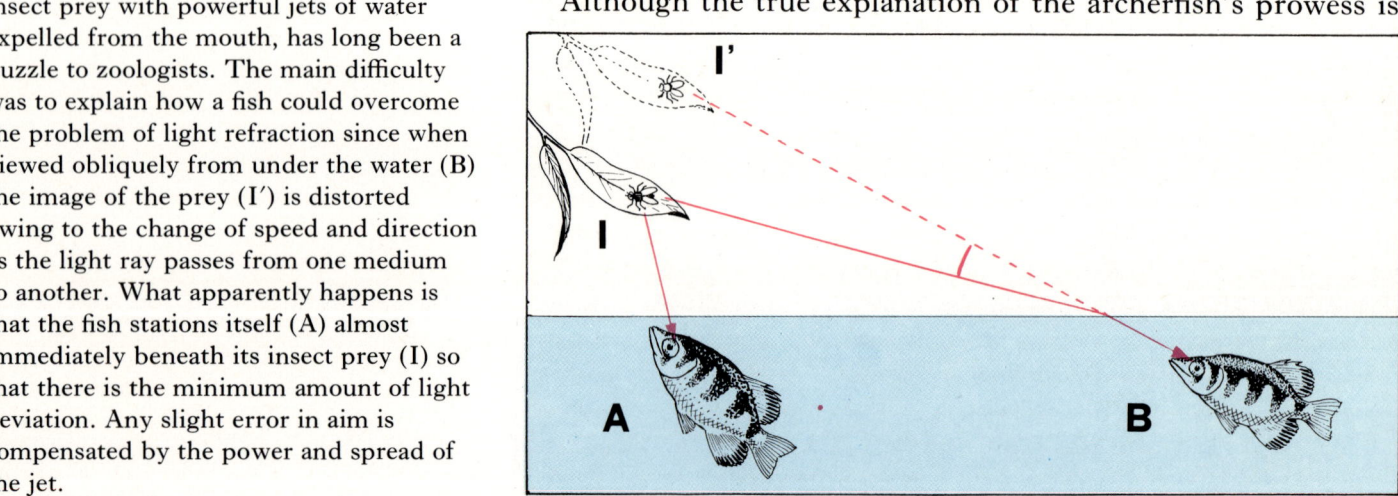

The archerfish, a characteristic mangrove swamp inhabitant, feeds principally on plankton despite its prowess in hunting insects. It seems to concentrate on the latter only when other food is scarce.

relatively uncomplicated, putting paid to the former assumption that the species must possess a higher intelligence quotient than the majority of fishes, one should not underestimate the way in which this unusual behaviour pattern is linked with the creature's anatomical and physiological development. Thus the eyes are very large, providing better binocular vision than is normal among fishes, and this facilitates the accurate gauging of distance. Furthermore, the unique technique of hunting practised by the archerfish has influenced the mental processes in the sense that it steadily improves its aptitude as it grows older and more experienced.

Nevertheless, not every aspect of the long-distance hunting capacity of the fish has been clarified. It would appear that such an elaborate predatory technique could only have come about in response to strong selective pressure, born of the need to survive. Yet, contrary to such reasoning, it seems certain that this specialised mode of hunting, nowadays at any rate, plays only a secondary role in the quest for food since, as already mentioned, the diet of the archerfish is mainly zooplankton. What is more, in other habitats, notably in the ocean, this gift is of no use whatsoever.

As more and more information comes to hand concerning the behaviour of this strange fish, it becomes obvious that there is no justification at all for the former assumption that it is endowed with superior intelligence. Thus it will frequently use the jet-propulsion technique in situations where it is either ineffective or wholly inappropriate. It is perfectly capable of catching an insect skimming the surface by leaping clear out of the water, like a trout, yet on other occasions it will direct a jet of water in the direction of an insect which is already floating and which could more easily be snapped up by mouth. Even

The archerfish can strike its insect prey accurately at a distance of up to 6 feet or more and practises its skill from a very early age. When young, however, it directs its jet of water in a somewhat random fashion, becoming more expert as it grows older.

stranger, and just as useless, is the habit of 'shooting' small objects and animals that are actually submerged, as can be seen when the archerfish contracts its gill cavity, stirring up, as it ejects the jet, a cloud of mud. It is hardly surprising that the jet of water expelled under such circumstances seldom finds its mark. What probably happens is that the fish spurts out its stream of water more or less indiscriminately in the vague direction of any moving object that arouses its curiosity, whether or not it is edible and no matter what the prevailing conditions.

In the breeding season archerfishes swim away from land out into the ocean, probably (though information is still inadequate) laying their eggs on the fringes of a coral reef. The young make the journey in the opposite direction back to the mangrove swamps and sometimes to river estuaries. The black stripes, separated by zones of iridescent yellow which flash like fluorescent lights, are already visible on their backs as they are in the adults. These bright colours may play an important part in intraspecific communication, enabling the fishes to recognise one another in muddy water where visibility is poor.

The amphibious dugong

An animal can only survive in a region of mangrove swamp if it is to some extent amphibious, able to lead an independent life both on dry land and in water. This is exemplified perfectly by two animals in particular—the mudskipper and the dugong. The former is a fish which can live like a terrestrial mammal and the latter is a mammal which can live like a fish.

The dugong (*Dugong dugon*) belongs to the order Sirenia and is undoubtedly one of the strange animals which helped to inspire the ancient belief in those mythical creatures, half fish and half-women, known as sirens, who with their singing lured sailors to their death. It is a peaceful, inoffensive creature and the fact that the mammae of the female are positioned on the chest does give it a vaguely human resemblance. Any further similarity is almost certainly fanciful. Although ancient books on zoology state that the animal also suckles her young, human fashion, supporting the baby with one of her flipper-like fore-limbs, this is not borne out by direct observation. In fact the only mammal belonging to this order which on one occasion has been seen suckling her offspring was completely submerged at the time, nor was she using her flipper as a support. This was a manatee, a related species, so that it is reasonable to assume that the dugong feeds her young in a similar manner.

The range of distribution of this mammal has shrunk considerably in recent times and even today, in those regions where it survives, the species is hunted mercilessly for its flesh and fat, and additionally because of the superstitious, pseudo-medical belief that certain parts of its body have miraculous restorative properties. As a result the dugong is practically extinct, its name listed in the IUCN Red Data Book; and although climatic changes might once have been partially blamed for its disappearance from some regions there can be no doubt that today man is the sole culprit. Its chances of survival seem to be very slim and it is ironic that the animal, after adapting to an environment where it had few land or water predators to fear, should have succumbed to the net, harpoon and gun.

The dugong is strangely built, the forequarters being much heavier than the hindquarters. Its forelimbs have been transformed into flippers, the hind limbs are absent and the paddle-like tail somewhat resembles that of a dolphin and other cetaceans, though not nearly as powerful. The eyes are kept moist with an oily secretion exuded by the preorbital glands and which seem to provide partial protection against salt water. The upper lip, playing a vital role in the feeding process, is horseshoe-shaped; and the male also has short tusks in the upper jaw. The teeth too are unusual, consisting of a few incisors and cheek teeth (for grinding) which are like those of an elephant.

Other anatomical features have helped the dugong to evolve in this environment, not only in the sea but also in the shallow coastal waters of the Oriental region. The skeleton is heavy and compact, providing valuable ballast, and the high position of the nostrils enables the animal to breathe almost without exposing the rest of the head at the surface.

The peaceable dugongs live in family groups and these are sometimes enlarged into herds, particularly in areas where food prospects are especially good. The main constituent of the diet is eel grass but this is supplemented occasionally by land plants growing near the shore. It is this habit of feeding in large groups which has made the alternative name of 'sea-cow' quite appropriate. The mammals may remain for a period of between one and ten minutes below the surface when feeding.

Under normal circumstances dugongs live harmoniously together with each male and female apparently forming a life-long relationship. Because of the general lack of reliable information about the reproductive habits of the species, however, this has not been positively confirmed. Mothers have frequently been seen in the company of two youngsters of different ages, one of them obviously having been born more than a year prior to the other.

Dugongs do not appear to mate at any definite time of the year for youngsters of all ages are to be encountered at different periods. The sexual act, like every other activity of the animal, takes place under water. Some authors claim that the male leaps out of the water as a preliminary to mating but Maurice Burton points out that the courtship rites, like everything else, are conducted in a tranquil, unspectacular fashion.

After approximately eleven months' gestation the female dugong gives birth to a single baby. The baby is born in the water and shortly afterwards floats up to the surface to take its first gulp of fresh air. There are still a few authors who put forward the suggestion that the birth takes place on dry land, but this seems extremely unlikely, given the fact that the dugong is much better adapted to life in the water and that the physiological processes, such as breathing, would be immeasurably difficult (considering the weight of the animal) on land. It does not seem possible that an activity as vital as giving birth could occur in such unfavourable circumstances.

DUGONG
(*Dugong dugon*)

Class: Mammalia
Order: Sirenia
Family: Dugongidae
Total length: 98–110 inches (250–280 cm)
Weight: 310–375 lb (140–170 kg)
Diet: vegetation, especially eel grass
Gestation: about 11 months
Number of young: one

Colour variable, often grey or brown. Massive and heavy with front part of body better developed than hindquarters. Pectoral mammae. Large head with broad, flat snout; upper lip horseshoe-shaped, with vibrissae; small eyes. Forelimbs transformed into flippers, paddle-like tail similar to cetaceans but less powerful. Dentition reminiscent of that of elephant; males possess short tusks.

CHAPTER 91

Fishers of the mangrove swamps

The adventitious roots of the mangroves might be said to anchor the land in the water. By their invasion of the sea these remarkable plants have created a new type of environment, the conditions of which are exactly right for a diversity of aquatic animals. Warmth, sunlight and shallow water combine to produce a strange and wonderful world suspended, as it were, between two elements, with a marvellously varied fauna composed, as usual, of primary consumers, predators and super-predators.

The hunters of the mangrove swamps – and we are concerned here not with predatory fishes but with birds, mammals and reptiles – have, from necessity, adapted to a very special mode of life which has, over an extended period, affected their body structure and modified their behaviour. Although most are land-based (at any rate for the vital purpose of reproduction) their prey is aquatic. Food in this habitat consists exclusively of fishes, crabs, amphibians and other water animals. The otter, the fishing eagle, the brown fish owl and the salt-water crocodile therefore have one thing in common; they all fish for their living. In fact the tangled roots, shady foliage and twisting trunks of the mangroves provide shelter for a greater diversity of fishing animals than in any other part of the world.

It is among the mangrove swamps of South-east Asia that we come across the only primate (apart, of course, from man) specialising in fishing – the crab-eating macaque, which, as its name suggests, feeds on crabs and other marine invertebrates. Most monkeys are either exclusively terrestrial or arboreal, but this species, though it lives in trees, dives, swims and catches its prey under water with consummate ease. It can remain completely submerged for several minutes if danger looms.

There is another primate living in this environment which is

Facing page : The Brahminy kite (*Haliastur indus*) is one of the several raptors of the Oriental region which feeds mainly on fishes and other aquatic animals, (*Haliastus*) operating in streams and rivers as well as on the fringes of the mangrove swamp.

also an expert swimmer and diver yet cannot be classified as a fishing animal in view of the fact that it feeds only on vegetation. This is the odd-looking proboscis monkey, which evidently regards the mangrove tree as a place of refuge and a valuable source of food. Both the crab-eating macaque and the proboscis monkey represent two stages in the gradual conquest of the watery element by the primates, but it is only the former which fully exploits all the food resources of its mangrove swamp habitat. It is probable that the species always had an instinctive inclination towards fishing; but undoubtedly the abundance of tree cover and the rhythm of the tides, which exposed extensive tracts of mud and sand that teemed with aquatic life, must have been vital determining factors in its subsequent evolution.

Favourable natural conditions similarly transformed the life habits of three other typical predators of the mangrove swamps – the fishing cat, the crab-eating mongoose and the otter civet. None of these animals, however, is so perfectly built for fishing as are the various species of otter that are active in the shallow waters of the swamps, notably the Indian small-clawed otter and the hairy-nosed otter. These sleek mammals, similar in appearance and behaviour to their European relative, are among the most successful colonisers of the mangrove swamps.

The coasts of the Oriental region offer exceptional opportunities too for a number of fishing birds. The osprey is the epitome of fishing expertise, diving from a great height into the water and emerging within seconds, clutching a prey in its talons. Although this raptor has a virtually cosmopolitan range it visits the mangrove swamps during the winter. There are three other specialised raptors found along the shores of tropical Asia that hunt in a similar fashion. Two of them, already mentioned, are the lesser fishing eagle, normally found along watercourses deep in the jungle, and the related grey-headed fishing eagle, usually inhabiting regions where the vegetation is less dense. The third is the white-bellied sea eagle, which, although lacking the extraordinary manoeuvring abilities of other birds of prey, feeds basically on fishes. In addition to these raptors there is one unusual member of the Strigiformes, the brown fish owl, which likewise finds its food in swamps, rivers and lakes.

Naturally there are plenty of other more common species making up the bird population of the coastal waters of Southeast Asia. They include herons, egrets, pelicans, cormorants, kingfishers and storks, all of which hunt fishes and other forms of aquatic life.

There is one formidable super-predator lurking in the waters of the mangrove swamp – the estuarine or salt-water crocodile. Although this enormous reptile is seen more often on land than in the water it is nevertheless extremely well adapted to both environments, including, as its name indicates, the sea. If it has one enemy in particular it is the monitor, a carnivorous member of the lizard family which devours the crocodile's eggs.

It is evident from this brief summary of a few characteristic species that the mangrove swamp, harbouring predators as massive as crocodiles and as agile as fishing eagles, is the home of

---

CRAB-EATING MACAQUE
(*Macaca fascicularis*)

Class: Mammalia
Order: Primates
Family: Cercopithecidae
Total length: 45½ inches (115 cm)
Length of tail: 20 inches (50 cm)
Diet: crabs, molluscs, sometimes small fishes; also vegetation
Gestation: 159–178 days
Number of young: one

Adults
Greyish hands, feet and tail; darker head; white patch between eyes. Tail long and muscular. Short beard in both sexes.

Young
Open eyes two hours after birth. Hair of head and back thicker than in adults.

an uncommonly wide range of animals. Yet despite the numbers involved competition between these species is obviated by the fact that there is more than enough food for all. The trees which serve as refuges and vantage points, and the salt waters that teem with life and assure the survival of an army of predators, make up an environment which surely reproduces, more closely than any other, conditions as they must have been millions of years ago in the primeval swamps where the earliest land animals evolved. In its way it is a wildlife paradise.

The crab-eating macaque

At high tide, when the waves ripple over the muddy banks of the mangrove swamp, a troop of monkeys watches patiently from the summit of a tree, temporarily cut off like an island, until the water recedes. They are crab-eating macaques (*Macaca fascicularis*) and they display no signs of anxiety or alarm, for they are skilled swimmers and, were they so minded, could easily make their way to another vantage point. But this mangrove tree suits them very well for the time being. In an hour or two the water will have retreated, exposing below them an expanse of wet, glistening mud literally swarming with all manner of appetising little animals, there for the taking.

Indeed their choice of this particular tree is not random or accidental but quite deliberate, dominating as it does the best feeding grounds. If necessary the macaques are prepared to defend this hunting territory energetically against intruders. Ridley has described a confrontation between a troop of crab-eating macaques and half a dozen langurs; and in the Udjung Kulon reserve on the island of Java Hoogerwerf witnessed

Stimulated by the abundance of prey to be found in and around the mangrove swamp, the crab-eating macaque has gradually developed fishing skills. Perching in a mangrove, it waits until the tide is low to come down to the mud flats where shellfish and crabs are to be found in great profusion.

An adult male crab-eating macaque surveys his territory from a tree.

Facing page: Crab-eating macaques (*above*) and langurs (*below*) often dispute a piece of territory but are sometimes found living harmoniously together in mixed troops.

a violent battle between these same two species for control of a tree or group of trees, in the course of which one young langur was actually killed. It is not uncommon, however, to find peaceful mixed troops of crab-eating macaques and langurs in the Indonesian reserve.

The number of individuals constituting a macaque troop is very variable, usually exceeding twenty, and comprises males and females of all ages. As a rule it is an exclusive society, closed to strangers, and fights against neighbouring troops of macaques are quite common. As with all species of social primates and the majority of zoological groups with communal habits, these monkeys spend much of their time grooming one another, not for hygienic reasons but because this has traditionally been a method of establishing mutual relationships and maintaining cohesion within the group. It is generally the females who groom the young and the males, but the latter often reciprocate at mating time.

This grooming activity has been recorded by Desmond Morris. He describes how a macaque may at intervals swallow tiny particles which it has removed from a companion's fur. As it carries these to its mouth it emits a series of tongue clickings as if it were savouring in advance what it is going to eat. When an individual is going to groom a companion the first movements it makes may be accompanied by this sound, even though it carries nothing to its lips. These sounds are of greater intensity than the 'functional clickings'. Generally the noises are made with the tongue against the teeth and are interrupted almost immediately to make way for the functional clicking during the rest of the grooming session. In some species they do not appear to be linked specifically with the grooming and may be emitted in other circumstances.

Apart from grooming, the gesture of submission which plays the most important role in establishing and maintaining class distinctions within the troop is the mating posture. The females, at any stage of the menstrual cycle, even when they are not in heat, as well as adult males and young, may, if menaced by a companion, adopt a position which cannot be distinguished from the normal posture of a receptive female presenting to a male. As is the case with baboons, this attitude is indicative of complete submission towards the animal that has provoked it. Nevertheless, the macaques are not as rigidly organised as baboons, and dominance of one individual over another does not play any significant role in everyday life. In the wild it is extremely difficult to distinguish dominants from subordinates.

When the tide recedes all the members of the macaque troop come down from the high branches of their tree to conduct a survey of their muddy surroundings, seeking crabs, snails and other small animals. They will not hesitate, if necessary, to enter the water (even those living some distance from the coast hunt freely in streams and rivers) and are always on the lookout for a fish or frog that may find itself trapped in a pool. Having satisfied their appetite the monkeys abandon the beach and head for the trees again – a reminder that they are, after all, essentially arboreal animals.

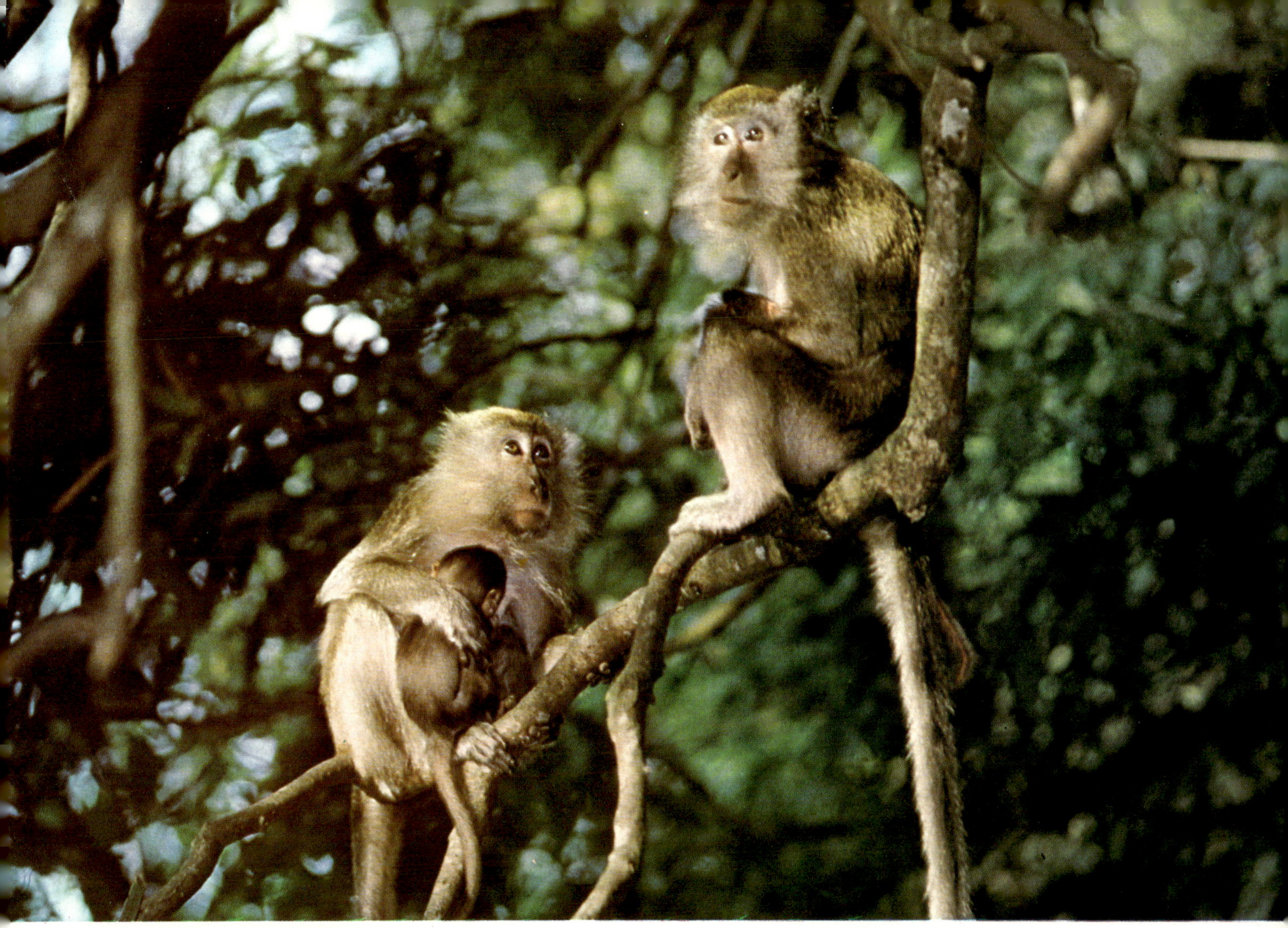

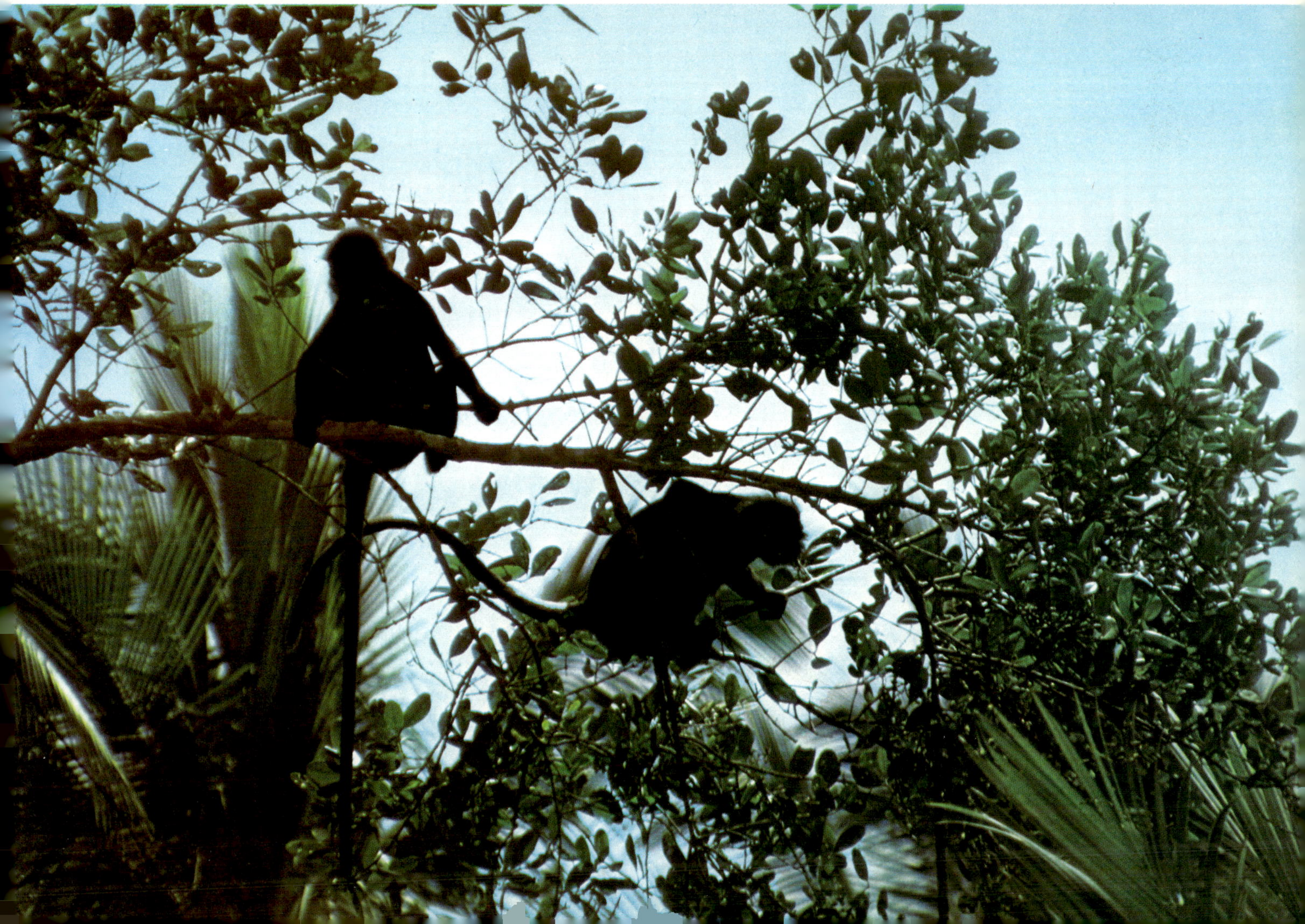

The proboscis monkey is a peaceful, indolent inhabitant of the mangrove swamps of Borneo. The long, pendulous nose of the male measures up to 7 inches but what purpose it serves is not known.

The sexual life of crab-eating macaques was studied for more than twenty years at the University of Jena in East Germany, first by Dr A. Spiegel and later by Dr J. W. Harms.

Among the females, as is the case with many other primates, the area surrounding the genital organs is naked and turns bright pink or red when the monkeys are receptive. This apparently occurs between the sixth and twentieth day of the menstrual cycle and normally lasts about five days.

During the entire period of rut the male remains in close attendance on the female who assumes a position on all-fours, presenting her raised hindquarters while her head rests on the ground. The macaques mate several times every day and when the sexual skin of the female fades in colour the male wanders off in search of another partner.

Gestation lasts 159–178 days, at the end of which time one baby is born. The mother immediately swallows the placenta and licks the baby clean. She then suckles the youngster for about eighteen months. The latter will be sexually mature between the ages of three and a half and five years.

The crab-eating macaque inhabits the greater part of Southeast Asia from Burma to Indo-China, the Malay archipelago and the Philippines. It is possible that man has helped to expand the size of its range for these trustful animals are easily tamed and were once taken along as mascots by fishermen in their boats, with the result that some managed to break loose and settled on remote islands where they subsequently multiplied.

In some villages in tropical Asia crab-eating macaques are regarded as sacred animals and their worshippers place offerings at the foot of trees regularly occupied by the monkeys.

Elsewhere, however, man is the principal enemy of the species, followed by crocodiles, pythons, wild dogs, eagles, leopards and a miscellany of smaller wild cats.

The grotesque proboscis monkey

One of the most popular characters of children's fiction is the wooden puppet Pinocchio, whose nose grows longer whenever he is naughty. Pinocchio might be said to have a counterpart in the animal world in the shape of the proboscis monkey (*Nasalis larvatus*). There is nothing very special about the appearance of the females and young of this species but the males are very singular individuals indeed. When a young male reaches puberty, at approximately seven years of age, his nose changes colour and begins growing longer. In old males it hangs down below the mouth like a trunk, measuring up to 7 inches in length. Sometimes, in order to feed, the monkey has to hold his nose aside.

What does this strange monkey eat? Although it evolved in a similar way to the crab-eating macaque and shared the same habitats, the proboscis monkey never took to fishing. It belongs in fact to the subfamily Colobinae and is theoretically a strict vegetarian, feeding chiefly on enormous quantities of leaves, above all those of the *Sonneratia* species of mangrove, flowers of nipa palms, and the fruits and shoots of various plants growing in mangrove swamps. Yet proboscis monkeys in the Frankfurt

Proboscis monkeys are vegetarians. Favourite plant species include the leaves of the *Sonneratia* mangroves and the fruits of the nipa palm.

PROBOSCIS MONKEY
(*Nasalis larvatus*)

Class: Mammalia
Order: Primates
Family: Cercopithecidae
Total length: male 26–30 inches (66–76 cm)
female 21½–23½ inches (54–60 cm)
Length of tail: 22–30 inches (56–76 cm)
Weight: male 35–49½ lb (16–22.5 kg)
Diet: leaves, flowers and fruits of mangrove swamp plants
Number of young: one

Adults
Back variable in colour, chestnut or brick red, belly grey, neck and sides of head pale; face naked. Cheeks bearded, forehead low, eyes wide apart, ears small. Trunk-like nose of male becomes soft and pendulous with age, up to 7 inches long. Nose of female shorter and pointed. Small laryngeal sac amplifies voice.

Young
Bluish face when born.

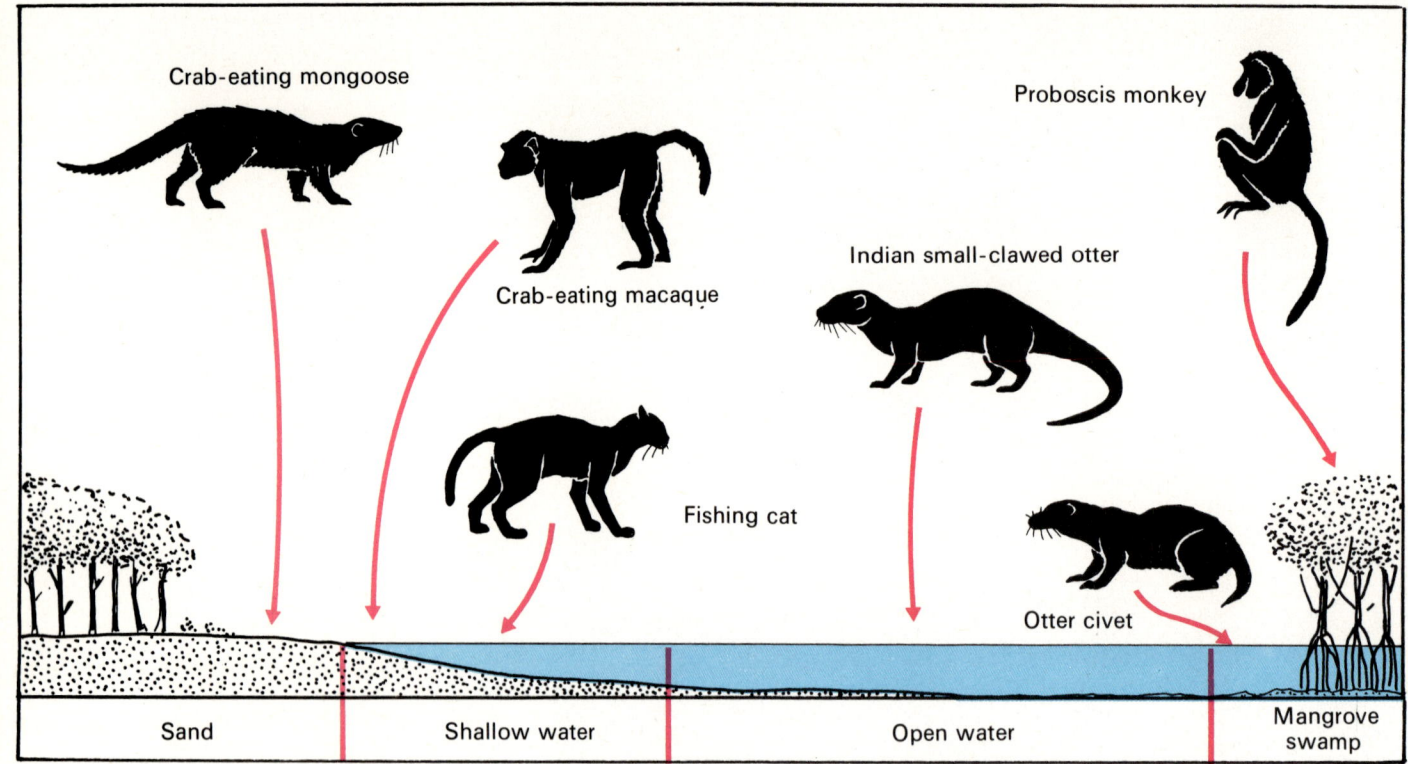

Crab-eating mongoose

Crab-eating macaque

Proboscis monkey

Indian small-clawed otter

Fishing cat

Otter civet

| Sand | Shallow water | Open water | Mangrove swamp |

Habitats of some of the typical fishing animals in and around mangrove swamps.

Zoo apparently eat mealworm larvae and grasshoppers as well, leading to the conclusion that just as crab-eating macaques vary their diet occasionally with fruit, leaves and flowers, so do proboscis monkeys with small invertebrates.

Large troops of proboscis monkeys, comprising twenty or more animals, are frequently to be seen sitting sedately in trees, yet if the need arises they display surprising agility, making leaps of up to 25 feet. The range of the species is much more restricted than that of the crab-eating macaque, for it is found only in Borneo, along muddy river banks, flooded areas of jungle and mangrove swamps. Living in such habitats, it is understandably a skilful swimmer and diver. William Beebe, head of the bird section of the New York Zoo, once saw a proboscis monkey swimming across the Rajang, one of Borneo's widest rivers. Alarmed by a sudden rifle shot, the animal vanished under the water, remaining submerged for almost half a minute, then reappeared, striking out again for the opposite bank.

There have been few detailed investigations into the biology and behaviour of the proboscis monkey in the wild and much remains to be discovered, especially concerning the development and function of the male's elongated nose. Some authors believe it to be a secondary sexual characteristic while others think that it may help to amplify the voice, for the male roars loudly and continuously, in contrast to the female who emits sharper and softer cries.

Proboscis monkeys are difficult to rear outside their natural surroundings, mainly due to the problem of obtaining a sufficient quantity of the type of vegetation peculiar to their mangrove swamp habitat. Up until 1959 no proboscis monkeys had survived for more than two and a half years in captivity, the record being held by one animal in Calcutta Zoo. But since then conditions

Otter civet
(*Cynogale bennetti*)

have improved a little and some of the animals have succeeded in breeding. On September 12, 1959, a baby was born to Penelope and Pinocchio in the zoo at San Diego, California. This was the first birth of a proboscis monkey outside South-east Asia, being watched with keen interest and described by Mr G. H. Pournelle, head of the zoo's mammal section. As the time drew nearer for the birth Penelope's customary high spirits flagged and she began to experience respiratory troubles. Her mate, Pinocchio, was clearly concerned for her welfare. In the course of their daily games he had followed her around the enclosure and treated her fairly roughly from time to time; but now he spent much of the day seated quietly by her side. He would even pick morsels of food out of their bowl and offer them to her.

At one stage he became noticeably mistrustful of the keepers and shortly before the birth he made it obvious that he was not prepared to tolerate their presence at all. After that nobody disturbed the couple. Eventually the baby was born, without any further complications, and Penelope proved to be an ideal mother. When born the baby's face was bluish (compared with the flesh colour of its parents). It soon showed signs of playfulness and would frequently approach Pinocchio to tweak his nose, pull his tail or otherwise interfere with his sleep. When the latter made it clear that he was not amused by such pranks the youngster sought refuge in its mother's arms.

Apart from the crocodile, the most deadly enemy of the proboscis monkey in Borneo is man. The Dayak people find the animal's flesh very much to their taste and the men go out hunting for it with blowpipes and poisoned darts.

Fish-eating carnivores

In the mangrove swamp almost all the carnivorous animals, large and small, resort to fishing at one time or another. The tiger and the leopard, both of them excellent swimmers, capture large fishes stranded in pools at low tide while the linsangs and civets prowl the mud banks for marine invertebrates. But whereas these form only occasional delicacies for the afore-mentioned carnivores, there are others which specialise exclusively in fishing. Among such hunters are the crab-eating mongoose (*Herpestes urva*), which wanders through the mud at dusk in search of crustaceans and molluscs, the fishing cat and the otter civet.

The fishing cat (*Felis viverrina*) is an inhabitant of Ceylon, India, Indo-China, Thailand, Sumatra and Java. Clearly it has adapted successfully to a variety of habitats but as far as zoologists are concerned it is a mystery animal. It measures over 4 feet long, tail included, but apart from this hardly anything is known about its biology or behaviour. H. C. Delsman, who studied the species for some time, wrote in 1930 that one individual he caught and kept in captivity only ate meat and would not touch fish. Although he later modified his opinion on this point his findings received support from J. J. Menden who claimed to have caught a fishing cat by baiting a trap with a chicken.

There is no doubt, nevertheless, that by and large this carnivore lives up to its name. On the coast of Pamanukan, Java, E. B.

The fishing cat is one of the most mysterious hunters of the Oriental region. Very adaptable, it is found, more often than not, in mangrove swamps but is sometimes encountered in mountain regions of India. It has frequently been seen catching fishes from the bank but its swimming and diving ability has not been recorded.

Like others of the family the smooth Indian otter is a lively, intelligent and inquisitive animal. At the least sound it sits up on its hind legs and attentively surveys the surroundings.

Facing page : The Indian small-clawed otter lives in river estuaries and mangrove swamps. It is an agile, skilful hunter which feeds principally, perhaps exclusively, on fish. The short claws are of no use for hunting, but in the opinion of some authors the hands are particularly sensitive.

Mulder once surprised a number of fishing cats lying in ambush for prey on the beach, pulling them out of the water with a swift movement of the paw and devouring them on the spot; and Pieters witnessed several similar occurrences in the same region. What is uncertain is whether the technique described by Mulder is characteristic or whether the cat normally dives completely into the water for its prey.

Reports have been submitted that the species also preys on kids, lambs and dogs, and that it has even been known to attack children lost in the jungle; but first-hand examination of excrement has revealed nothing more incriminating than the remains of shellfish, crabs' claws and rats' hairs.

On Java the fishing cat lives near sea level in mangrove thickets and on sandy beaches. In India, however, it has been sighted at altitudes of around 5,000 feet. But in none of its known habitats has any useful information been forthcoming about its breeding habits.

The otter civet (*Cynogale bennetti*) at first glance looks much like an otter but a certain distinguishing feature is its short, slender tail. An inhabitant of Indo-China, Malaya, Sumatra and Borneo, the animal is almost always found close to water, either on flooded plains or in mangrove swamps. Well adapted to a semi-aquatic existence, the nostrils and ears of the otter civet are sealed automatically (like those of the otter) as it dives into the water. But in some respects the animal emulates the true civets; thus when pursued it will seek refuge high in a tree and not in the water.

The otter civet is a comparatively slow swimmer for its feet are not webbed and the short tail is of very little use for propulsion or steering. Nor can it make sudden, rapid turns in the water to chase its prey. On the other hand, the position of the nostrils makes it possible for the animal to stay completely submerged for fairly long periods and to lie silently in ambush for any birds and small mammals that happen to stray down to the water for a drink. Nevertheless, food consists in the main of fishes, crabs, snails and the like, together with fruit.

Little precise information is available about the reproductive behaviour of this species. Ernest P. Walker states that the female gives birth to a litter of two or three babies and reports having seen, in Borneo, a mother accompanied by young during the month of May. The animals have three glands, on the skin close to the genital organs, secreting an odorous substance.

The true otters are naturally the best equipped of all the fishing animals of the mangrove swamps. There are several species but all are characterised by a long, powerful tail, and nostrils and ears that are sealed at the moment of diving. Broadly speaking, their habits are identical to those of the common otter of Europe and other parts of Asia. But the Indian small-clawed otter (*Amblonyx cinerea*), which greatly resembles the clawless otter of Africa, differs to some extent from its other relatives. In India it is regarded as a typical mountain animal since it is normally found frequenting highland streams and rivers; but elsewhere, especially in eastern coastal areas, it inhabits river estuaries and mangrove swamps. On Java the small-clawed otter is seen both

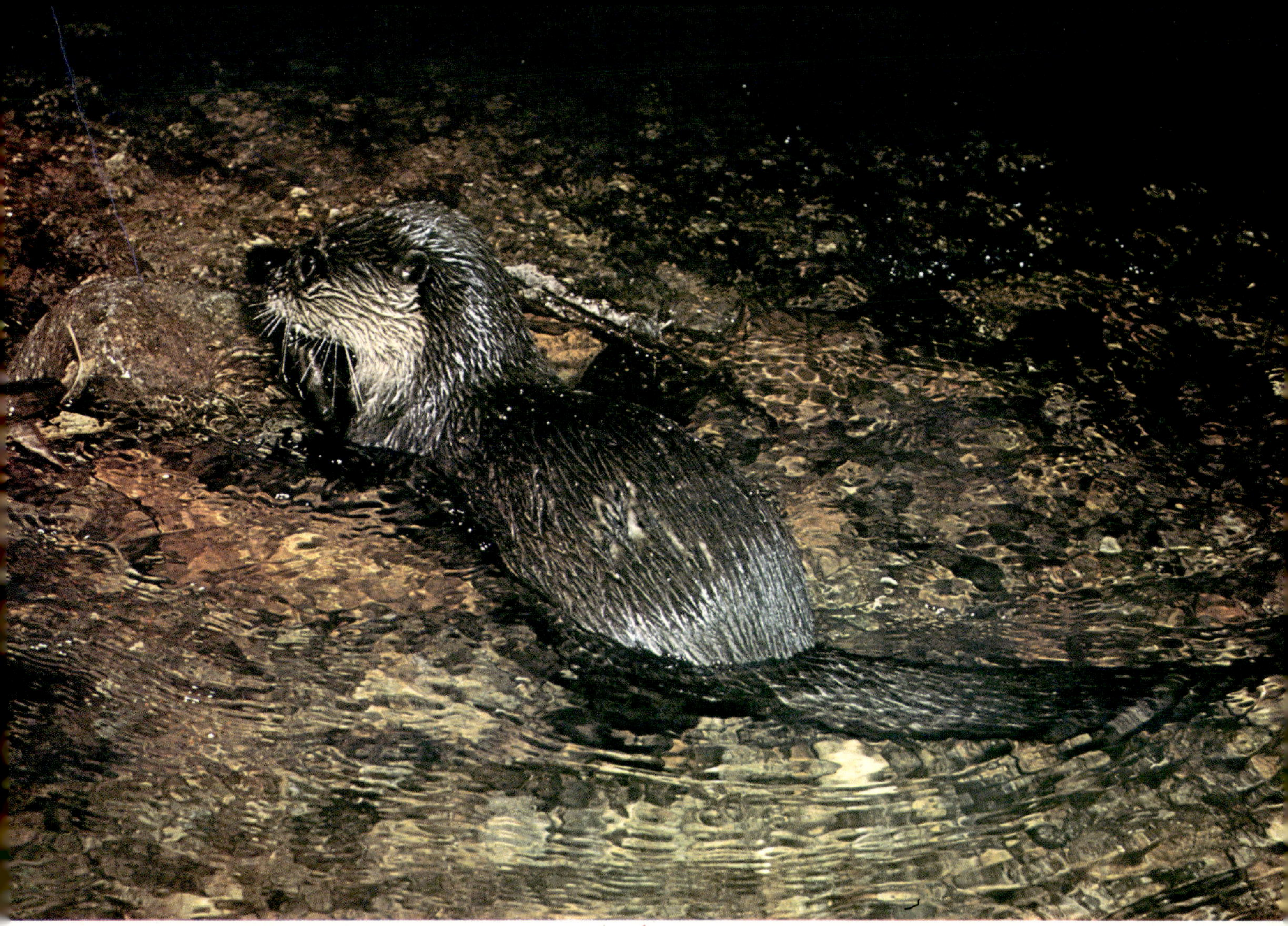

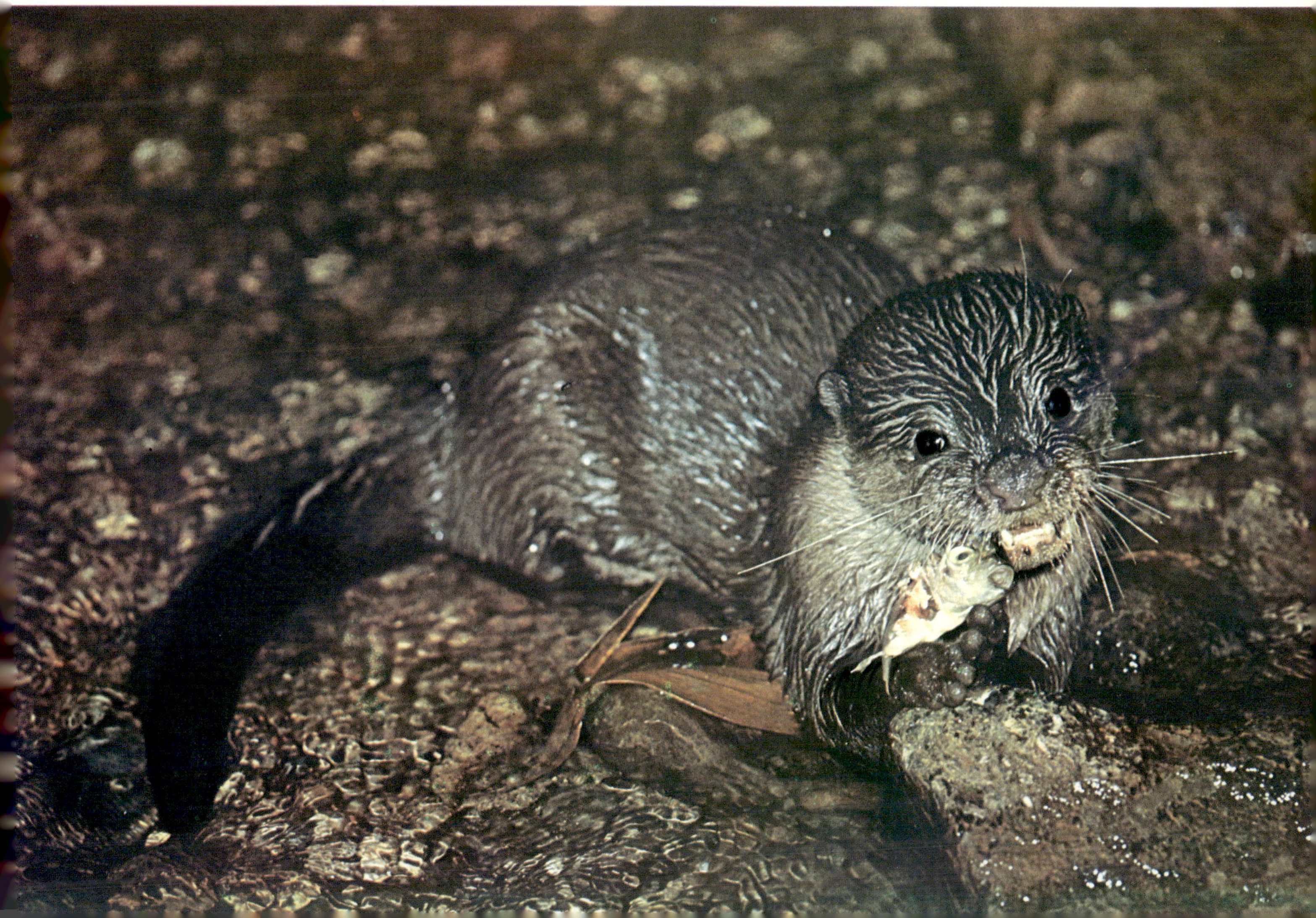

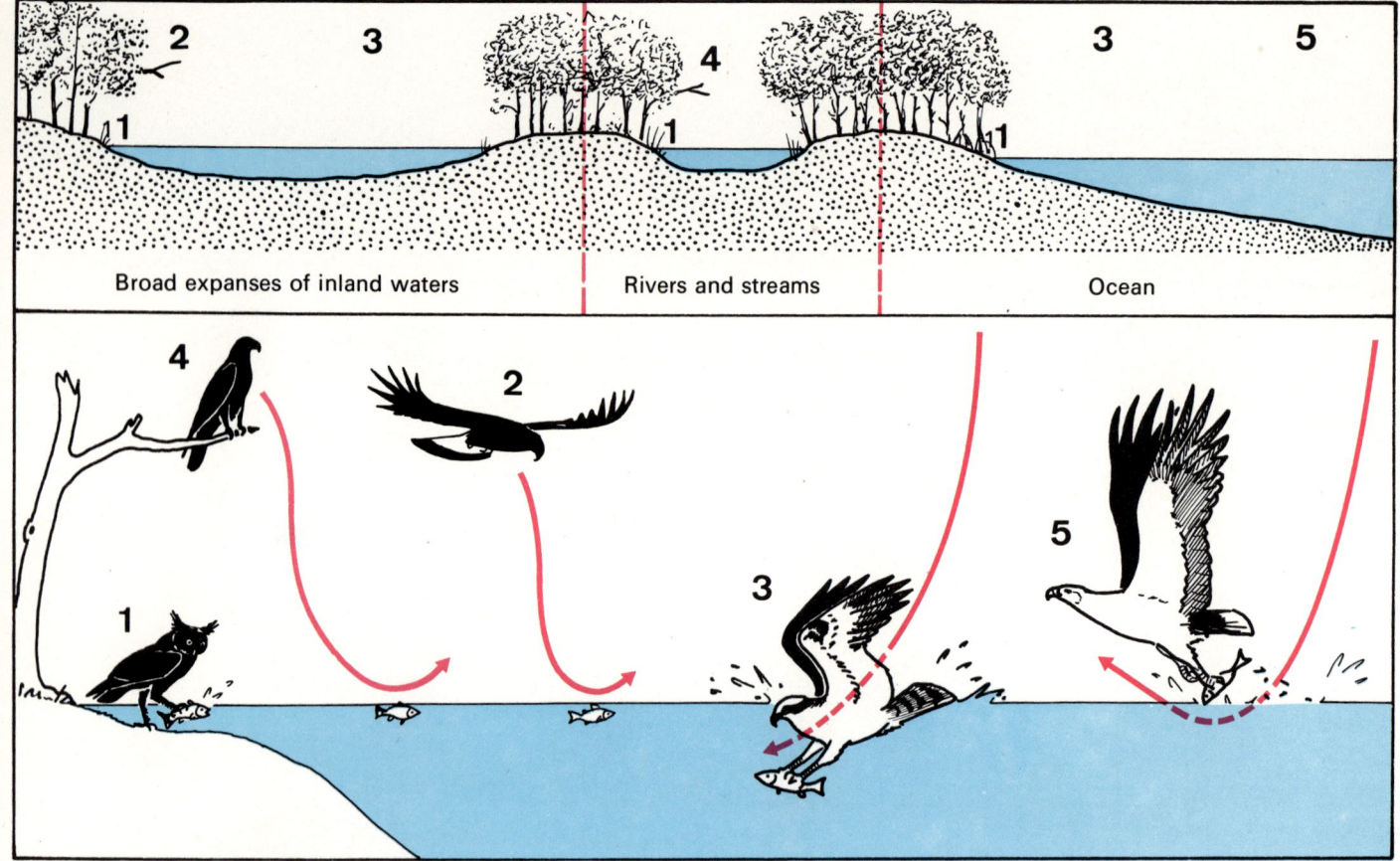

Hunting grounds and fishing techniques of various birds of prey frequenting the mangrove swamps of the Oriental region: 1. Brown fish owl. 2. Grey-headed fishing eagle. 3. Osprey. 4. Lesser fishing eagle. 5. White-bellied sea eagle.

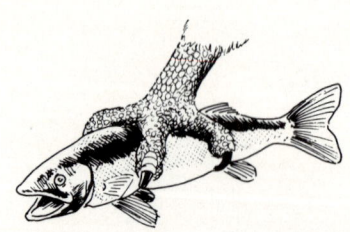

The osprey has extremely strong toes and claws. The fourth and outer toe can be turned to the rear so that a powerful grip can be taken on the prey.

at sea level and higher at altitudes of up to about 7,000 feet. As a rule this species lives and hunts in groups. Little bands of from four to twelve individuals are seen more frequently than single animals or pairs. As the name suggests, the claws of the forelimbs are very short and this, according to R. I. Pocock, is linked with a heightened sensitivity of touch. Certainly, like other otters, the animal is very playful, especially in captivity. Pocock tells how he once saw a small-clawed otter toying with a marble and handling it with as much dexterity as a juggler with a cricket ball.

Some authors claim that the small-clawed otter feeds on crabs and snails to a greater extent than fishes but this is contradicted by others who state that only fish is eaten.

Much larger, and confined to coastal regions, the hairy-nosed otter (*Lutra sumatrana*) is sometimes described as a sea otter although it has little in common with the true sea otter (*Enhydra lutris*). In some respects it resembles the Indian small-clawed otter but it does not manipulate objects with anything approaching the same measure of skill as the latter species.

The smooth Indian otter (*Lutra perspicillata*) is similar to the hairy-nosed otter but its pelage is paler and of uniform colour. Not much is known about its behaviour in the wild but on the basis of studies carried out on a pair in captivity it is possible to say that the animals mate in the water, that gestation lasts 63 days and that the one or two babies of a typical litter are born blind, their eyes opening ten days later. They are as large as the adults after a year though not sexually mature for two more years. In captivity the animals have lived for up to fifteen and a half years. This is about the same as has been recorded for several species of other.

In conclusion, one should mention that the common otter (*Lutra lutra*) is found throughout temperate Asia and also in many parts of tropical Asia.

The osprey: king of fishing birds

The osprey or fish hawk (*Pandion haliaetus*) is a familiar bird of prey with a wide global distribution, to be found in all types of watery habitat where fish is plentiful and conditions are suitable for its specialised hunting technique. The species breeds more regularly in the northern hemisphere although some birds do nest in Australia and the adjacent islands. The northern populations leave their breeding zones in the autumn and fly south to warmer regions for the winter. Normally it is only during the winter months, therefore, that this raptor is seen in the mangrove swamps, on the river banks and around the shores of South-east Asia. Very occasionally, however, pairs are sighted in these habitats during the spring.

The osprey is undoubtedly the most specialised of the fishing birds of prey and its anatomy bears all the signs of being moulded for this purpose alone. The feet are large and powerful, with naked tarsi terminating in short toes and long, hooked claws. The fourth or outside toe of each foot is particularly mobile and can be directed backwards so as to form a pair with the rear toe, providing a strong pincer grip in conjunction with the two front toes. The plumage is thick and compact, blunting the impact when the bird hits the water. The head is small, the bill short and only slightly curved, and the feathers of the neck form a small crest when erected.

When at rest the dark brown facial marks are a sure means of identification, but the bird is even easier to recognise in flight, when the brown back is seen to contrast strikingly with the white underparts.

The osprey spends much of the day perched motionless on a branch, interrupting its rest intermittently to clean its feathers with its beak. The moment it straightens up for action, however slight the movement, terrified ducks and coots in the vicinity take wing while ducklings scuttle for shelter among the water plants. There is no need for these birds to panic or hide for the raptor has no intention of attacking them, being interested solely in fishes.

The hunting territory of this large bird of prey is extensive, frequently more than five miles. Surveying its domain from a height of between 30 and 150 feet, the osprey's flight is slow and graceful, the regular, rhythmic beating of the long wings punctuated from time to time with elegant glides. The head moves incessantly from side to side as the glittering yellow eyes scour every square inch of water below. As soon as it sights prey the raptor hovers immediately above and makes its final preparations for attack. With the victim clearly pinpointed in a patch of clear water, the osprey folds its wings, flexes its muscles and dives, either straight down or at an angle, as swiftly as any falcon. Just before touching the water the wings are reopened, the head comes up and the feet with their deadly talons are extended. Then it breaks the surface, feet first, remaining submerged for only a couple of seconds, time enough to grasp the fish and alter direction again. With a series of powerful thrusts of wings and tail the raptor is soaring upwards,

In the course of flying over the clear, tranquil waters that constitute its fishing grounds the osprey will suddenly sight prey (1). Half-closing its wings the raptor dives (2) and just prior to making contact with the water will extend its feet (3). Breaking the surface feet-first, the osprey needs only a second or two to snatch up a fish (4).

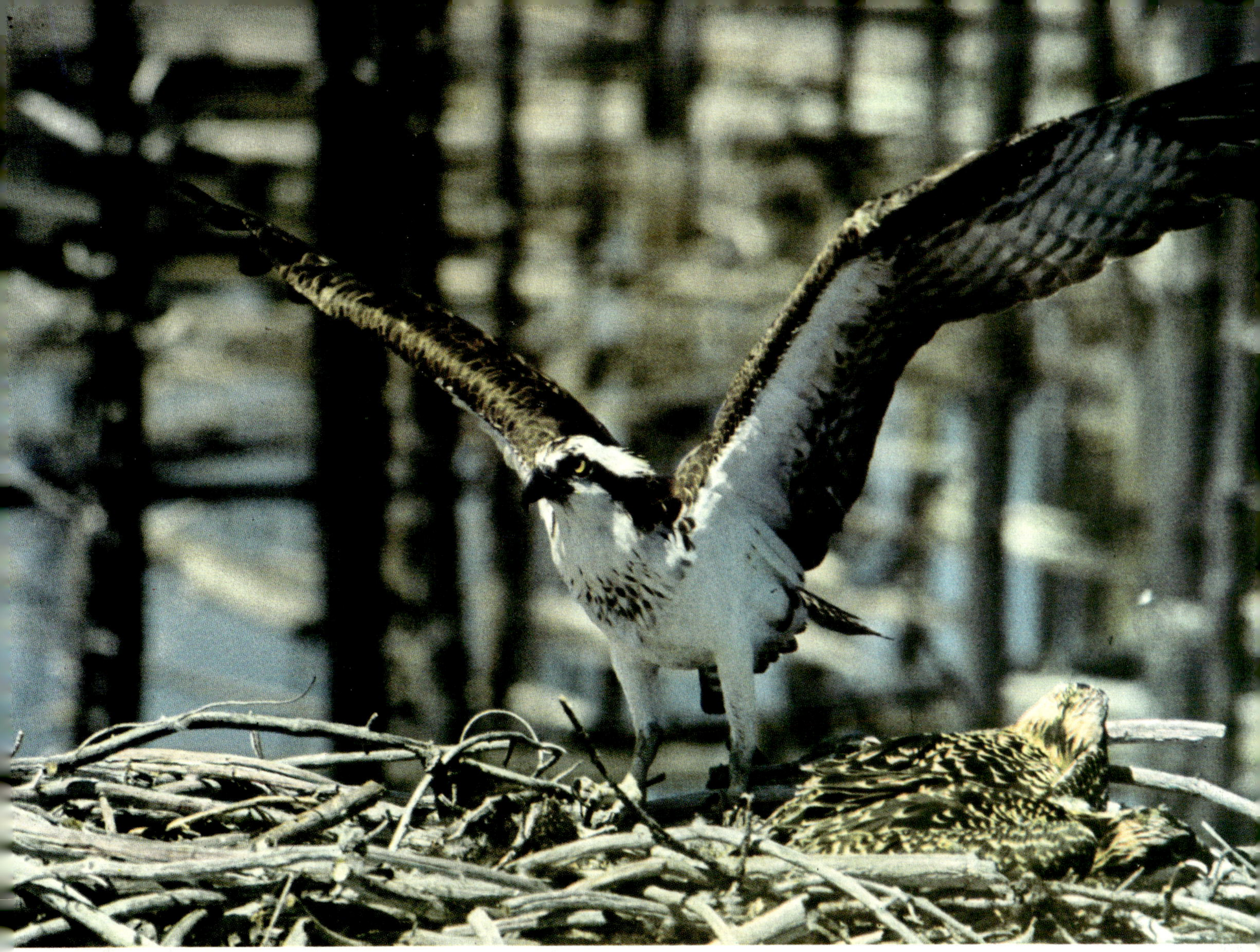

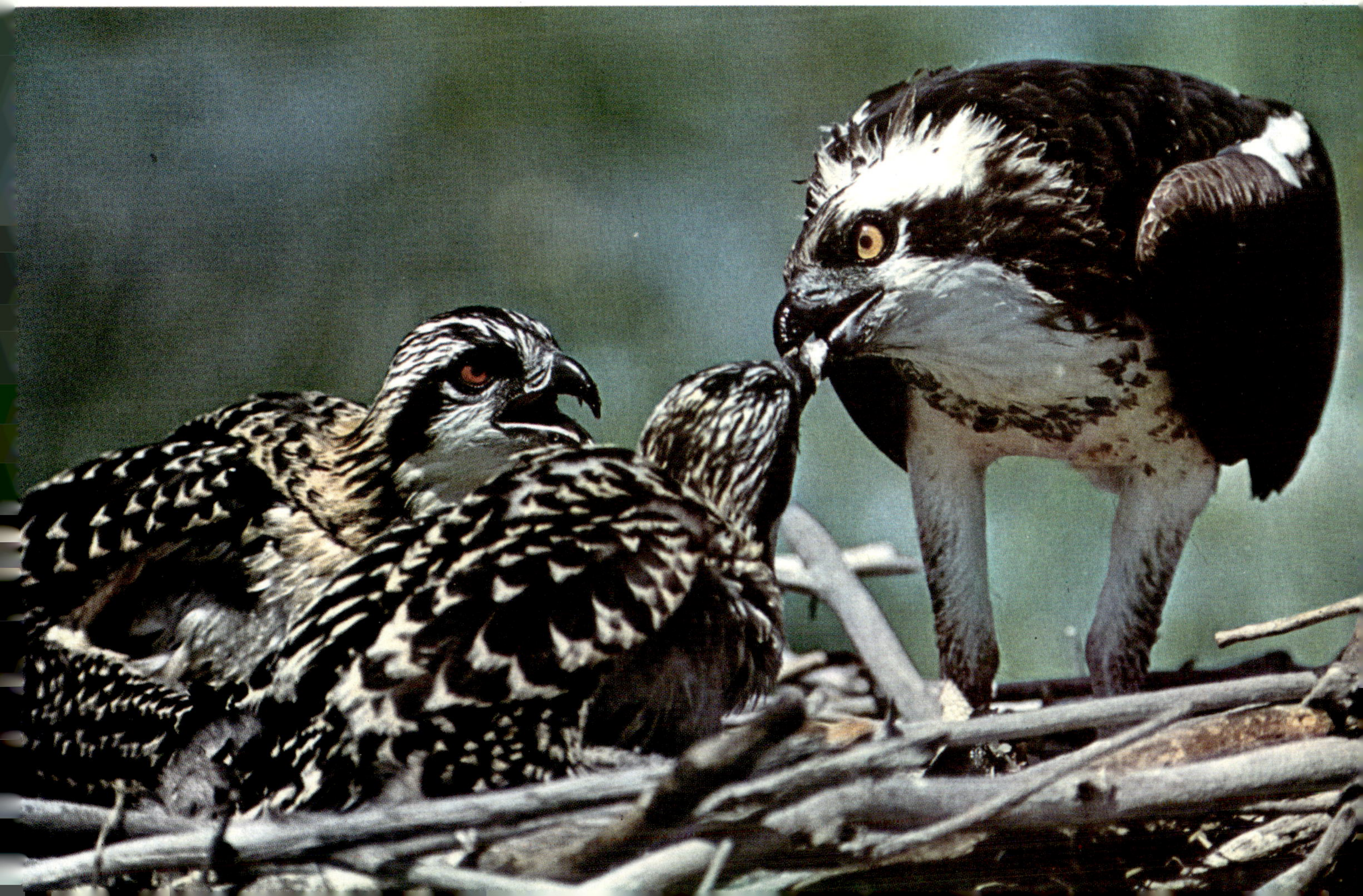

Preceding pages : As soon as a pair of ospreys return to their breeding grounds they start building a new eyrie or repairing the old one. The young hatch after an incubation of about five weeks and are fed for approximately two months, by which time they have acquired the rudiments of hunting. Soon they will be independent enough to leave the nest.

leaving a thin trail of bubbles in its wake. Two toes of both feet are used to grip the victim in an unbreakable stranglehold. The fish is never eaten in mid-air but dies slowly as the osprey wings its way back to the shore. The claws are not relaxed even when the bird settles down for its meal and considerable skill is needed to retain its balance on the branch while feeding.

Although the osprey's fishing expeditions are not always successful, Leslie Brown is probably right in saying that nine out of ten attacks achieve their objective. What sometimes happens is that the osprey may grab hold of prey which is too strong and failing to relax its grip in time will be drowned. Moll reports a number of such incidents from Mecklenburg in East Germany. The raptors were dragged far below the surface by 7 lb pike – roughly twice the weight of the attacking birds.

As a rule ospreys will hunt only medium-sized fishes, understandably ignoring those that are too small, firstly because they are difficult to catch and secondly because they do not offer a sufficient quantity of protein, not even compensating for calorie losses entailed by the attack. In examining 496 prey Schnurre found that 42 fishes measured 3–6 inches, 352 measured 6–12 inches and 102 measured 12–23 inches. According to this naturalist the favourite species in northern Germany are pike and barbels, followed by bream, perch and carp. Farther south, along the coast of Morocco, mullet are the main victims. Surveying the contents of several ospreys' nests Moll found that 87 per cent of prey consisted of bream, 7 per cent of tench, 2 per cent of perch and 4 per cent of other unidentifiable species. Moll also claims that the osprey requires at least 12 ounces of food a day, but Przygodda puts it higher, pointing out that the bird will take at least two 1 lb fishes daily. Probably the real amount lies somewhere in between, the discrepancy perhaps being explained by the fact that the former figure represents the basic minimum need and the second the average quantity caught in a single day.

It is quite obvious, given the specialised hunting methods of the osprey, that it can only operate successfully when conditions are suitable. The majority of failures occur when the water is rough or turbid, obscuring the raptor's clear view of its prey. Nor can it risk launching attacks in regions with dense tree cover for it must have space in which to manoeuvre. This is why the bird is almost always found near large rivers and lakes and in coastal zones. It is true that winter finds it frequenting mangrove swamps but only where the trees are not too closely packed together.

Fishermen often accuse the osprey of depopulating rivers and marshes but this is patently ridiculous. Under normal conditions no predator, however large its appetite, could possibly consume all the food at its disposal. Furthermore, since the osprey, like most birds of prey, concentrates its attacks on the feeblest individuals or most abundant species (in some areas, as already mentioned, it attacks pike which are themselves likely to create more havoc among local fish populations) it plays a useful, not destructive, role. By removing weak, sick and moribund fishes that tend to swim near the surface, the raptor helps to keep

---

OSPREY
(Pandion haliaetus)

Class: Aves
Order: Falconiformes
Family: Accipitridae
Total length: 23½–28½ inches (60–72 cm)
Wing-length: 18–20 inches (45–51 cm)
Wingspan: 59–75 inches (150–190 cm)
Weight: 2¾–3¾ lb (1.3–1.7 kg)
Diet: fish, almost exclusively
Number of eggs: usually 3, rarely 2 or 4
Incubation: 35–38 days

Adults
Colours alike in both sexes, female slightly larger than male. Upper parts dark brown, white below. Streaked brown band across chest; yellowish tints on flanks and beneath wings. Head pale with small crest, faintly striped with dark brown. Dark brown band through eye and running down nape. Long, pointed wings; powerful grey legs, unfeathered. Short, strong toes, covered with scales, terminating in long, hooked claws. Cere grey-blue; iris yellow.

Young
Grey down on back, creamy white on underparts. Yellow feet.

streams and rivers clean and this in itself encourages the development of healthy stock. It is worth emphasising once more that it is not inoffensive birds of prey such as the osprey that are responsible for the declining fish populations of our rivers. Man with his toxic effluents, his herbicides and his insecticides, is the sole culprit, the true destroyer of his own freshwater environment. In North America the careless and indiscriminate use of DDT has in fact proved catastrophic to the local communities of ospreys, the majority of the contaminated birds having become sterile.

Having completed its meal the osprey scrupulously cleans its bill and claws and indeed spends much of its time at this activity. When, for some reason or other, conditions do not favour fishing it may feed on young birds, small mammals or frogs. Thus Zarudny points out that on the Asiatic steppes when rivers are too muddy the raptor catches frogs, susliks and other rodents, none of which are easy victims.

Early life of an osprey

As often as not the female osprey precedes her mate to the nesting site which the pair occupy year after year. As soon as they have both arrived the birds busily begin to build the eyrie or to repair the old one, the result being an enormous construction of dead branches situated at the top of a tree (in Europe, usually a conifer). They may, however, nest on the ground, particularly on the steppe and in sparsely wooded areas. Other nesting sites include steep rock ledges overlooking the sea.

The male, who at all other times is silent, soars up into the sky to perform his courtship display, emitting loud, continuous cries. He will climb to well over 1,000 feet above the eyrie, sometimes holding a small branch in his talons, which is his manner of announcing to possible rivals that he and his mate are already in residence. He then settles on the nest or on an adjacent tree and allows his partner to soar aloft in her turn, keeping his eyes firmly fixed on her, wings hanging on either side of his body like a cape, tail lowered and spread wide. After performing her own aerial acrobatics the female rejoins the male who circles around her until she raises her tail and lowers her head, thereby signifying her readiness to mate. Sometimes he offers her a fish prior to coupling. The sexual act is accompanied by soft cooing noises.

Although they normally live in pairs these birds may, under favourable conditions, come together at this time of year to form breeding colonies. Paul Géroudet has described how on the American coasts ospreys in their hundreds will build their eyries within 30–100 feet of one another. At no time, and especially today, when the species is thinly distributed, has this size of colony been reported from Europe. In northern Germany, thanks to an abundance of food and stringent protective measures, one may come across perhaps ten nesting sites or eyries to the square mile, never less than 300 feet apart, but even this is rare. As for Australia, Leslie Brown indicates that there are not more than one pair of birds to each five-mile stretch of coastline,

The white-bellied sea eagle frequents coasts and estuaries. Sometimes it fishes by gliding low over the water but more often watches for prey from the bank.

although small colonies may be formed on some Pacific islands.

About two weeks after her arrival at the nesting site the female lays her eggs. The normal number is three, but occasionally two or four. They are whitish with brown and grey spots. Incubation commences as soon as the first egg is laid, the female assuming the main responsibility but being relieved at intervals by the male, usually two or three times a day, when he brings her food. The father may thus spend three or four hours of daylight assisting in the incubation whereas for the remaining part of the day and the entire night it is the female who will be found guarding the eggs.

Incubation lasts for five weeks or a few days longer. The father continues to carry fish to the nest but it is the mother that actually feeds it to the eaglets. She cares attentively for her brood for about six weeks (the period may be slightly shortened or extended according to the amount of food consumed by the young) and then the eaglets gradually learn to lead an independent life. When they are two months old they leave the eyrie but frequently fly back to collect scraps of food left by their parents. They quickly acquire the rudiments of the adults' fishing technique, the first lesson being to snatch dead fishes deliberately dropped into the water by the parents. Once capable of doing this they soon become expert in capturing live prey and no longer need to be fed. Less than two months after they leave the nest they are ready to abandon family territory and to migrate to warmer climes.

White-bellied sea eagles, black kites and crows often attack ospreys, trying to snatch their prey as they fly back to the roost, and sometimes succeeding in the attempt.

The white-bellied sea eagle

The sea eagles of the genus *Haliaeetus* are large, heavy birds of prey, characterised by a powerful bill, long, hooked talons and hard pads underneath the toes which help to secure prey. There is nothing very remarkable about their fishing technique, which cannot compare for skill and agility with that of the osprey; nor are they noted for unusual courage or speed. In fact they supplement a fish diet with reptiles, mammals and birds.

The white-bellied sea eagle (*Haliaeetus leucogaster*), with a wide range extending along the coasts of India and Ceylon to the southern shores of China, and islands eastwards to New Guinea, Australia, Tasmania and the Bismarck archipelago, is the nimblest and most elegant of the tribe. It is a particularly noisy bird, frequently heard at dawn and dusk. Although sometimes sighted on the banks of rivers and lakes quite a distance from the sea, it rarely wanders far from the coast. In Borneo, for example, it has never been seen much more than 10 miles inland from the shore.

The hunting grounds of the white-bellied sea eagle are sometimes restricted to an area not more than a mile in diameter but are usually much more extensive. The raptor may either patrol this territory from the air, swooping low over streams and rivers, or simply perch on a dead branch (the same one every day),

---

WHITE-BELLIED SEA EAGLE
(*Haliaeetus leucogaster*)

Class: Aves
Order: Falconiformes
Family: Accipitridae
Total length: 27½ inches (70 cm)
Wing-length: male 20½–23 inches (52–58 cm)
female 23–23½ inches (58–60 cm)
Wingspan: male up to 71 inches (180 cm)
female up to 86 inches (218 cm)
Weight: up to 6¼ lb (2.8 kg)
Diet: basically fish
Number of eggs: 2–3
Incubation: 51 days

Adults
Back, wing and tail coverts dark grey; tail greyish, white at tip; wings black, contrasting strongly with lighter colours of rest of body. Iris dark; cere grey-blue; feet and toes yellowish.

Young
Covered with white down at birth.

waiting for a fish to appear at the surface. It then scoops up the prey with its claws but, unlike the osprey, never dives into the water. Sea snakes seem to be favourite victims and some authors claim that these, together with fishes, constitute 90 to 95 per cent of the eagle's diet. This is probably an exaggeration, however, for in addition to aquatic animals the raptor eats the eggs of sea birds as well as small mammals and young crocodiles. Occasionally it will also feed on carrion and, as previously mentioned, will not hesitate to snatch a dying fish from the talons of an osprey.

The courtship ritual of this species is in no way spectacular. Both birds simply fly at some considerable altitude, emitting shrill cries. Nest building is a shared activity, the eyrie being situated either on a rock ledge or, more frequently, in a tree, at a height of 50–100 feet.

The female lays two or three eggs–white with tiny black specks–and incubates them alone for 51 days, beginning as soon as the first is laid. As a rule only one eaglet will survive and although it will be ready to leave the nest about two months after hatching it continues to pay frequent visits to the parents for at least six months, mainly for the purpose of increasing its food supply. In favourable seasons the parents may succeed in rearing two or even three eaglets but this is exceptional.

The brown fish owl

There is one nocturnal bird of prey living in the Oriental region whose habits are insufficiently recorded but which, because of the rich opportunities offered, specialises in fishing, unlike any other related species. The raptor in question is the brown fish owl (*Ketupa zeylonensis*) and had it not adopted this form of hunting it is doubtful whether it would have survived. As large as an eagle owl, if not somewhat larger, this interesting bird ranges widely across Asia from the countries of the Mediterranean to northern Manchuria and China. Some authors distinguish four different species but others regard them merely as subspecies or geographical races.

Outwardly the brown fish owl resembles other members of the family for it has large orange eyes, a strongly curved bill, long tufts of feathers (misleadingly called 'ears') on either side of the head, and sketchy facial discs. The feet may be covered with feathers but the thickness of this layer varies from one region to another. In fact the owls inhabiting tropical countries have naked tarsi; those of the Himalayas and mountains of China have a few feathers scattered over the front of the legs but a dense covering on the backs; and the legs of those living in north-western Asia are completely feathered. The wings of the brown fish owl are slightly modified in that they are rather shorter and more rounded than those of other nocturnal raptors (suitable for manoeuvres in wood and forest); and the long, powerful and sharply curved claws have a keen cutting edge. The pads on the underside of the toes are also scaly, thus helping to retain a grasp on slippery prey.

Fishes and crustaceans (especially freshwater crayfishes of

The Brahminy kite ranges widely through the Oriental region from India to Australia and the Pacific islands.

Brown fish owl
(*Ketupa zeylonensis*)

the species *Astacus schrenckii*) form the basis of the owl's diet but other items include small mammals, lizards, frogs, insects, birds up to the size of pheasants, and, occasionally, carrion. Two fishing methods are employed depending on the surroundings. If the water is shallow the owl simply wades in and waits for its prey to appear, rather in the manner of a heron, the difference being that it uses its claws and not the bill for harpooning the victim. If the river is deep it stations itself on a branch overhanging the bank and dives, feet extended, on any fishes swimming close to the surface.

Because of its specialised hunting methods this owl is seen on the ground much more frequently than other raptors, and its comings and goings are often clearly shown by its footprints in the sand or mud.

The day is usually spent high in a tree, concealed in the foliage, for like other Strigiformes it is likely to be tormented by diurnal birds of prey. In Borneo, during the late afternoon of a winter day in 1952, Smythies saw a brown fish owl being attacked by a peregrine falcon. The owl normally sets out in search of food towards dusk although some authors believe that it is less nocturnal by habit than most of its relatives. Certainly its behaviour seems to vary from region to region – hardly surprising when one considers the extent of its Asiatic range.

Very little is known of the breeding habits of this bird. The nest is generally built in a tall tree but the owl appears to settle fairly often in nests abandoned by other species. In some areas it has been found nesting on the ground or among rocks. The female lays from one to three eggs, usually two, between February and April although nestlings have also been observed in November.

Birds of the river

Although the Oriental region abounds with animals which depend for survival on water, there are not a large number of birds native to the rivers of South-east Asia. Most of the species encountered on the banks of streams and rivers are also common to Africa and Palearctic Eurasia. Among the many birds nesting and breeding on the banks of forest streams, in flooded terrain and on the shores of mangrove swamps are ducks, plovers, ringed plovers, godwits, curlews, sandpipers, purple gallinules, herons, storks and cormorants.

It is in large river estuaries and on other open expanses of water that one often sees the Asiatic skimmer (*Rhynchops albicollis*), closely related to the African species. Although as graceful as a tern the shape of the bill is unmistakable. The lower mandible is much longer than the upper one and this is how the skimmer catches its prey, gliding low over the water and hardly getting its feathers wet.

Herons and egrets that feed on fishes, amphibians and molluscs are numerous in these parts. Many of them, notably the grey heron, the purple heron, the little egret and the great white egret, are familiar to European ornithologists, but others, such as the reef heron (*Demigretta sacara*) are characteristic only of the shores washed by the Indian and Pacific Oceans.

The purple gallinule (*Porphyrio porphyrio*), rare in Europe, is much more common in Asia and often seen flocking above rivers. About the size of a moorhen or coot, this bird is more handsome than either, with gleaming purple and turquoise plumage and red frontal shield, bill and feet.

Closely packed rows of cormorants (*Phalacrocorax*) are often

Facing page and above : Among the many riverine birds of South-east Asia, but not exclusive to areas of mangrove swamp are (*left to right*) ducks, darters, cormorants and painted storks.

The flooded areas around the large river estuaries of India are a paradise for numerous species of water birds, including marabous, painted storks and egrets (*above*) and purple gallinules (*below*).

seen sunning themselves, wings spread wide, on trees bordering rivers and seacoasts. These well known fishing birds are not seeking warmth but drying their plumage, because the secretions of the uropygial glands at the base of the tail are not sufficient to waterproof the feathers. So these graceful underwater hunters are compelled to spend some time on land drying out.

Close relative of the cormorant is the Indian darter (*Anhinga anhinga melanogaster*), another bird with a long, flexible, serpentine neck which fishes in the same way as its African counterpart.

Black kites also fish the rivers of the Oriental region as they do in Europe and Africa. Pfeffer once saw a dozen of these raptors attentively watching the activities of a group of pied kingfishers (*Ceryle rudis*). Every time one of the small birds emerged from the water carrying a fish in its bill it was attacked by a kite and robbed of its prey.

Finally there are the pelicans, fishing in groups in shallow water, often far from the sites of their breeding colonies.

Super-predator of the mangrove swamps

The oceans of the world harbour many powerful predators. Ever since man began to sail the seas terrifying tales have been circulated, some with a grain of truth, others born of pure invention, about man-eating sharks, killer whales, giant squids and enormous sea serpents. Yet, strangely enough, very few stories have been told of one 'monster' which really does deserve its fearsome reputation, a sea animal famed for its unpredictable temper and tremendous strength. This is the estuarine or salt-water crocodile (*Crocodylus porosus*) which, in certain circumstances, is indeed a man-eater.

The salt-water crocodile, despite its name, is just as content to live in the fresh waters of lakes and rivers as in the sea; but it is the only member of the crocodile family to venture far from the shore. Salt-water crocodiles, for example, have made their way to the Fiji and Solomon Islands; and one individual reached the Cocos Islands in the Indian Ocean, more than 600 miles from the nearest land. No crocodilian, in fact, has a wider distribution for its range extends from the east coast of India to northern Australia and the Philippines, and probably includes the southeastern tip of China. Most favoured habitats are rivers, irrigation canals, flooded plains, beaches, marshes and mangrove swamps.

The reptile is the largest of living crocodilians. Although its average size as 12–15 feet Dr Schmidt indicates a maximum length of 23 feet. Reports of individuals measuring more than 30 feet have not been substantiated.

Because of its size and the frequency of its appearances in heavily populated areas, the salt-water crocodile is much feared. In the Sundarbans reserve, with its chain of jungle-covered islands formed by the deltas of the Ganges and the Brahmaputra, it is considered even more dangerous than a shark. Official estimates show that in one district in Ceylon fifty-three persons were killed by these crocodiles over a period of twenty-five years. Villagers who habitually bathe in rivers and streams inhabited

by these reptiles protect themselves by building wooden stockades but even then it is quite a common occurrence for smaller crocodiles to wriggle through the barricade. Dr Deraniyagala tells how one 10-foot individual, penetrating such a stockade, killed two people in the space of a few minutes.

It was probably a salt-water crocodile which attacked the well known American entomologist Philip J. Darlington Jr when during the Second World War he was working on a health project in New Guinea. Arthur Loveridge, describing this incident in his book *Reptiles of the Pacific World*, tells how Captain Darlington was astride a tree trunk collecting mosquito larvae from a swamp. He had lowered his test tube into the water when he noticed the snout of a crocodile below. In his haste to get away from the spot Darlington slipped and fell into the swamp. He recovered quickly but as he scrambled to his feet the crocodile, about 10 feet long, grabbed his arm and tried to drag him deeper into the water. Although he himself was 6 feet tall and weighed about 175 lb Darlington struggled, lashing out blindly with his feet. One lucky blow must have found its mark in the reptile's belly, however, for it momentarily loosened its grip, allowing Darlington to reach the bank and crawl to safety. Only then did he realise how gravely he had been hurt, his left hand being badly gashed, one finger fractured, the elbow crushed and the muscles and ligaments of the right arm torn.

Nevertheless the villagers of South-east Asia often have their revenge. Thus in the Irrawaddy delta the crocodile is hunted for its flesh and hide, with ducks and dogs as bait.

The huge estuarine or salt-water crocodile is found in lakes, rivers and swamps as well as in the sea. It frequently lurks among shore vegetation for prey.

The salt-water crocodile is the largest living crocodilian and is the super-predator of the mangrove swamp. Feeding on all kinds of animals, it will, in certain situations, attack and kill humans.

When it is time to breed the female begins to build her nest. Her first task is to find a suitable site where there is no risk of the eggs being swept away by tides or flood waters. Sometimes it is concealed in the vegetation but quite often it will be in an open place, fully exposed to the sun. The materials include leaves, branches and grass, and it seems likely that she uses her powerful tail to rake everything together close to the site, then carries the scraps of vegetation in her mouth and treads them down with the forefeet.

When completed, the nest measures 5–7 feet in diameter and is $1\frac{1}{2}$–$2\frac{1}{2}$ feet deep. When the crocodile has gathered all the necessary materials (the preparation may take a week) and made numerous journeys to and fro, the nest is completely exposed in the centre of a trampled mass of vegetation extending up to 25 feet in all directions.

The salt-water crocodile normally lays 30–50 eggs and this is fairly general throughout its range. In eastern Java, 35–38 is the usual figure, but in the western part of the island Kopstein discovered four nests containing 52, 54, 56 and 72 eggs respectively. The size of the eggs is also variable – ranging from 2·9 x 1·8 inches to 3·9 x 2·2 inches – and they each weigh between $2\frac{1}{2}$ and 5 ounces.

Some naturalists believe that the female mounts guard over the eggs until they are incubated but others deny this, asserting that she does not look after them to any extent. All are agreed, however, that she does visit the nest from time to time in order to keep away monitors, monkeys, birds and small carnivores.

Another theory—and it remains no more than that because there have not been enough direct observations to confirm it—is that the reptile sprinkles the eggs at intervals, thus keeping them moist and fresh, by shaking her wet tail over them.

The mass of vegetation piled up by the crocodile around and over the eggs gradually decomposes and this produces the warmth that is necessary for embryo development. On Sumatra the temperature inside the nest has been measured at 28°–32°C, a little lower than the outside temperature. In Borneo, on the other hand, Witkamp has taken a reading of 33·5°C, evidently somewhat higher than the outside temperature. Deranyagala, studying the habits of the salt-water crocodile in Ceylon, states that the vegetation serves to keep the nest temperature uniform at 32°C during the day and that this goes down a couple of degrees at night. He has also incubated crocodiles' eggs artificially and these hatched in 96 days. Under normal conditions, however, incubation does not take more than two and a half months.

The baby crocodiles measure about 12 inches at birth and feed primarily on insects. Hompes, who has reared a number of them, says that they grow very slowly. At a year they are about 19 inches long, at two years 24 inches and at three years barely 30 inches. Delsman states that the species is mature at two years but this seems unlikely, given their small size at such an age. Recent work on Nile crocodiles by H. B. Cott, of Cambridge University, shows that the species procreates only when twenty years old, at which age individuals measure approximately 8 feet.

Salt-water crocodiles are carnivores, normally hunting aquatic prey (fishes and crustaceans); but those reptiles that habitually lurk in waterholes cannot survive unless they attack and kill large herbivores that visit these sites for drinking.

These crocodiles are without doubt the most formidable predators of the mangrove swamp and in the ocean they have only one natural enemy—the shark.

Pirate of the mangrove swamp

We in the West, particularly if we have visited the Mediterranean region, are fairly familiar with the tiny lizards that bask on walls and scamper about in the dust or sand. Thus our first glimpse of a monitor, the only other huge reptile commonly found in the mud of the mangrove swamp, would probably lead us to conclude that here was a giant lizard, perhaps the last surviving descendant of a long line of prehistoric dragons capable both of climbing and swimming. But a closer look at the reptile as it follows the track of a prey might raise doubts, for this is evidently no ordinary lizard. The forked tongue is more characteristic of a snake's and it uses it in a similar way, picking up odour particles from the ground and transferring them to Jacobson's organ, situated in the roof of the mouth. This organ, described in a previous chapter dealing with the Gila monster of North America, helps the reptile to identify substances, probably by a combination of taste and smell.

The monitor has other features in common with snakes. It swallows its victims without crushing them with its jaws; and its

The female salt-water crocodile uses her powerful tail to collect branches and leaves which she piles up to form a nest. In the view of some, but not all, authors she also keeps the nest and eggs moist and cool by splashing water over them.

ESTUARINE OR SALT-WATER CROCODILE
(*Crocodylus porosus*)

Class: Reptilia
Order: Crocodilia
Family: Crocodylidae
Total length: up to 23 feet (about 750 cm)
Diet: animals of all kinds, from crustaceans to large herbivores
Number of eggs: 30–50
Incubation: 2½ months

A horny crest juts out from the eye region and extends to the centre of the snout which is two and a quarter times as long as it is wide. There are no plates on the back of the head but the reptile's back is protected by 6–8 rows of longitudinal plates and 16–18 rows of transverse plates. There are also raised plates on the outsides of the hind feet. Adults are olive-green with black markings on the head; young are lighter in colour. Toes of the hind feet are webbed, but those of the forefeet are linked only at the base.

The tongue of the water monitor is forked, like that of a snake, but the reptile is classified as a lizard, related to smaller and less sinister species of the Palearctic region. The tongue is furnished with papillae, sensory protuberances which, in conjunction with Jacobson's organ, enable the monitor to locate prey by taste and smell.

Facing page : The Indo-Malayan water monitor is a characteristic inhabitant of mangrove swamps, although often found in rivers several thousands of feet above sea level. An excellent swimmer, it spends most of its time in water; but it can also climb sloping banks and clamber up onto fallen tree trunks.

tail, unlike those of many true lizards, does not grow again after being cut off.

Thus the monitor, although classified among the Sauria (lizards) displays certain characteristics akin to those of the Ophidia (snakes). Indications are that the family is very ancient and related to extinct reptiles of the age of dinosaurs.

The typical mangrove swamp forest representative is the Indo-Malayan water monitor (*Varanus salvator*), a huge reptile that measures up to 8 feet long. It is no exaggeration to call it a pirate for as often as not its food consists of eggs filched from turtles and salt-water crocodiles. Like related species in Asia and Africa, however, it is a carnivore, although not, as was long believed, exclusively so. It is now known to consume large quantities of fruit and other vegetable matter. In fact the diet is very varied, according to the habitat, and may include crabs, snails, eggs, fledglings, wild and domesticated birds, mammals of all sizes, and even carrion.

The reptile is frequently encountered in coastal regions but has also been sighted in rivers more than 3,000 feet above sea level. Not only is it an excellent swimmer but it is also an amazingly skilful climber, clambering up steep banks, sloping tree trunks and rock faces with remarkable agility. The mangrove swamp is therefore an ideal habitat. In water it propels itself, like the crocodile, with powerful side-to-side flicks of the tail, the limbs being kept close to the body. The stout feet are furnished with long claws which are useful for climbing and for scooping holes in the ground, especially at breeding time.

Nobody has calculated precisely the length of time that one of these reptiles can remain under water; but its African counterpart, the Nile monitor (*Varanus niloticus*) can stay submerged for more than an hour so that it is reasonable to assume that the Asiatic species can do likewise.

The water monitor does not begin to stir until some time after the sun has risen, when its body temperature has risen sufficiently high. At night it remains hidden in the vegetation along the water's edge, either in a natural hollow or in a depression that it has dug for itself with its powerful claws.

In the breeding season the female scoops out a larger hole, up to 5 feet deep, in which she eventually lays her eggs. The number will vary between seven and thirty, depending both on the age of the individual and the region.

Although the hide of the water monitor is not so highly valued as that of the crocodile, it is nevertheless eagerly sought. According to Dammerman some 350,000 skins were exported from Indonesia (then the Dutch East Indies) in 1926; and Vorstman confirmed that in the year before that 400,000 skins were sent abroad from the island of Java alone. It is quite obvious that the species has suffered greatly from such depredations over the decades. Today hunting is strictly regulated and some countries have totally prohibited trading in the hides. Stringent measures have been taken in those areas where, albeit belatedly, the ecological value of the monitor has been appreciated.

Although there are other Asiatic monitors, the water monitor is the most typical species inhabiting the mangrove swamps.

CHAPTER 92

The dry lands of South-east Asia

In our survey of the Oriental region we have travelled through the immense jungles of the South-east Asian mainland and off-shore islands, tracked the course of great rivers and cut our tortuous way through the mangrove swamp forests. The over-riding impression thus far has been of dense, luxuriant vegetation – a mass of greenery extending high up into the mountains and down to the coasts. But although it is true that a vast part of the region does consist of jungle and woodland, in some areas the landscape takes on a very different aspect. There are, for example, broad tracts of desert, steppe and savannah.

In central Indo-China (those lands making up the huge peninsula stretching from the Bay of Bengal eastwards to the South China Sea) the typical form of tree cover is not that of the jungle or true savannah but a type of open woodland or grassy steppe, largely composed of Dipterocarpaceae – tropical trees with characteristic winged fruits – which do not grow densely nor to any great height. Elsewhere the rather monotonous face of the countryside is relieved by scattered acacias, palms, figs and jujubes. Only along the banks of streams and rivers do these formations take on the look of a forest. This combination of wood and grassland is of course an ideal habitat for a variety of large herbivores – gaur, banteng, kouprey, elephant and deer – and, correspondingly, for their traditional predators, including leopards and wild dogs.

In Indo-China and the islands of the Malay archipelago – lands subject to the influence of the monsoons – one does not find any steppes or deserts. Such biomes are characteristic of the Indian subcontinent to the west. Here there are two principal arid regions. The first is the Indus plain in the north-west which is mainly desert or semi-desert; the second is the Deccan triangle,

Facing page : On the Indian subcontinent the cobra is responsible for thousands of human deaths every year. Yet it is not the most dangerous of the world's snakes and renders a service of a kind in destroying other venomous snakes.

a vast plateau rearing up from the Indo-Gangetic plain, flanked by three mountain ranges. The highest peaks are to be found in the Western Ghats (Anai Mudi and Doda Betta are almost 9,000 feet); the mountains of the Eastern Ghats and the peaks of the Vindhya-Satpura range to the north do not exceed 4,000 feet. The average height of the Deccan tableland itself varies from 1,300 to 3,000 feet.

The incidence of rainfall in the Deccan is variable. The western part is strongly influenced by the monsoons and covered by rain forests, whereas the eastern zone south of the Godavari River is barricaded by mountains and receives from 20–30 inches (sometimes less) of rainfall annually, conditions that give rise to a typical landscape of acacias, jujubes and other poorly developed thorny plants, alternating with broad stretches of grassland, devoid of all tree cover. The alternation of areas of scrub and grass creates conditions highly favourable for wildlife of all descriptions. In addition to a multitude of rodents the Deccan is populated by herbivores such as gazelles, four-horned antelopes and nilgai, while among typical predators are mongooses, hyenas, foxes, jackals, wolves, leopards and loepard cats.

The Great Indian Desert

The most arid terrain on the Indian subcontinent lies east of the Indus River and extends for about 90,000 square miles, mainly in the Indian state of Rajasthan but partly in Pakistan. It is known as the Thar Desert or Great Indian Desert and is one of the driest and most desolate regions in the world, a continuation of the desert zones of Iran and Afghanistan. The parched soil receives a minimal amount of annual rainfall (5–15 inches) and the few human inhabitants of the area spend the greater part of the day sheltering from the broiling sun in rudimentary underground dwellings framed by the ruins of forts that once marked the ancient caravan routes.

There are hardly any signs of life in this empty wasteland and it must seem inconceivable to the rare desert traveller that there was ever a time when the landscape looked different. Yet in fact this desert is not very ancient. Palaeontologists investigating the region have turned up fossil remains of animals dating from the middle of the Pleistocene epoch. These include water buffaloes, rhinoceroses and elephants, all of them species which depend on an abundance of vegetation and fresh water. Nor is it necessary to go back that far to plot the first emergence of desert. Only a few thousand years ago this region was much wetter than it is today and was thickly covered with trees and plants. But as the climate gradually became hotter most of the original rivers disappeared, the water table dropped and the lush vegetation died. Yet the process was a slow one, for even when the armies of Alexander the Great invaded India in 326 B.C. they used wood from the trees growing in these now-vanished forests for the construction of rafts.

Nevertheless the Thar Desert is very different from the Sahara. There are, for example, no immense expanses of stony terrain which give parts of the immense African desert an even

more inhospitable and lifeless appearance. The Great Indian Desert is sandy throughout, so that despite the extreme scarcity of water some plants and trees do manage to grow here and there.

The comparatively recent origin of the Great Indian Desert and the fact that it adjoins the arid lands of Iran and Afghanistan have enabled certain plants and animals from the Palearctic region to penetrate it. According to a survey by the Indian zoologist Prakash 56·4 per cent of the mammals of this area are of Palearctic provenance, 41 per cent come from the Indo-Malayan region and only 2·6 per cent can be regarded as endemic species. Because of this varied distribution some scientists have preferred to include the desert in the Palearctic region proper whereas others count it as part of the Oriental region. This is the sort of problem that may arise wherever there is a transitional zone separating two main zoogeographical regions, for the frontiers are seldom rigidly marked or wholly impenetrable.

Animals of the Indian Desert

As is the case with all the world's arid regions, the Great Indian Desert, despite first appearances, is not entirely devoid of life. For the experienced zoologist it is no less interesting than a forest, a lake or a prairie. Of course, compared with such biomes, the desert contains far fewer species of plants and animals but what is lacking in numbers is amply compensated by the extraordinary adaptations displayed by the representative fauna and flora in a struggle for survival in extreme conditions. In some

The Indian Desert is less arid and desolate than the Sahara and parts of it are covered with acacias, jujubes and other thorny plants. It is of comparatively recent origin for a little over two thousand years ago the invading armies of Alexander the Great were still able to gather wood for raft-building in what had previously been a flourishing forest.

The small Indian mongoose is an agile and courageous carnivore, specialised for killing cobras. It can operate equally efficiently in thick undergrowth and open desert.

cases these adaptations are of an evolutionary nature and exclusive to one particular type of plant or animal; in other instances they are characteristic of an entire group. Thus almost all the mammals of the desert are pale in colour or at any rate lighter in hue than related species in wetter regions. This feature serves a dual function. In the first place it is a well attested fact that a light-coloured pelage absorbs less heat than a dark one—an important factor in the regulation of an animal's body temperature. Secondly, it is much simpler for an animal of this type to blend with its surroundings in a landscape where colours in general are pale and blurred.

These laws apply alike to hunter and hunted. A leopard concealed behind a thorn bush, keeping watch on a nearby herd of browsing gazelles, would be detected immediately by its victims if its coat were dark; and once deprived of its customary prey the carnivore would surely die of slow starvation. Centuries of natural selection have therefore decreed that the leopard of the Indian Desert possesses a much paler coat than its jungle relative —an adaptive process that has enabled the species to survive.

The herbivores of the Great Indian Desert have evolved along parallel lines. Thus the desert species of Indian antelope, also known as the blackbuck (*Antilope cervicapra*) is very pale in colour. This beautiful animal with spiralling horns is one of the swiftest and most agile animals in the world, its leaping ability on a par with that of the graceful African impala. As a result it can outdistance all its natural enemies apart from the cheetah. In bygone days the blackbuck and the cheetah prospered side by side, each a match for the other in grace, beauty and speed. Both were magnificently fashioned by nature to live in this inhospitable environment and the harmonious equipoise of predator and prey would have continued had it not been for human intervention. Today the handsome Indian antelope is a rare desert inhabitant while its traditional enemy the cheetah is virtually extinct.

By a stroke of irony it was man's clumsy attempt to profit from this natural relationship and to reproduce it in artificial circumstances that was to result in the decimation of both species. The blackbuck had for many centuries been a prized hunting trophy and the chief agent of its destruction proved to be the cheetah, now domesticated and trained for that single purpose. Oriental princes would venture out into the desert on horseback, their cheetahs on the leash. Once a herd of the antelopes were sighted the sleek cats were let loose and over short distances they demonstrated their hunting prowess to deadly effect.

As time went by the antelope population dwindled. Less predictably, the numbers of cheetahs also went down steadily because they would not breed in captivity. The position became so desperate that envoys were despatched to Africa to replace the species and it was not long before the cheetah was beginning to be scarce in the Dark Continent also.

The advent of firearms completed the sad sequence of events. On a winter night in 1947 the last three cheetahs recorded in the Great Indian Desert were caught in the glare of a truck's headlamps and killed on the spot. It was hoped that they might have formed part of a litter and that an adult breeding pair had possibly

been able to survive, but these hopes were soon dashed. Since then there have been rumours of the odd cheetah roaming the desert but no real grounds for supposing the stories are true.

Another animal of the Indian Desert threatened with extinction is the ghorkar (*Equus hemionus*). Only about 800 of these wild asses are reckoned to remain.

Fortunately not all the wild animals of the desert are in such extreme straits as the three species already mentioned. The nilgai or blue bull (*Boselaphus tragocamelus*), a large bovine-like antelope regarded in India as a sacred animal, is, principally for that reason, still abundant in the central and north-western parts of the peninsula; but the Indian gazelle (*Gazella gazella*) has suffered a decline in numbers.

Silent death

Ancient legends and cold contemporary statistics bearing on the various causes of human deaths in modern India and Pakistan both bear witness to the fact that snakes have played and continue to play an important role in human affairs. Perhaps if one were to take a poll among a cross-section of Europeans to ascertain the single image which evoked for them all the mystery and exoticism of the East, a high percentage would select the traditional snake-charmer of the market and bazaar. For people who have never visited these parts there is something undeniably fascinating and sinister about the figure of the swarthy, white-turbaned peasant, head bent low over a basket, from which emerges the head of a snake, swaying in time to the rhythm of the strange melody played on his pipe. It does not require any expert knowledge to recognise the shape of the cobra, immediately identifiable by its dilated neck which swells into the form of a hood. But what few outsiders realise is that the cobra does not necessarily need this musical stimulus to perform what is erroneously regarded as its 'dance'. This is a perfectly natural posture for the snake, as from the moment it is born the young

One of the most typical inhabitants of the desert regions of southern Asia is the ghorkar, now unfortunately threatened with extinction. Some 800 animals are estimated to survive, all of them in the Little Rann of Kutch, an arid region in north-western India about 2,000 square miles in area.

Symbol of Eastern exoticism, the snake-charmer of the Indian market and bazaar. will gladly demonstrate his skill for the benefit of the tourist. Contrary to popular belief the fangs of the cobra are not removed for the snake is a sacred animal in Hindu mythology.

cobra, sensing itself endangered, instinctively rears up to assume the characteristic position which is a clear warning of its intention to attack. Even at such an early age the snake will be prepared to confront an animal much larger than itself.

The most common species in South-east Asia is the Indian cobra (*Naja naja*), which measures, on average, 5½ feet. Considerably larger, measuring over 12 feet in length, and the only species that builds a nest in which to lay its eggs, is the king cobra or hamadryad (*Ophiophagus hannah*). The female hamadryad heaps up leaves and rotting wood to construct a nest, using her tail as a hook to draw it together. Then she coils herself in the middle and revolves her body, making a chamber for the eggs.

Although the venomous cobras are capable of killing humans they sometimes perform a useful service by disposing of other types of poisonous snake. No less deadly than the cobras, for example, are the kraits of the genus *Bungarus*. The latter, however, only constitute a real threat when they are accidentally disturbed, perhaps by being stepped on in the half-light of dusk when they are gliding over the ground in search of prey. In the daytime these brightly coloured and patterned snakes bury head beneath body in the event of disturbance, raising the tip of the tail to give warning of their presence.

The number of deaths from snakebite and the prominent publicity given to such occurrences have given many people the impression that cobras, with their efficient poison apparatus, must be the most deadly snakes anywhere in the world. Dangerous they certainly are, but not more so than the African mambas and the coral snakes of the North American deserts. Like the latter reptiles cobras belong to the group of Proteroglypha—snakes whose venomous fangs are situated in the front part of the mouth. The fangs are grooved, the poison from the venom sac flowing along these grooves (which in the more highly evolved proteroglyphs are in the form of closed channels) and thus into the wound. The cobra's poison-injecting mechanism is therefore much more effective than that of many snakes although it is compelled to retain a grip on its victims to allow the venom time to penetrate the tissues. In this respect it is not such an efficient killer as the viper or rattlesnake (solenoglyphs) in which the fangs are erected preparatory to striking and the venom sac simultaneously compressed by the temporal muscle so that the poison is driven under pressure through a closed channel into the wound. There is no need for the snake to hold on to its prey for the venom is immediately injected and the death process irrevocable, even though some time may elapse before it takes its full effect, and the prey may wander away prior to dying.

Like the poison of all venomous snakes that of the cobra is composed of various noxious chemical substances. In common with other proteroglyphs the venom consists predominantly of neurotoxins which have a swift effect on the central nervous system and cause serious, often fatal, disorders to the lung and heart activities. The bite itself is not necessarily very painful but the result—failing a timely application of an antidote—is paralysis or death.

To sum up, cobras, although not the most dangerous of all reptiles, are responsible for many deaths on the Indian sub-continent. The high percentage of fatalities is due to the markedly aggressive temperament of these snakes and the high population density in the areas where they are most active. Nor does it help matters that so many people are in the habit of walking everywhere barefoot.

The fers-de-lance

The Asiatic fers-de-lance of the genus *Trimeresurus* are also venomous snakes but at least their bites are not fatal to humans. These snakes possess a small opening on either side of the head between eye and nostril, each consisting of two cavities separated by a membrane. It was long thought that these pits had some connection with the reptile's senses of smell and hearing and that their function was to receive low-frequency vibrations of the air. Later it was demonstrated experimentally that they are organs that are acutely sensitive to infra-red rays. Invisible radiations of this nature are emitted by all warm-bodied creatures. Tests with the snakes showed that even when deprived of sight, smell and taste a fer-de-lance was still capable of detecting a heat source, in this case a glowing light bulb, and striking out at it instinctively with great precision.

Obviously this sensitive mechanism is of great utility in the wild, enabling the reptile to locate and kill small animals in complete darkness. The membrane separating the double cavities of each pit in fact contains some 500–1,500 miniscule receptors

The snakes of the genus *Spalerosophis* have a wide range of distribution extending from India across North Africa to the Atlantic.

to the square millimetre, recording temperature variations as slight as 0·002°C. In a way, therefore, the pit and its membrane might be likened to an eye and its retina, and since the range of the two pits partially overlaps one could say that the fer-de-lance uses a stereoscopic 'vision' to locate its prey in the dark.

The remarkable egg-swallower

Although it is the venomous snakes which, by virtue of the fact that they are potential killers, attract the most attention and comment, there are large numbers of snakes on the Indian subcontinent which are absolutely harmless yet still of great interest to zoologists for a variety of reasons.

One of the most spectacular of these inoffensive species is certainly the egg-eating snake (*Elachistodon westermanni*). It is true that many snakes include eggs in their diet but only six species—five in Africa and this one alone in Asia—feed on eggs almost to the exclusion of everything else. Because of this highly specialised diet the snake has no need of venom; in fact its teeth are weak and situated far back in the mouth.

To compensate for this the egg-eating snake possesses a special mechanism for breaking the shell and digesting the liquid content of an egg in the shape of a series of 'teeth' ranged along the length of the esophagus, the foremost of which are particularly sharp and cutting. They are not true teeth but well-developed, enamel-covered projections of the vertebrae which project through the rear wall of the esophagus. In addition, the mouth and neck can be dilated to an incredible size, enabling the snake to swallow eggs of considerable dimensions.

The egg-eating snake is much more active at night than by day and is capable of locating its food by means of smell. When it discovers a nest in the grass or among the branches of a tree it begins by making a scrupulous examination of the eggs with its tongue so that it can gauge their condition and estimate their size. If they seem suitable the snake selects one of them and winds its body round it to get a firm grip. Then it opens its huge mouth as wide as possible and starts to swallow the egg whole. When the egg reaches the esophagus and comes into contact with the 'teeth' contracting muscles prevent it from moving back towards the mouth or travelling farther towards the stomach. The snake then raises its head and swivels its neck back and forth so that the 'teeth' can go to work like a saw and slit open the egg. The shell is propelled in the direction of the blunter back 'teeth' which crush it flat and in this state it can be spat out. At the same time the closing of a special valve at the entrance to the stomach prevents the edible portion of the egg being expelled along with the ground-up shell.

The snake-killing mongoose

The efficacy of the cobra's venom as well as the speed and suddenness of its attack are sufficient to keep most predators at bay. A direct confrontation with a cobra demands on the challenger's part an unusual combination of courage and agility,

Soon after it is born the baby cobra rears up defensively in the face of danger and is capable of inflicting a venomous bite. The venom of the adult snake is stronger, however, and all the more toxic when the cobra can maintain a grip on its victim.

plus a measure of immunity or resistance to the toxicity of the poison. These are precisely the qualities possessed by the mongoose, a small carnivore specialised for killing venomous snakes.

The legendary bravery of this animal has been immortalised by Rudyard Kipling in his *Jungle Book*, where the mongoose appears under the name of Rikki-Tikki-Tavi, doing battle against a pair of cobras, Nag and Nagaina, and killing them.

The tendency of a mongoose to attack and kill snakes is purely instinctive and not the result of upbringing and learning. Tests conducted with a mongoose in captivity in which a plastic replica of a cobra was placed before the animal resulted in the latter launching a violent attack against the make-believe reptile, even though it had never seen a real cobra.

There are six species of mongoose on the Indian peninsula but only two of these, the Indian grey mongoose (*Herpestes edwardsi*) and the small Indian mongoose (*Herpestes auropunctatus*) are inhabitants of desert regions.

Despite its reputation the mongoose does not hunt snakes to the exclusion of other animals, and its diet also includes insects, lizards, frogs and rodents. Nevertheless, it is for its snake-killing ability that it is famed and a battle between the slender little mammal and its hereditary enemy cannot fail to be a highly spectacular affair.

As soon as the mongoose catches sight of the cobra the hairs of its body bristle so that it appears to be twice the normal size – an immediate psychological advantage. It then circles the snake, trying to provoke it into attacking. This is essential, for unless the cobra rears its body the mongoose cannot dart in for a mortal blow. The invariable reaction of the cobra is to assume the usual menacing posture, front part of the body erect, mouth agape, hood spread. As it strikes the mongoose leaps rapidly to one side. This happens again and again, with the mongoose avoiding each thrust of the fangs. After a succession of bold advances and equally swift withdrawals on the part of the little animal, the snake gives way to exhaustion and sinks to the ground. This is precisely the moment chosen by the mongoose to spring to the attack in its turn. It will never risk a direct assault on the cobra's head or neck until the reptile's strength is completely sapped. Patience, persistence and courage eventually reap their reward. The outcome is almost always predictable and the only occasions when the mongoose fails to make a kill is when the cobra decides to flee rather than fight.

The secret of the mongoose's success is the swiftness of its reflexes – that complex process whereby the eye records information, the brain relaying it back to the nerve centres which immediately stimulates muscle action. Neither a snake nor a human is capable of reacting to a given situation anything like as rapidly. A fight between a mongoose and a cobra (and such a spectacle is often artificially staged as a tourist attraction) brings this fact home dramatically. What is especially astonishing is the way in which the mongoose always appears to be one step ahead of its furious adversary, anticipating every movement in time to take evasive action. But everything occurs so quickly that the finer details are blurred and the chances are that the spectator will

Following pages : In doing battle with its traditional enemy the mongoose tempts the cobra into attacking first. Time and time again the little carnivore retreats and advances, skilfully avoiding the venomous fangs and slowly exhausting its adversary. When the snake is no longer capable of defending itself the mongoose sinks its teeth in neck or head and leaves its enemy dead on the ground.

be unable to see exactly what is happening, especially to appreciate the subtle manner in which the mongoose plans the attack, conserving its energy while steadily wearing down the opponent, teasing and provoking rather than risking a premature assault, then leaping in for the death blow at the strategic moment when the enemy is at its mercy.

The fantastic speed of the mongoose is a vital weapon but the thickness of its coat also affords some protection. Although not immune to the cobra's venom the mongoose seldom succumbs to a random fang thrust. For the poison to have any lasting effect by finding its way into the blood system the snake would have to bury its fangs deeply in the mongoose's flesh and this is a rare occurrence.

The beneficial action of the mongoose is not confined to its snake-destroying activities for it is just as efficient a killer of rodents. It was for this reason that groups of them were introduced into various countries in attempts to control the local rat populations, notably in Hawaii, Jamaica and other islands. Unfortunately this proved to have the opposite effect to that intended, as has so often been the case with animals introduced from abroad. After eliminating the rats the mongooses went on to create havoc with other local animal species, causing far greater ecological upheavals than had existed in the first place, and posing unparalleled problems for the authorities. One unfortunate consequence of the mongoose invasion which occurred in Jamaica was the complete disappearance of a rare native bird of prey, the burrowing owl.

Crossing of the ways

The Indian subcontinent is a meeting point for a multitude of migrating birds. Geese and ducks from the tundra, cranes of marshes and steppes, raptors of the forests, little bustards, houbara bustards and geese from arid lands, storks, peregrine falcons, lapwings, sandpipers, curlews and a host of other birds too numerous to list make use of the identical flight paths every year as they wing their way south to their winter quarters. The majority of them complete their journeys in India but others treat it as a brief halting post on the way to Africa.

Above and facing page : The Indian subcontinent is a winter assembly point for many birds migrating from the Palearctic region. Among these are three species of cranes and they are joined by another, pictured in these photographs, the sarus crane. The latter is, however, a sedentary species which breeds in the Oriental region.

Sarus cranes form life pairs and are popularly looked on in the Orient as symbolising married bliss. Like the storks of Europe they often construct nests on public buildings.

The great Indian bustard, once found in large numbers in the Great Indian Desert, is now a threatened species, listed in the IUCN *Red Data Book*.

The most spectacular of these winter visitors are undoubtedly the cranes, here represented by three different species—the common crane (*Grus grus*), the demoiselle crane (*Anthropoides virgo*) and the Siberian crane (*Grus leucogeranus*). In addition there is one native species which nests in South-east Asia, the sarus crane (*Grus antigone*), one of the most familiar and characteristic inhabitants of the Oriental region. It is a superb bird, standing 5 feet high, its greyish plumage flecked with red on head and neck. In the Far East the species enjoys the same kind of popularity and measure of protection that the white stork does in Europe. Like the latter it sometimes builds its nest on a tall building or tower without fear of disturbance. Male and female form lifelong pairs and the sarus crane is widely regarded in the East as the symbol of married happiness and mutual loyalty. It has been reported that when one of the birds dies its companion remains near the site for several weeks, uttering melancholy cries, and then wanders off for a while on its own until it eventually dies of grief.

The great Indian bustard (*Ardeotis nigripes*) is another typical inhabitant of arid regions but, unlike the sarus crane, it does not enjoy the same degree of local protection, so that its future is in jeopardy. The adult birds are comparatively easy to supervise because of their sedentary habits, but when the young migrate southwards in the winter they are completely at the mercy of unscrupulous hunters who are quite oblivious of the fact that this is a rare species and that in shooting the birds they are breaking the law.

The distribution of the houbara bustard (*Chlamydotis undulata*) extends westwards as far as the Canary Islands but here in the dry lands of the Oriental region its numbers are steadily on the decline. In the southern parts of its range the species is sedentary but the northern communities migrate in winter. Each year flocks have assembled in the Asian deserts, there to be massacred by falconers. After playing havoc with this and other species in their native lands hunters would spill across the border into the Indian Desert to continue their sport. In one year they killed some 600 bustards but the scale of the destruction so alarmed the Pakistani authorities that they eventually prohibited all further activities of this kind.

The houbara bustard, with its distinctive black and white head feathers, is midway in size between the great and little bustard. It lives on rocky and sandy plains and on tree-steppes. Like other members of the family it is also often found on the fringes of cultivated land.

Food consists both of vegetable and animal matter–fruit, shoots and bulbs in addition to lizards, beetles, grasshoppers and crickets. In summer, however, the birds, particularly the young, show a clear preference for insects. During the breeding season the males fight for possession of the females, then court the favours of the latter by following them about, neck extended, crest erected and wings half-opened, but without uttering the slightest sound.

The female houbara bustard lays from one to four (usually three) eggs at regular intervals of twenty-four hours in a small hole, at least half a mile from the next nesting site. She only begins incubating them when the last egg is laid and receives no help from her mate. When engaged in this activity she is extremely difficult to detect, so perfectly do the colours of her plumage match the surroundings. Should she be alarmed she will wait until the last possible moment before abandoning the eggs. Meanwhile the males wander around in small groups. The nestlings are nidifugous and consequently may be seen scampering about soon after they hatch, quickly learning how to take refuge in the undergrowth at the least strange noise. While she is rearing them the mother displays extraordinary courage and resorts to all manner of daring manoeuvres to protect the brood, often deliberately attracting the attention of a predator to herself and away from her young.

Around mid-August the houbara bustards begin assembling in groups of seven to nine individuals, including adult males and females and young. These groups lead a nomadic existence for a while and then head southwards.

In addition to bustards other characteristic birds of the Indian subcontinent include partridges, quails and francolins; but the real feature of interest of birdlife in this region resides in the number of raptors to be found in the north-western part of the peninsula, some of which are easily identifiable, even by a layman, others recognisable only to an experienced ornithologist. Mingling with familiar European birds of prey such as vultures, eagles, kites, buzzards and other species are raptors exclusive to the Oriental region. Quite frequently it is simply a matter of Palearctic birds which are wintering in India but just as often it will be a subspecies whose relationship to better known races will not be evident at the first casual glance. Thus in winter particularly, the Indian subcontinent is a principal assembly point and crossroads for innumerable Palearctic birds of prey on migration.

The honey buzzards

The range of the honey buzzards of the genus *Pernis* extends from the Iberian peninsula in the west to the Philippines in the east, and from Siberia in the north to New Guinea and adjacent

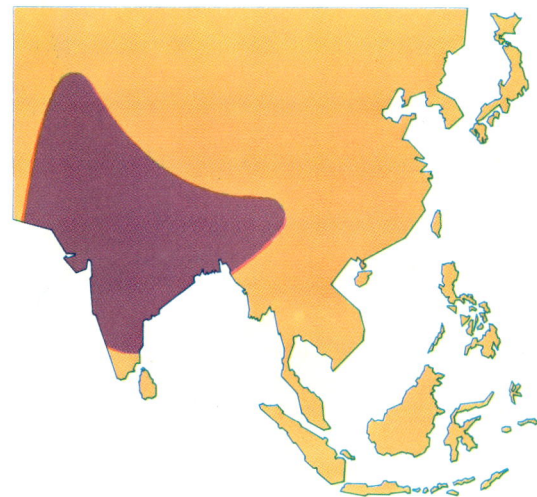

Geographical distribution of the laggar falcon.

LAGGAR FALCON
(*Falco jugger*)

Class: Aves
Order: Falconiformes
Family: Falconidae
Wing-length: male 12–13 inches (30–33 cm)
female 12½–14½ inches
(32–36.5 cm)
Diet: birds, small mammals, lizards, insects
Number of eggs: 2–5, usually 3 or 4

Grey-brown above with whitish band around upper part of head; belly white; flanks sometimes flecked with brown.

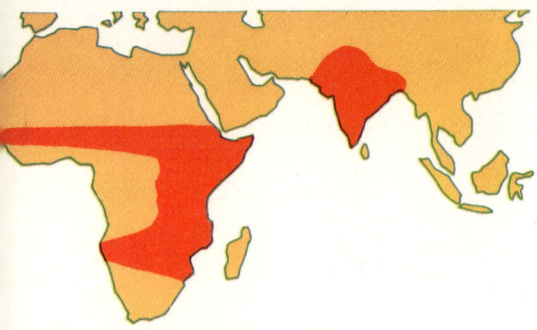

Geographical distribution of the red-headed falcon.

island groups in the south. Misleadingly described in some countries as falcons they belong to the Perninae, a subfamily of the kites. Although similar in appearance to the buzzards of the genus *Buteo* they are distinguishable by a slenderer silhouette, longer, narrower wings and the bar patterns of the tail.

The diet does not consist of honey, as the bird's name might imply, but of wasps and bees, together with their eggs and larvae. Such nests are often found in the undergrowth so that the raptor spends much time perching on a branch or walking about on the ground. Once a nest is discovered the honey buzzard sets about filching the contents, risking an attack by an angry swarm of wasps or bees determined to expel the intruder. The stings of the insects, however, seem to make no impression on the dense plumage and the raptor is further protected by scale-like feathers around the eyes and at the base of the bill. So the honey buzzard normally continues unhurriedly to rummage with its feet for the insects and larvae.

Although nest-pillaging is a frequent activity the honey buzzard sometimes supplements its diet with ants, grasshoppers and other insects. It also hunts small mammals, snakes and frogs as well as feeding on fruit, nestlings and birds' eggs.

Honey buzzards are migrants and may often be seen in flocks. European varieties normally fly southwards to central and southern parts of Africa, but those living farther east spend the winter on the Indian subcontinent and in South-east Asia. The former groups are faced with the problem of crossing the Mediterranean. Because it cannot benefit from the rising thermal currents characteristic of hotter climes the raptor, like other soaring birds, crosses by way of the Straits of Gibraltar or the Bosphorus, spending a minimum time over the water.

The laggar falcon

An inhabitant of India, Baluchistan and Afghanistan, the laggar falcon (*Falco jugger*) is the counterpart in the Oriental region of the lanner falcon of Africa and the saker falcon of the Asiatic steppes. In habit, however, it bears a closer resemblance to the former than to the latter.

The laggar falcon is a sedentary species which may be seen at any time of year patrolling its territory, perching on a rock or a tree and hunting in company with its mate. Although not so swift in flight as the peregrine falcon it is nevertheless a versatile and mobile predator, capturing its victims either on the wing or on the ground.

The diet of the laggar falcon comprises birds, small mammals, lizards and insects, in varying proportions. Thus in certain regions the raptor displays an obvious partiality for birds whereas elsewhere it concentrates on insects.

When male and female hunt together they combine their efforts skilfully, one partner skimming the ground to flush out prey, the other hovering at altitude and diving steeply down at any moving animal. Occasionally the laggar falcon feeds on carrion but in common with other falcons it cannot properly be classified as a scavenger.

RED-HEADED FALCON
(*Falco chicquera*)

Class: Aves
Order: Falconiformes
Family: Falconidae
Wing-length: male 7½–8 inches (19–20 cm)
　　　　　　female 8¾–9¼ inches (22–23 cm)
Diet: mainly birds, occasionally bats and lizards
Number of eggs: 3–5

Adults
Head, neck and moustachial streaks russet; breast white; upper parts and tail pale blue-grey; belly, wing coverts and scapulars black-barred; rectrices also have black band and white tip.

Young
Head darker than adult; underparts russet.

In the course of their nuptial display laggar falcons embark on a series of unusually rapid flights across their territory, the male often bringing his mate offerings of food. In February – give or take a few days on either side – the female lays three or four eggs on a rocky ledge, at the summit of a tall tree or in an abandoned nest of a bird of the crow family.

Incubation of the eggs is undertaken in rotation by both parents although the mother devotes more of her time to the activity. During this period and while the young are being reared the father hands over most of the prey he catches to his family, his partner ripping up and sharing out the food.

The red-headed falcon

The red-headed falcon (*Falco chicquera*) is a very beautiful little raptor whose distribution range includes the Indian subcontinent and Africa south of the Sahara. It is to be found in deserts, steppes and tree-covered savannahs and appears to be especially drawn to areas where palms and acacias grow.

After selecting a suitable stretch of territory both male and female make a systematic exploration of their hunting grounds as if to familiarise themselves with all available natural resources. The birds often hunt together, assisting and complementing each other in quite remarkable fashion. They are amazingly brave and will attack prey much larger than themselves. Unlike other falcons they often hunt in dense undergrowth although these are not ideal surroundings for such manoeuvres. Although lizards are sometimes caught, prey, consisting mainly of birds but also bats, are usually taken on the wing.

On the Indian subcontinent the sarus crane enjoys the same kind of protection as is afforded to the white stork in Europe. The birds are often found in flocks in the marshes, mangrove swamps and rice paddies of southern Asia.

CHAPTER 93

The vanishing animals of steppe and desert

The steppes and savannahs that separate the Great Indian Desert from the monsoon and tropical rain forests are inhabited by a number of animals which do not to any large extent depend upon water and which obtain adequate liquid nourishment from tough, fibrous grass, thorny plants and stunted trees.

Visibility in these wide open spaces is not interrupted by dense tree cover and such surroundings are therefore ideal for those herbivores that rely only on fleetness of foot to escape from predators. These animals are not merely fast runners but also possess the keenest sensory perceptions—whether sight, smell or hearing—so that they receive timely warning of danger. Once alerted, the remarkable group alarm system comes into operation and necessary evasive action is immediately taken.

The animals concerned are artiodactyls—even-toed ungulates—and among the most characteristic species of these dry regions, which despite meagre food resources offer a high degree of security, are nilgai, blackbucks, four-horned antelopes and crested wild boars.

The nilgai: bull or deer?

In the latter half of the 18th century the German-born zoologist Peter Simon Pallas, then Professor of Natural History at the Imperial Academy of Science in St Petersburg, made a six-year journey across Asia. To one animal he met on his travels he gave the name *Boselaphus tragocamelus*. *Bos* is Latin for bull and *elaphus* for deer, and the description certainly fits the animal in question. We know it today as the nilgai or blue bull and zoologists since Pallas' time have found it difficult to agree on its proper classification. Does it belong to the Bovidae or Cervidae?

Facing page: Most of the large herbivores that once roamed the Asiatic steppes and savannahs in large numbers are nowadays under threat of extinction. One exception is the nilgai or blue bull which, partly because it is a sacred animal in India, is still fairly abundant.

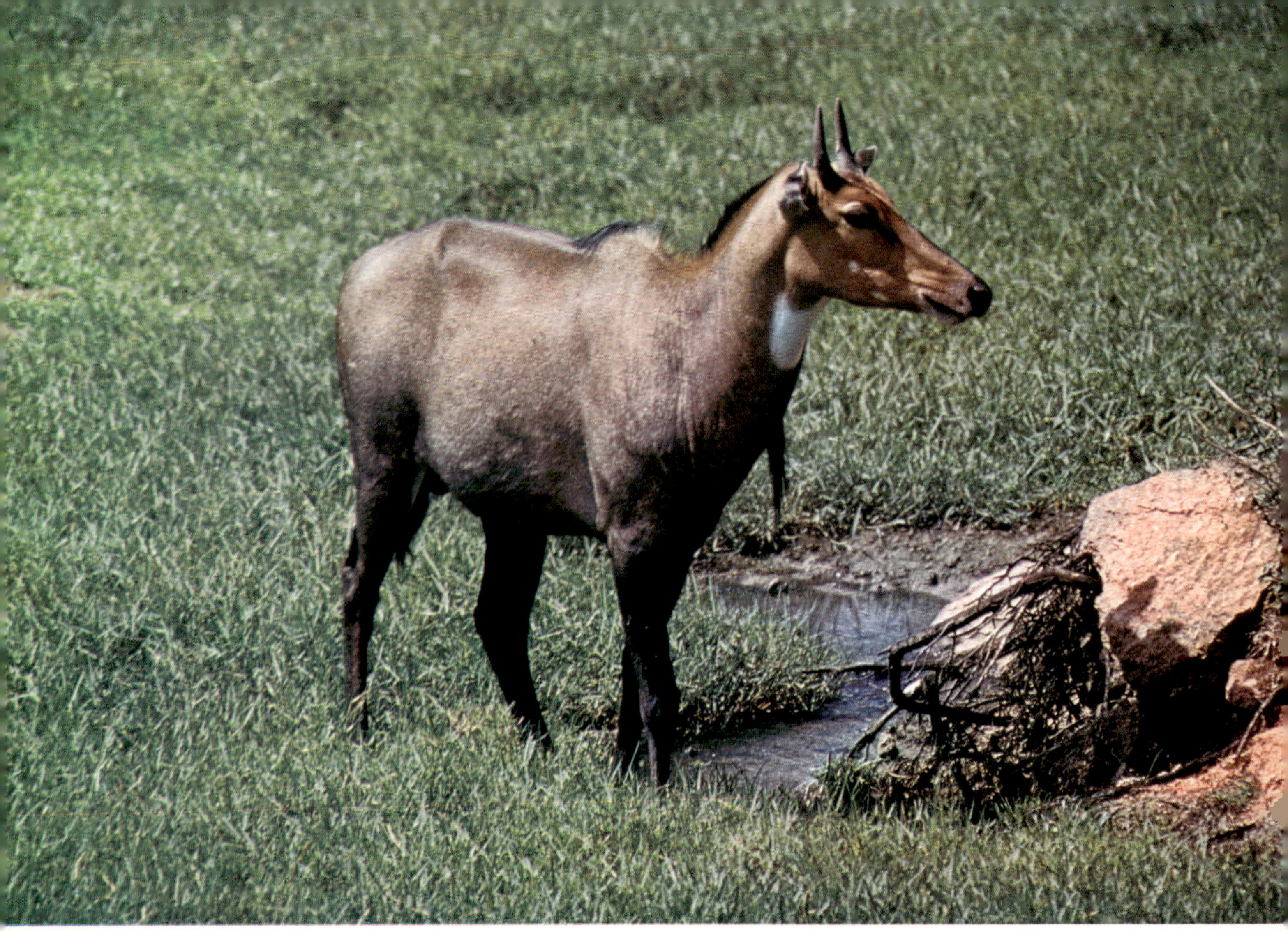

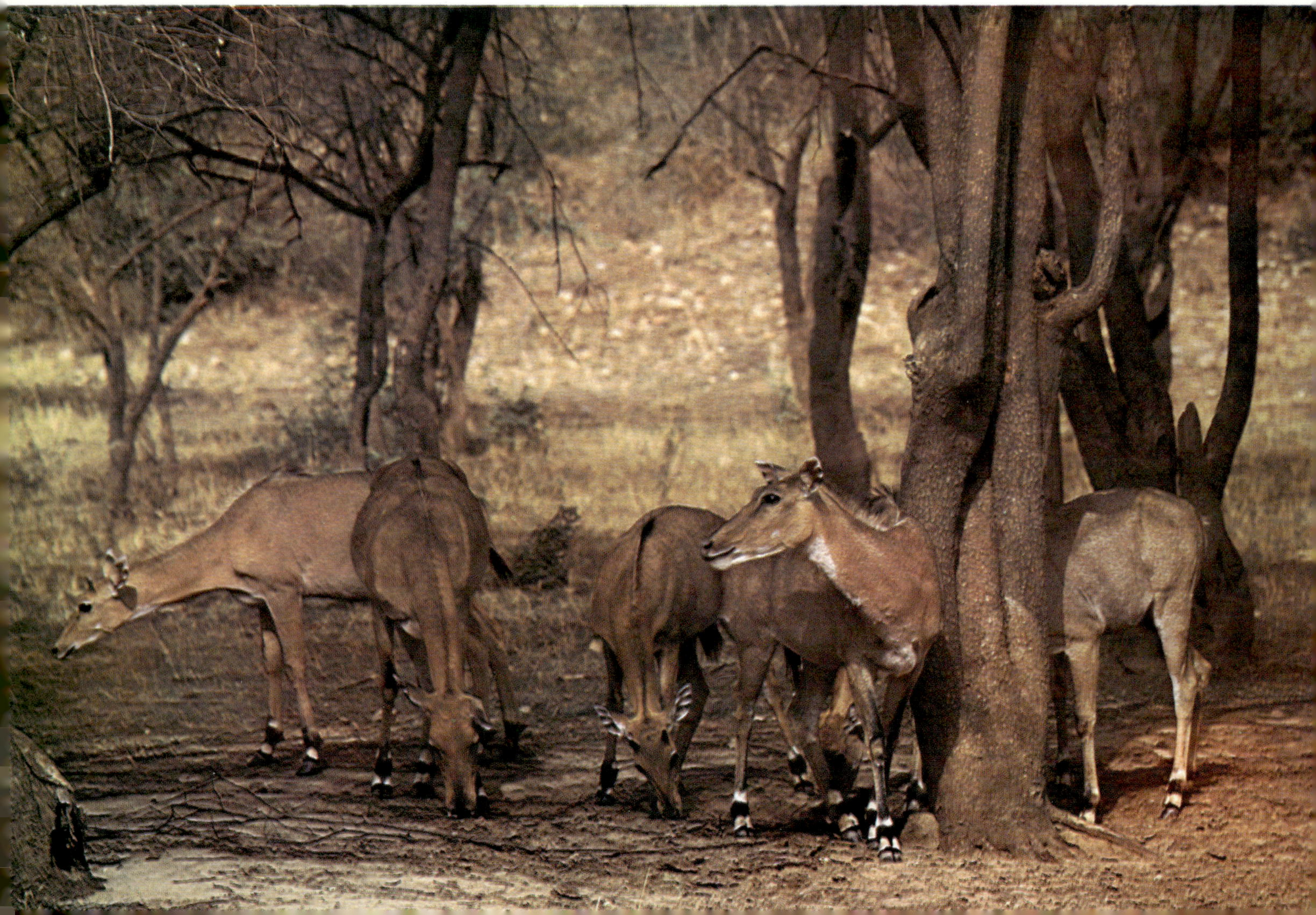

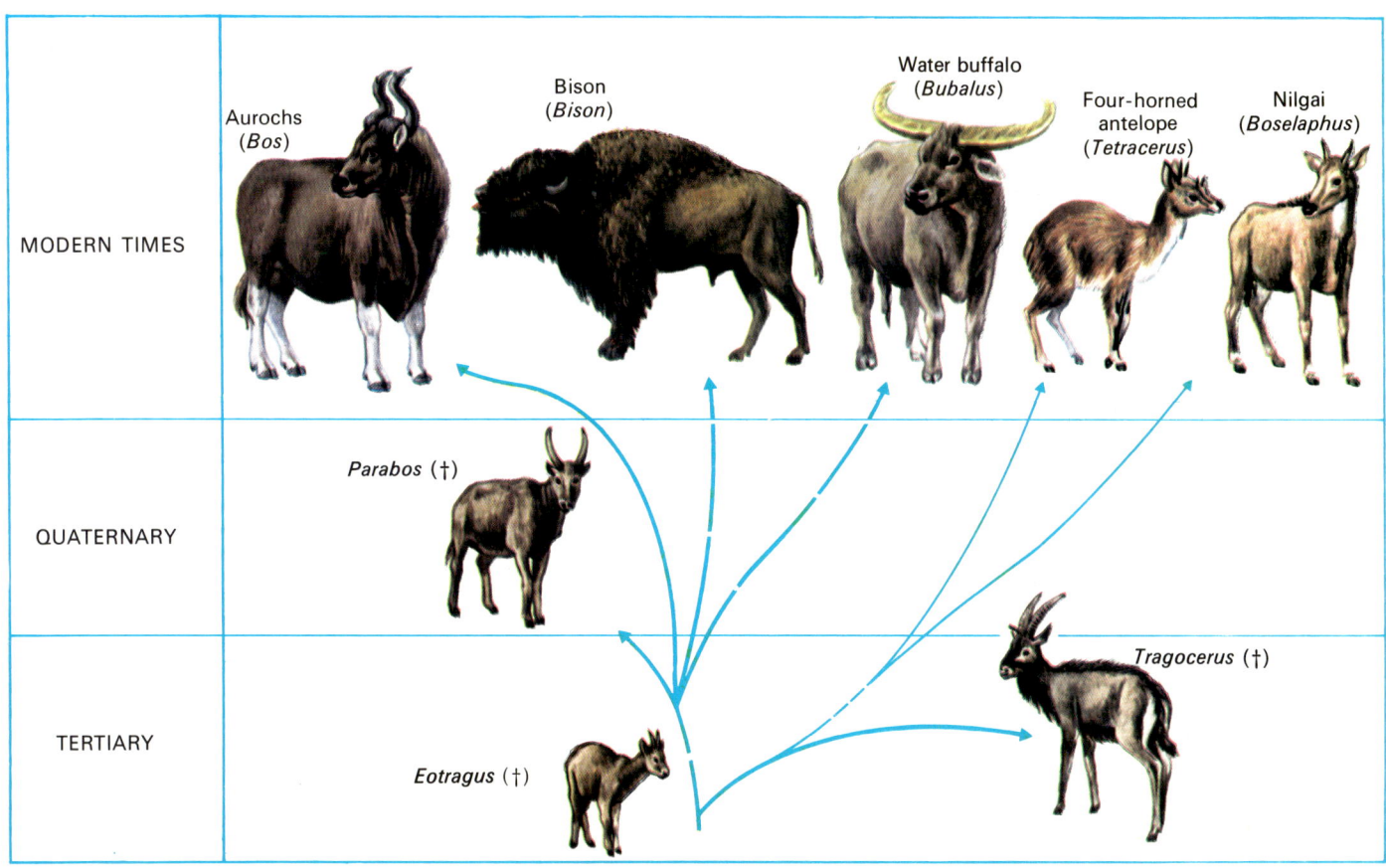

Despite appearances to the contrary, the nilgai, with its massive, well-muscled frame, its short and powerful neck, and its long, sturdy feet, is not a deer. It is in fact a true antelope, related not only to the equally curious four-horned antelope of the Oriental region but also to the harnessed antelopes, kudus and elands of Africa.

Thanks to recent investigations by the palaeontologist Pilgrim we now know that both the nilgai and the four-horned antelope have evolved very little during the past fifteen million years, that is, since the Miocene age. Not only do they strongly resemble their prehistoric artiodactyl ancestors but they are closely related to buffaloes and other forms of cattle, all descended from the same ancient stock. Among common ancestral features are the structure of jaws and teeth: and the white marks on the cheeks of the antelopes are also present in the anoa, smallest of all wild cattle.

A privileged species

The alternative name blue bull is derived from the coat colour of the male nilgai—not really blue but slate-grey mingled with dark brown. The coat, however, does give off gleaming bluish reflections when the sun shines directly on it. The colour of the female is light brown and the sexes are further distinguished by the fact that the male is larger, that it has short horns (they are not more than 8 inches in length and look slightly ridiculous in relation to the size of the animal's body) and also a lock of black hair hanging from the front of the neck.

The nilgai and the four-horned antelope are primitive Bovidae which bear a strong resemblance to their ancestors of the Tertiary period more than ten million years ago. Also descended from the latter—as this chart shows—are the now-extinct aurochs, the bison and the water buffalo.

Facing page: The male nilgai (*above*) possesses short horns and a distinctive tuft of black hair on the neck. The females (*below*) have no horns and their coat colour is paler. Whereas the males tend to be solitary the females roam about in small groups together with immature animals.

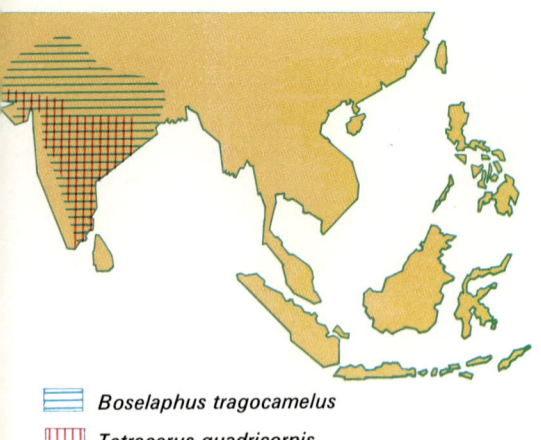

Geographical distribution of the nilgai (*Boselaphus tragocamelus*) and the four-horned antelope (*Tetracerus quadricornis*).

NILGAI
(*Boselaphus tragocamelus*)

Class: Mammalia
Order: Artiodactyla
Family: Bovidae
Length of head and body: 71–83 inches (180–210 cm)
Length of tail: 18–21 inches (45–53 cm)
Length of male's horns: 8 inches (20 cm)
Height to shoulder: 47½–59 inches (120–150 cm)
Weight: up to 440 lb (200 kg)
Diet: grass, tender shoots, fruit
Gestation: about 245 days
Number of young: 1–2
Longevity: 15 years

Coat of male slate-grey mingled with brown, with lock of black hair on neck. Hornless female has light brown coat. Both sexes have white rings above hooves and white patch on throat.

FOUR-HORNED ANTELOPE
(*Tetracerus quadricornis*)

Class: Mammalia
Order: Artiodactyla
Family: Bovidae
Length of head and body: 39½ inches (100 cm)
Length of tail: 4¾–6 inches (12–15 cm)
Length of rear horns: 3–4 inches (8–10 cm)
Length of front horns: 1–1½ inches 2·5–3·8 cm)
Height to shoulder: 23½ inches (60 cm)
Diet: grass, tender shoots
Gestation: 7–8 months
Number of young: 1–3

Russet coat, turning yellow with age; belly, inside of legs and muzzle white; prominent dark brown line on lower part of each foot. Male unique among Bovidae in possessing four conical, pointed horns, one pair in normal position, other pair farther forward, above eyes.

Small groups of females and young males may sometimes be seen grazing or drinking together, whereas adult males tend to be solitary, occasionally forming little bands which wander to and fro in search of food. The females often deposit small piles of excrement in the centre of group territory, in the same manner as rhinoceroses. Many zoologists believe that the function of these dung heaps is the same, namely to mark territorial bounds and in the case of females to indicate to the reproductive males their state of sexual receptivity.

Mating occurs most frequently in March or April but since young have been observed at all times of the year it seems probable that there is no fixed breeding season. As is the case with other Bovidae rival rutting bulls engage in ritualised fights for possession of a herd of cows. The contestants kneel on their forelegs and butt each other with their foreheads. After a series of violent but hardly dangerous clashes the weaker animal retires while the victor claims his harem.

Gestation lasts eight or nine months and at time of birth the calves are pale brown like their mother. They remain with her, often in company with elder siblings, until they are eighteen months old. By then they are sexually mature and ready for an independent life. The young bulls, which have now taken on the blue-grey coat colour of their elders, stray apart from the herd to adopt a solitary existence.

Nilgai feed on a mixed diet consisting of leaves, tender shoots, fruit and large quantities of grass which they crop by sinking to their knees. This is because the neck is so short that they would be unable to reach the ground from a standing position. They feed twice daily, at dawn and at dusk, spending the rest of the day in the shade, sheltering from the intense heat. Water holes, provided they are not dried up, are visited regularly but during periods of drought the nilgai can go for some time without drinking, obtaining liquid from vegetation and from the morning dew.

Predators of the species include the tiger, the leopard and the Asiatic lion, for an adult nilgai represents more than 200 lb of solid meat for a hungry carnivore. But it is not caught that easily. Ever on the alert, the herbivore raises its head at the slightest unaccustomed sound and if there is good reason for being alarmed makes off. Its gait–a mixture of gallops and leaps–may not be exactly graceful but is undeniably effective even across rough ground.

The nilgai is in a sense a privileged species inasmuch as it has not suffered greatly at the hands of man. In India it is a sacred animal so that hunting is out of the question. In Pakistan it is not much threatened for more practical reasons, for the animal's short horns and rather ungraceful lines do not place it among the upper ranks of hunting trophies. Consequently the nilgai is one of the few artiodactyls of the Oriental region which is not in imminent danger of extinction; quite the reverse. In 1936 the authorities even had to organise a cull in the Chhanga Manga reserve, in what is today Pakistan, to reduce the numbers of these animals which were at that time threatening the livelihood of other herbivores in the area.

The reason why such extreme measures have to be contemplated is that the numbers of nilgai can no longer be controlled in the natural way by predators. Indiscriminate hunting of carnivores has decimated them throughout their range. Those tigers and leopards that have managed to escape the massacre are not numerous enough on their own to check the growing nilgai population and it is unfortunately left to man to deal with the problem of restoring the natural equilibrium that he has disturbed.

Just the same, by virtue of the fact that the nilgai enjoys so much protection and is therefore in the habit of straying onto cultivated land, it is annually becoming more difficult to find the animal in its original wild surroundings, except in nature reserves.

The antelope with four horns

The four-horned antelope (*Tetracerus quadricornis*) is far more elegant in build than the nilgai and much nearer to the popular conception of a typical antelope. But it too possesses one curious feature – four conical, pointed horns, not two, sprout from its head. The first pair measure 3-4 inches long and are situated in the normal position, but the other pair, farther forward immediately above the eyes, are about half that size.

There has been no shortage of ingenious, not to say highly fanciful, explanations by scientists of this curious phenomenon. Many authors point out that there are probably local variations of horn structure since some individuals apparently develop only a single pair of horns. The Belgian zoologist S. Frechkop has put forward one of the more plausible theories, suggesting that the rear pair are equivalent to the horns carried by other Bovidae whereas the front pair may be analogous to the single central bony boss on the forehead of a giraffe, but in this particular case, in double form.

When male nilgai reach sexual maturity they leave the herd in which they have grown up and go off to lead a solitary life, only returning to join the females during the breeding season.

Because of its short neck the nilgai is forced to kneel down on its forelegs to crop grass or feed on other forms of ground vegetation.

The beautiful, fleet-footed blackbuck or Indian antelope is one of the imperilled species of the Oriental region. For centuries the males have been hunted for their splendid horns.

Four-horned antelope
(*Tetracerus quadricornis*)

Despite its distinctive horn structure this antelope, seen from a distance, may be mistaken for the hog deer (*Axis porcinus*). Yet although similar in outline, the two species are easily identified when in motion for the four-horned antelope is a swift, rather ungainly galloper whereas the hog deer is slower and considerably more graceful.

The four-horned antelope has more or less the same geographical range as the nilgai, although found a little farther west. It too prefers open terrain and woodland clearings, likewise feeding in the early morning and when the sun sets; but, in contrast to its relative, it requires a larger supply of drinking water. Unless it has a regular source of water it is in dire straits and in times of drought the situation may be critical. To find water the four-horned antelope may be obliged to travel many miles and its journeys sometimes bring it to the very borders of the jungle (where tigers and leopards lurk) or to streams and wells in the neighbourhood of villages. Its treatment will then be a matter of pure chance for although its has nothing to fear from Indian peasant, villagers in Pakistan hunt it.

The male of the species is normally a solitary animal and it is only briefly, from time to time, that it may form part of a small group with companions. The females, like nilgai, herd together and deposit their excrement in the centre of their territory to demarcate boundaries.

The antelopes mate during the rainy season and ritual fights between rival males are frequent. Gestation lasts about eight months, which is quite a long time considering the fairly small size of the animal.

The four-horned antelope has unhappily been hunted remorselessly in the past, so that nowadays the population is very small and the species is in imminent danger of extinction. Attempts have been made to rear very young individuals in zoos, but although the animals are sociably disposed they have not acclimatised to such surroundings. One antelope lived for four years in the London Zoo, others failing to live that long.

The swift and supple blackbuck

The blackbuck or Indian antelope (*Antilope cervicapra*), one of the world's fastest-moving animals, is an inhabitant of open ground where the absence of bushes and trees enables it to exercise to the full its running and leaping abilities.

Indian antelopes are gregarious animals, forming groups of from fifteen to fifty individuals, each led by an old, experienced male. These groups often band together to make up herds comprising several hundreds of animals. Every year the leader of the group expels a number of young males who have meanwhile reached sexual maturity and the latter assemble with similarly situated males from other groups to form bachelor herds. These antelopes roam about at random until, by proving their worth in ritualised combat, they succeed in winning the favours of a harem of receptive females.

The blackbuck feeds on all kinds of vegetation – especially grasses – during the coolest hours of the day, retiring into the shade to rest and ruminate when the sun is high. Although this would seem to be a peaceful enough existence the calm is sometimes shattered by the females responsible for protecting the herd. Easily alarmed by the slightest sound, even a breath of wind, these sentinels begin leaping high in the air, like impalas, to warn their companions of looming danger. The signal is passed down the line and within a matter of seconds all the antelopes in the vicinity are snorting and jumping about in their turn. Order is soon restored as the entire herd gallops thunderously off across the plain.

The male Indian antelope is particularly handsome, elegantly built and with a magnificent pair of divergent, spiralling horns. The coat is blackish-brown with white underparts and white patches on the muzzle and white rings around the eyes.

The antelope figures prominently in ancient Hindu mythology and astrology but its recent history has been sad, especially since the introduction of firearms. The species has virtually disappeared from India and in neighbouring Pakistan its plight is almost as grave. Conservationists hope that by protecting the survivors in national parks and nature reserves the species may yet be saved; and experiments in introducing the antelopes to other countries also provide grounds for cautious optimism that one day it may be possible to restore them to their traditional habitats on the Indian subcontinent. On the Texas prairie, for example, a herd of about 5,000 blackbucks has been built up in a comparatively short period.

In February or March – the breeding season – male blackbucks fight for the females. Neither animal is badly hurt in the course of these ritualised contests and the loser leaves the victor to claim the harem.

Adult male

Young male

Female

Sexual dimorphism is very marked in the blackbuck. The female has a paler coat than the adult male and carries no horns. The young male is distinguished by his horns and his coat resembles that of the female.

Disputed species

During the last one hundred and fifty years or thereabouts a number of zoological expeditions have gone out to India to study and describe the biology and behaviour of the country's wildlife. Much valuable information has been provided, frequently based on first-hand observations, but inevitably there has been a good deal of confusion and error as well. This is understandable, bearing in mind the difficult conditions under which such scientific enquiries often had to labour—inadequate transport facilities, poor channels of communication, untrained personnel and so forth. As a result a number of animals, new to science, came to be described with rather more imagination than accuracy, and were often linked with species to which they were not really akin. Conversely, some scientists allotted species status to animals which later proved to be mere subspecies. Two cases in point were the Indian gazelle and the crested wild boar; and in fact controversy rages over their classification.

The Indian gazelle (*Gazella gazella*) is regarded by some authors as a species in its own right and by others simply as a

subspecies of the Dorcas gazelle (*Gazella dorcas*). This is not the place to take sides or to argue the comparative merits of the two theories. Suffice it to say that the animal is admirably built for running and jumping and that like related species on other continents safety depends on quick reflexes and speedy flight. Gregarious by habit, the gazelles form herds in the driest parts of the desert but not even isolation has fully protected them from man's destructive enterprises. Like blackbucks they have been mercilessly hunted for centuries and today there are all too few of them left.

The other animal whose correct scientific identity is still disputed is the Indian or crested wild boar, a heavy animal with a mane of black bristles. Like other wild pigs elsewhere, it can be counted upon to defend itself fiercely against predators. Many authors describe it as a characteristic species of the Oriental region (*Sus cristatus*); some consider it to be the same animal, apart from a few distinctive details, as the European wild boar (*Sus scrofa*). Others classify it as a subspecies of the latter (*Sus scrofa cristatus*).

The hunting grounds of this boar, as of related members of the

The blackbuck is an agile, lively animal with remarkable leaping powers that are on a par with those of the African impala. Its speed, combined with extraordinary stamina over long distances, ensure that it is seldom caught by predators. Its only serious rival over a short course used to be the cheetah, now thought to be extinct in Asia.

272

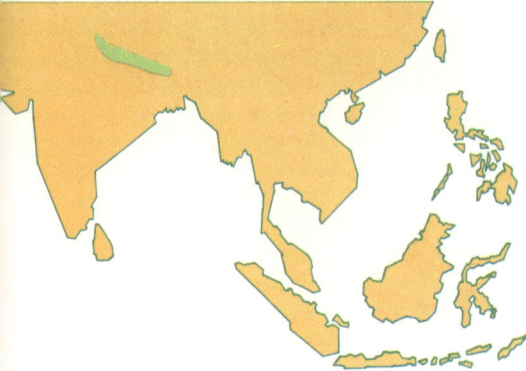

Former geographical distribution of the pygmy wild boar.

PYGMY WILD BOAR
(Sus salvanius)

Class: Mammalia
Order: Artiodactyla
Family: Suidae
Length of head and body: 19½–25½ inches (50–65 cm)
Length of tail: 1¼ inches (3 cm)
Height to shoulder: 10–12 inches (25–30 cm)

Habits little known of this species, which is believed extinct. Female has three pairs of teats. Skull measures only about 5¼ inches from tip of nasal bone to ridge of occipital bone.

Facing page (above): The crested wild boar is encountered in various habitats, including forests and marshes. (Below) The rare Indian gazelle is only found on open plains and in desert and semi-desert regions.

family, are extensive. It is as likely to be encountered in dry zones with hardly any tree cover as in dense forest, feeding on whatever vegetation is available—shoots, fruit, roots and so forth. It is also to be found in swampland, where it spends much time rolling in the mire.

There is one small representative of the Suidae, nevertheless, which is—or was—truly typical of the Oriental region, not being present in any other part of the world. This is the pygmy wild boar (*Sus salvanius*). Unfortunately, this species is feared extinct, although certain authors maintain that it is still to be found in Nepal. This boar differs from other members of the genus *Sus* not only as a consequence of its diminutive size (its total length does not exceed 27 inches) but also because the female possesses only three pairs of teats, as against the customary six of related species. Since the boar is to all intents and purposes extinct zoologists have been unable to make a proper study of the animal's behaviour in the wild, and therefore nothing is known of its habits.

The wild boars inhabiting the frontier regions of India and Pakistan are, like wild cattle, affected by local religious beliefs and customs, but in their case the position is reversed. For whereas the local Hindus eat their flesh (it was the British who instructed them in the noble sport of pig-sticking) the Moslems regard them as unclean animals; and since pork is for them a forbidden food, not only do Moslems refrain from hunting the boars but they also avoid any contact with them.

Population control and natural selection

Before man appeared on the scene to make his meddling presence felt in the savannahs, steppes and deserts of the Oriental region a flourishing herbivore population was kept under control by the hunting activities of predators. As happens in all biomes where there is no human interference, the laws of natural selection operate by eliminating weak, wounded and sick individuals so that no species overflows its habitat and exhausts the available food resources.

Man's arrival and his need to adapt himself to life in an inhospitable corner of the earth soon threatened to upset the natural balance and in time he succeeded in decimating the populations of plant-eaters and carnivores alike. The peaceful herbivores were the obvious first victims and uncontrolled hunting quickly reduced their numbers to an alarming degree. Denied their traditional prey the wild dogs and cats of the region starved to death; those that survived were in such a weakened condition that they too ended as sporting trophies.

Those species that for one reason or another were not hunted—the nilgai, for example—increased uncontrollably, all the more because their former predators had now vanished. This expansion was often detrimental to other herbivores which found themselves driven into a confined space where there was not enough food for them all to survive.

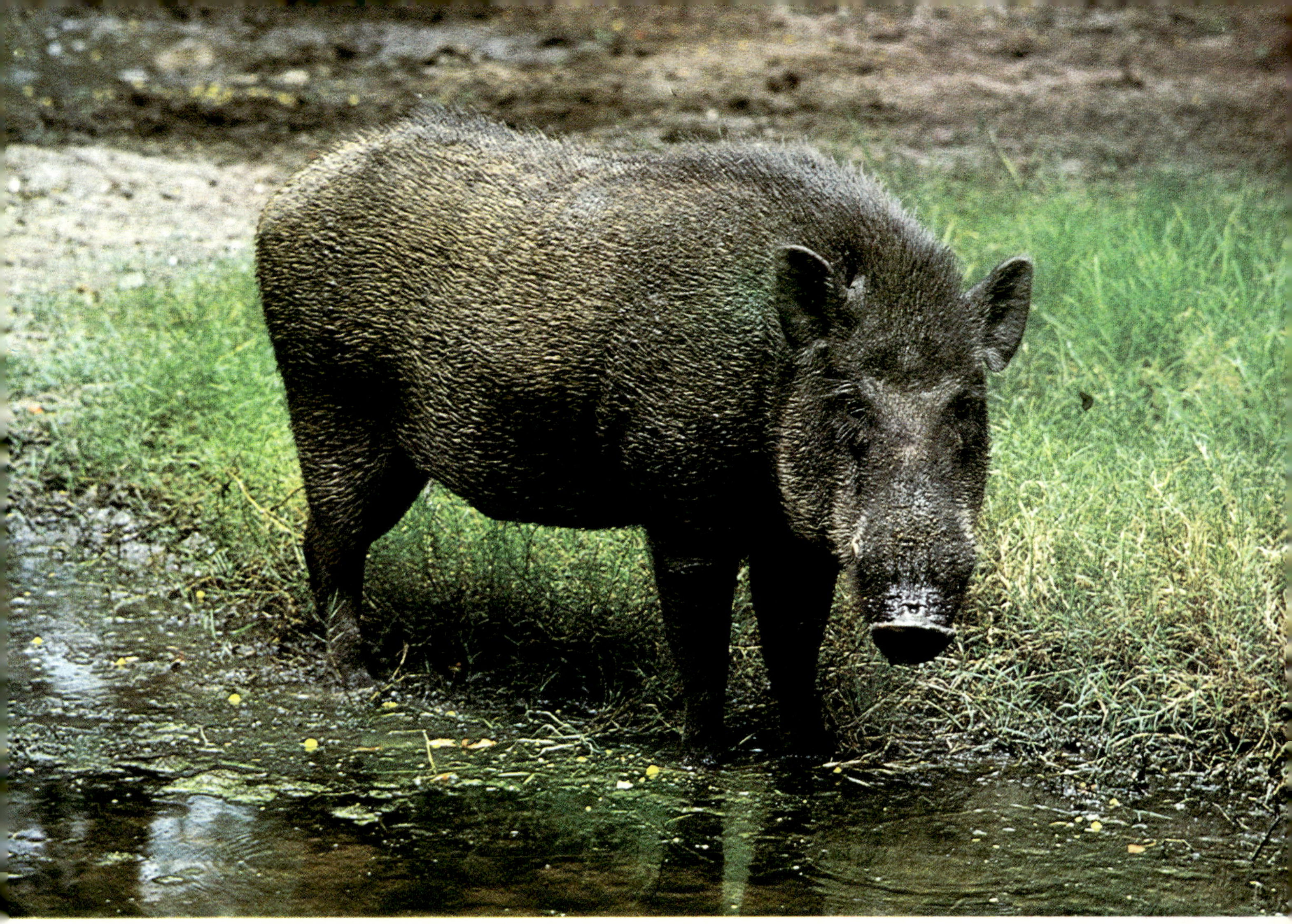

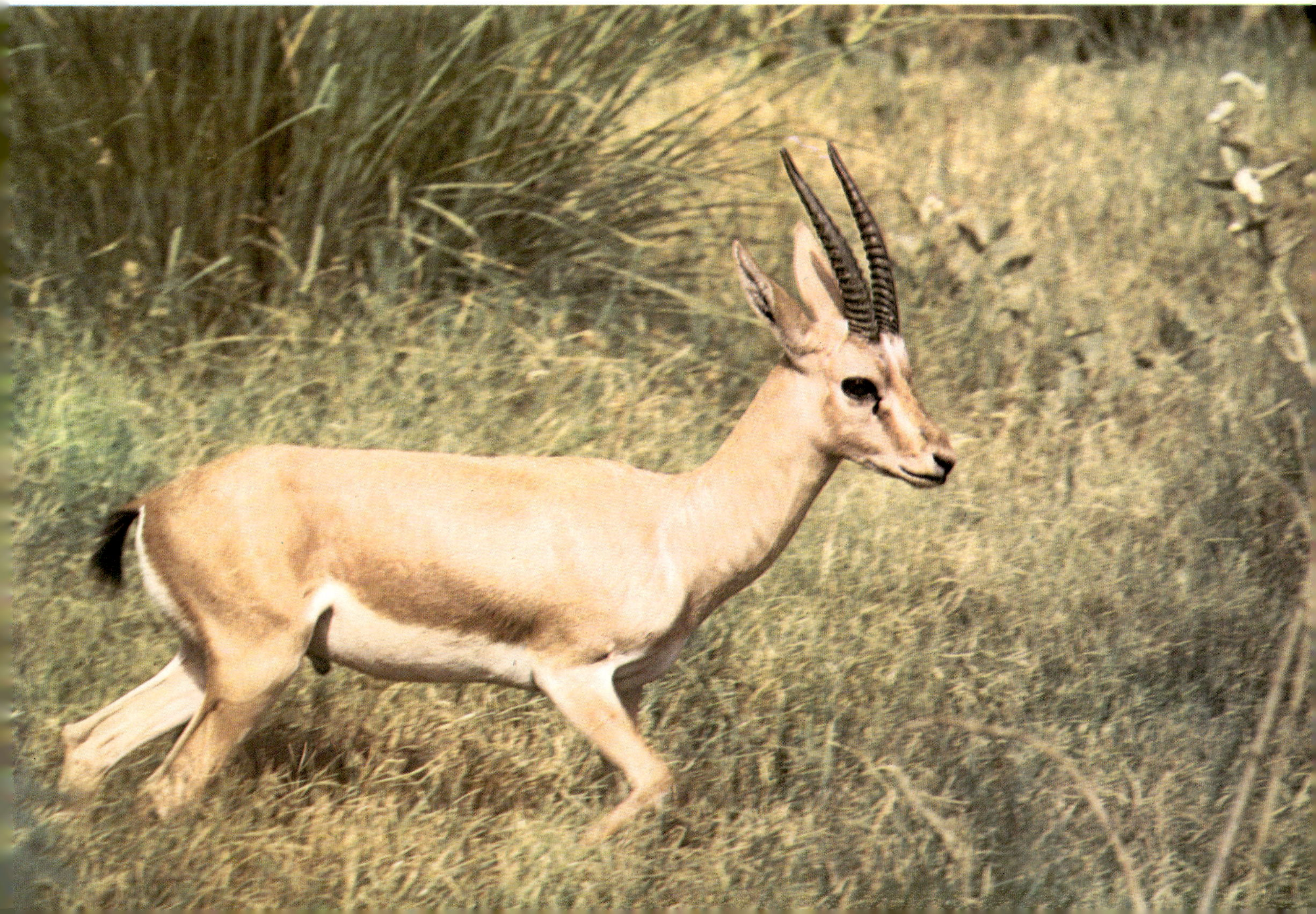

274

| DESERT AND STEPPE | SAVANNAH | JUNGLE | RIVER |

Indian gazelle

Blackbuck

Kouprey

Water buffalo

Anoa

Nilgai

Muntjac

Chevrotain

Gaur

Four-horned antelope

Axis deer

Great Indian rhinoceros

Sambar

Wild boar

Sumatran rhinoceros

Javan rhinoceros

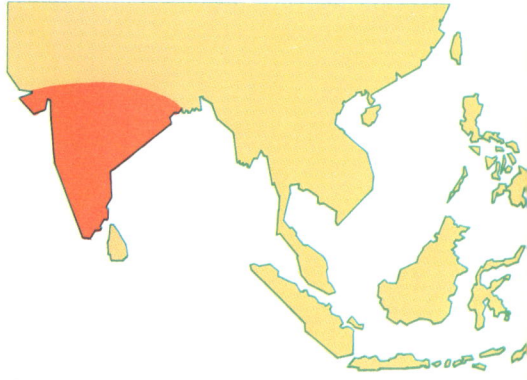

The Asiatic lion: a king in exile

One of the predators of the Oriental region which has suffered most grievously from the direct and indirect consequences of man's invasion of its territory is the Asiatic lion (*Panthera leo persica*), a relative, of course, of the better known African lion (*Panthera leo leo*).

Although nowadays one of the rarest carnivores in the world, the Asiatic lion looks back on a long history. The Greek authors Herodoitus and Aristotle, as well as the Greek general Pausanias, all mention the existence of large numbers of the animals in Macedonia, Thrace and Thessaly around the year 480 B.C., the period which marked the defeat of the army of Xerxes I, king of Persia. Oppian, the poet and naturalist, and Apollonius of Tyana, five or six centuries later, likewise refer to the wild lions which roamed the deserts of Syria, Armenia, Babylonia, Persia and Arabia. So there can be little doubt that a couple of thousand years ago there still flourished a sizeable population of lions whose range extended from India to southern Europe.

The Asiatic lion has also featured importantly in the mythology and religious symbolism of many peoples. Thus the monstrous Nemean lion strangled by Hercules in the course of his twelve labours was placed by Jupiter among the gods of Olympus. In punishment for profaning the shrine of Aphrodite, the lovers Hippomenes and Atalanta were transformed into lions and sentenced to draw the chariot of Cybele, mother of the gods. Samson ripped out the throat of a lion, and Daniel emerged unscathed from the lions' den.

Lions were widely portrayed in classical art. In honour of Leonidas I, the Spartan king who commanded the small force of Greeks in their famous victory over the Persians at Thermopylae, a statue of a lion was erected. The Persians in their turn depicted the beast in various forms in bas-reliefs. Later artists faithfully reproduced the animal's features in their own sculpture and painting; and the reason that such representations differ in some details from our familiar conception of the lion, as seen in zoos and circuses, is that these works of art depict Asiatic, not African, lions. This long artistic heritage proves conclusively that the present much diminished population of Asiatic lions is indeed descended from the same ancient stock.

According to the reports of British hunters of the 19th century, there were obviously large numbers of lions on the subcontinent at that time, constituting a challenge that the 'sportsmen' in question met by proceeding to massacre them as quickly and efficiently as possible. Around the year 1850 one colonial gentleman boasted of having killed 300 lions single-handed during his sporting career; soon afterwards one of his neighbours managed to slaughter 80 more in a three-year hunting spree. Other equally unedifying accounts make it clear that these must have been the last declining years of the Asiatic lion.

The principal difference between the Asiatic lion and its African relative is that the former has growths of hair extending to the joints of the forelegs and along the belly, together with a tuft of hair (larger than in the African lion) at the tip of the tail.

Geographical distribution of the blackbuck.

INDIAN ANTELOPE OR BLACKBUCK
(*Antilope cervicapra*)

Class: Mammalia
Order: Artiodactyla
Family: Bovidae
Length of head and body: $39\frac{1}{2}$–$47\frac{1}{2}$ inches (100–120 cm)
Length of tail: 4–7 inches (10–17·5 cm)
Length of male's horns: 18–27 inches (45·6–68·5 cm)
Height to shoulder: $25\frac{1}{2}$–$33\frac{1}{2}$ inches (65–85 cm)
Weight: 82–88 lb (37–40 kg)
Diet: grass, tender shoots
Gestation: 180 days
Number of young: 1–2
Longevity: 15 years

Blackish-brown coat, white underparts and inner legs; white patches on muzzle, lower jaw and around eyes. Male possesses divergent, spiralling horns. Females and young light brown.

Facing page: Distribution, according to characteristic habitat, of some of the most common ungulates of the Oriental region.

African lion
(*Panthera leo leo*)

Asiatic lion
(*Panthera leo persica*)

The Asiatic lion differs from its better known African relative in certain physical details. Although its mane is shorter, hair extends to the forelegs and the belly. The tuft of hair at the tip of the tail is also somewhat larger.

Facing page : Once abundant throughout southern Asia and south-eastern Europe, the Asiatic lion was so mercilessly hunted that it is now an imperilled species. Between two and three hundred survivors are today protected in the Gir Forest reserve. The tranquil expression of this lioness with her cubs seems to testify that here, in its last refuge, the species is safe.

As a cub, the coat of the Asiatic subspecies is also less mottled. Furthermore, the mane of the male Asiatic lion is shorter. The zoologist E. P. Gee was of the opinion that this last feature may be due to climatic contrasts between Africa and the Indian subcontinent and to the fact that hunters in their heyday preferred to shoot only those lions with splendid manes, with the result that breeding was confined to the less imposing survivors.

Many zoologists have argued that the Asiatic lion must have been the oldest of the super-predators of the Indian subcontinent, preceding the tiger which appears to have penetrated the region from the north-east during the last Ice Age. But the theory advanced until recently that the two carnivores must have engaged in a life-and-death struggle for supremacy surely lacks validity. Not only are their habitats different but so are their patterns of behaviour. The tiger lives in areas of dense forest whereas the lion is a creature of the exposed open spaces. The tiger is a nocturnal animal while the lion is much more active by day. Furthermore, the tiger is a solitary hunter, cautious and easily frightened. The lion is gregarious, living and hunting in groups, bold almost to the point of impudence.

To declare, as some authors have done, that the reason for the disappearance of the Asiatic lion must have been the superior strength and stamina of the tiger is quite ridiculous. Apart from the fact that their hunting grounds seldom overlapped, it makes no sense to assume that in any hypothetical contest between predators of more or less equal size, weight and muscle power, the tiger should always have come out on top. The truth is more likely to be found in the contrasted temperaments and behaviour patterns of the two animals. Because the lion was usually found by day in the open, and in company, it was a much easier target for the sportsman's gun, and it was this vulnerability which was undoubtedly the principal reason for its dwindling numbers.

The last surviving Asiatic lions are nowadays protected in India's Gir Forest wildlife reserve on the Kathiawar peninsula. The sanctuary consists of broad open stretches interspersed with zones of dense forest and covers 527 square miles. Official statistics gave the number of lions in the reserve in 1950 as 240, and in 1955 as 290. Latest figures suggest that the size of the population has again dropped to around 250. In theory the lions would seem to have plenty of food—wild boars, nilgai, axis, hog deer, sambars and blackbucks. But in fact these animals are not all that abundant and the lions frequently kill domestic livestock, which poses serious problems for the administrators. The local shepherds regard this as a small price to pay in return for grazing their flocks, but in times of famine when the lions are, figuratively speaking, snatching vital food from the mouths of the local

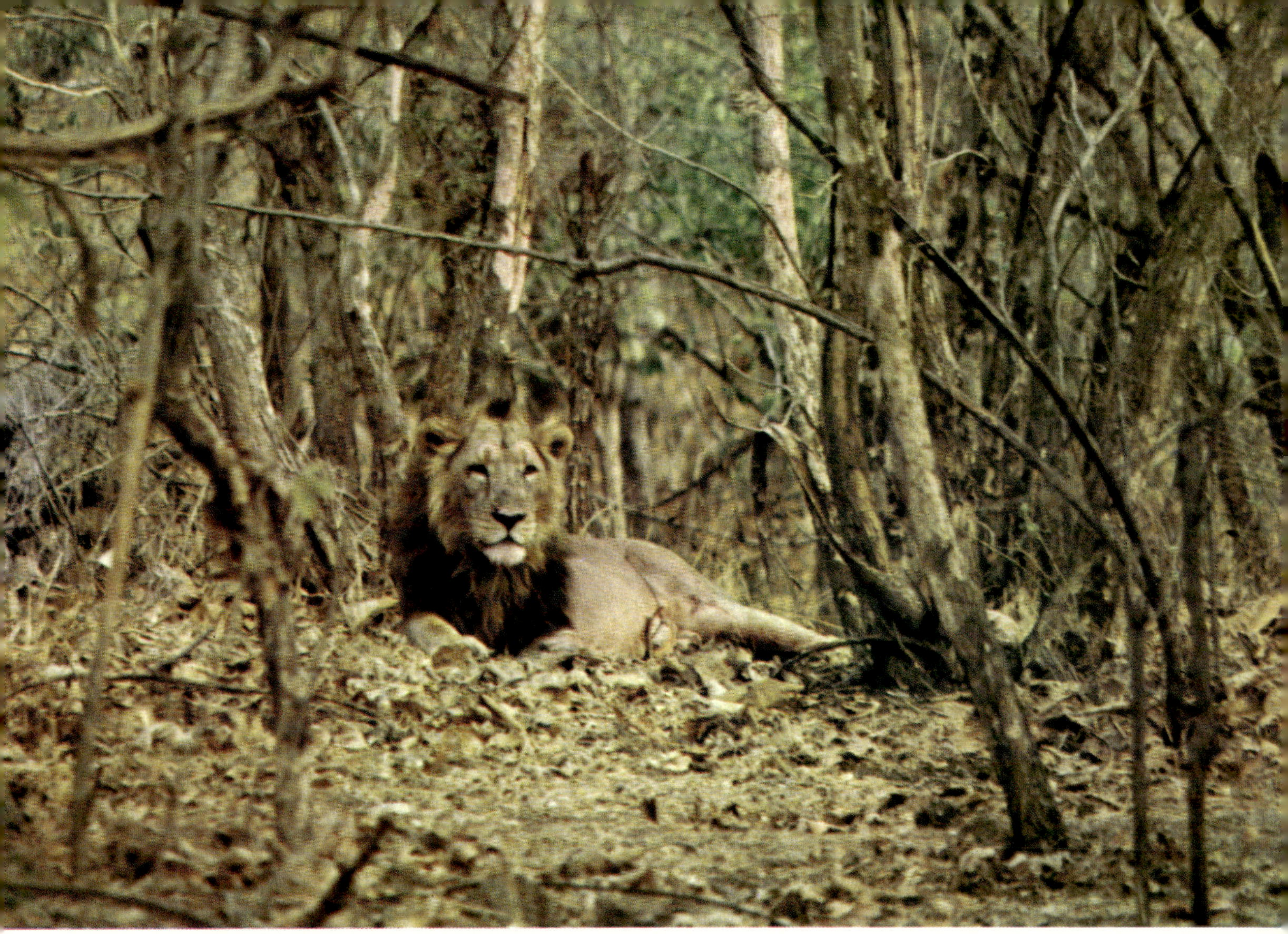

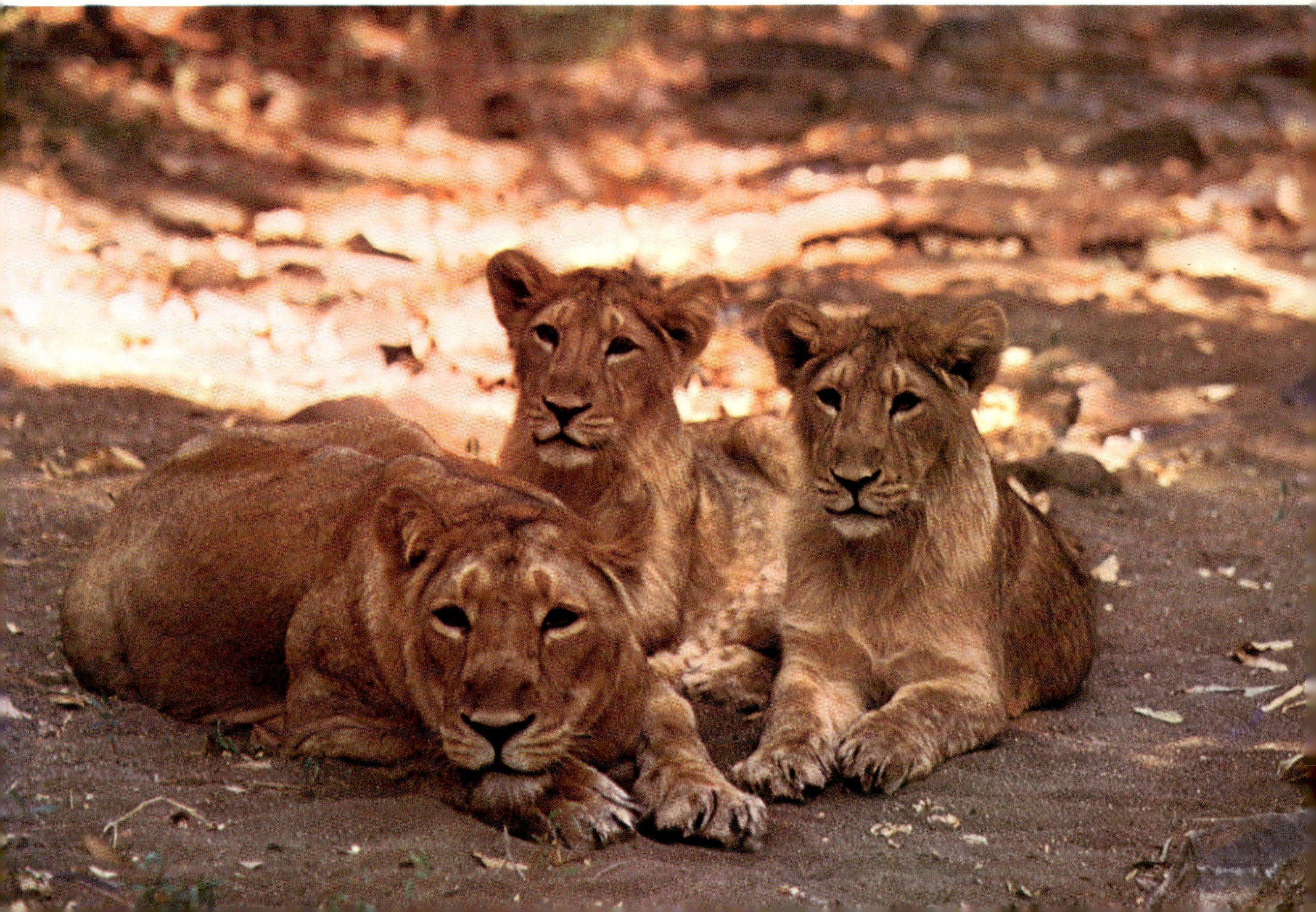

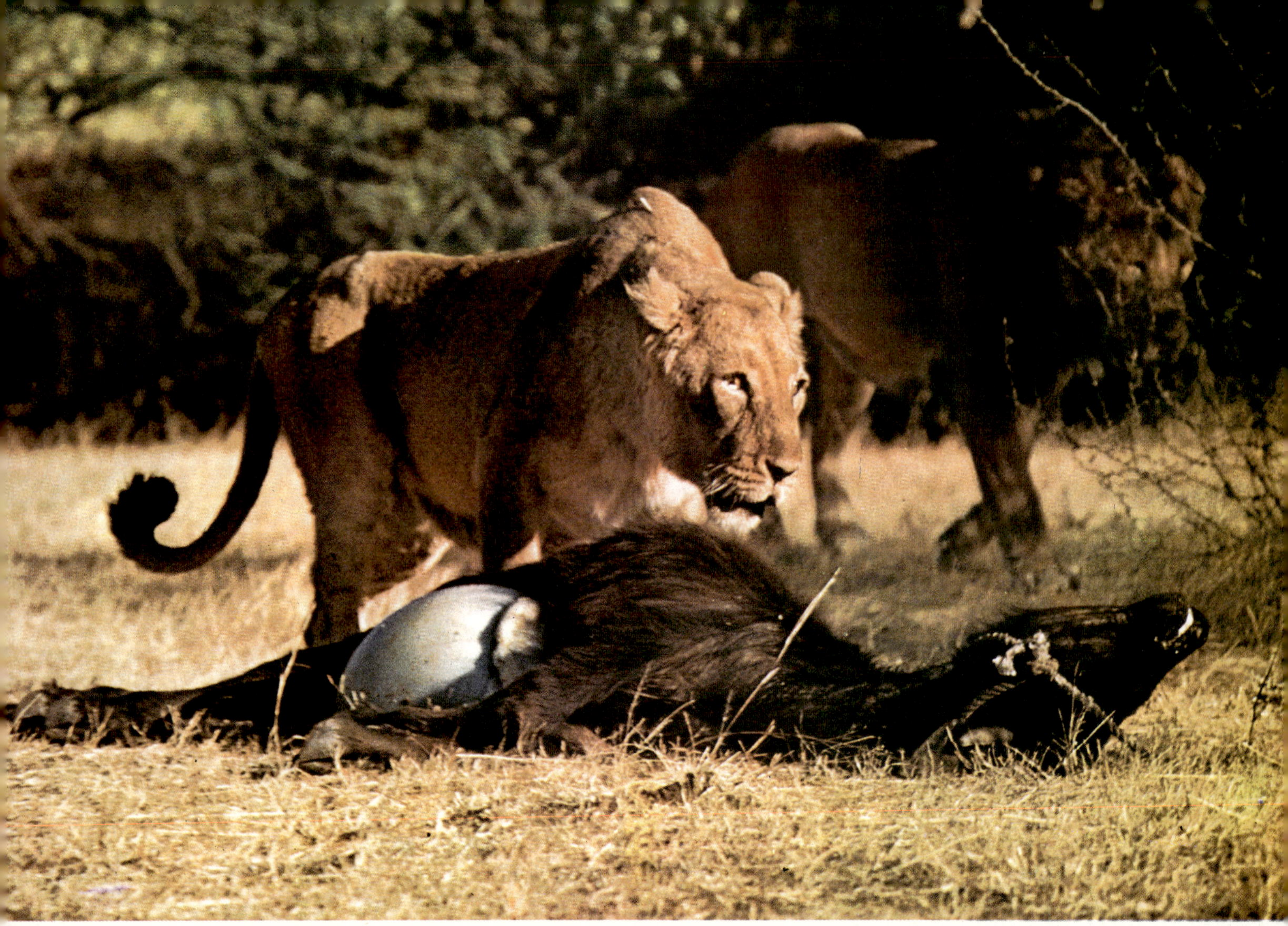

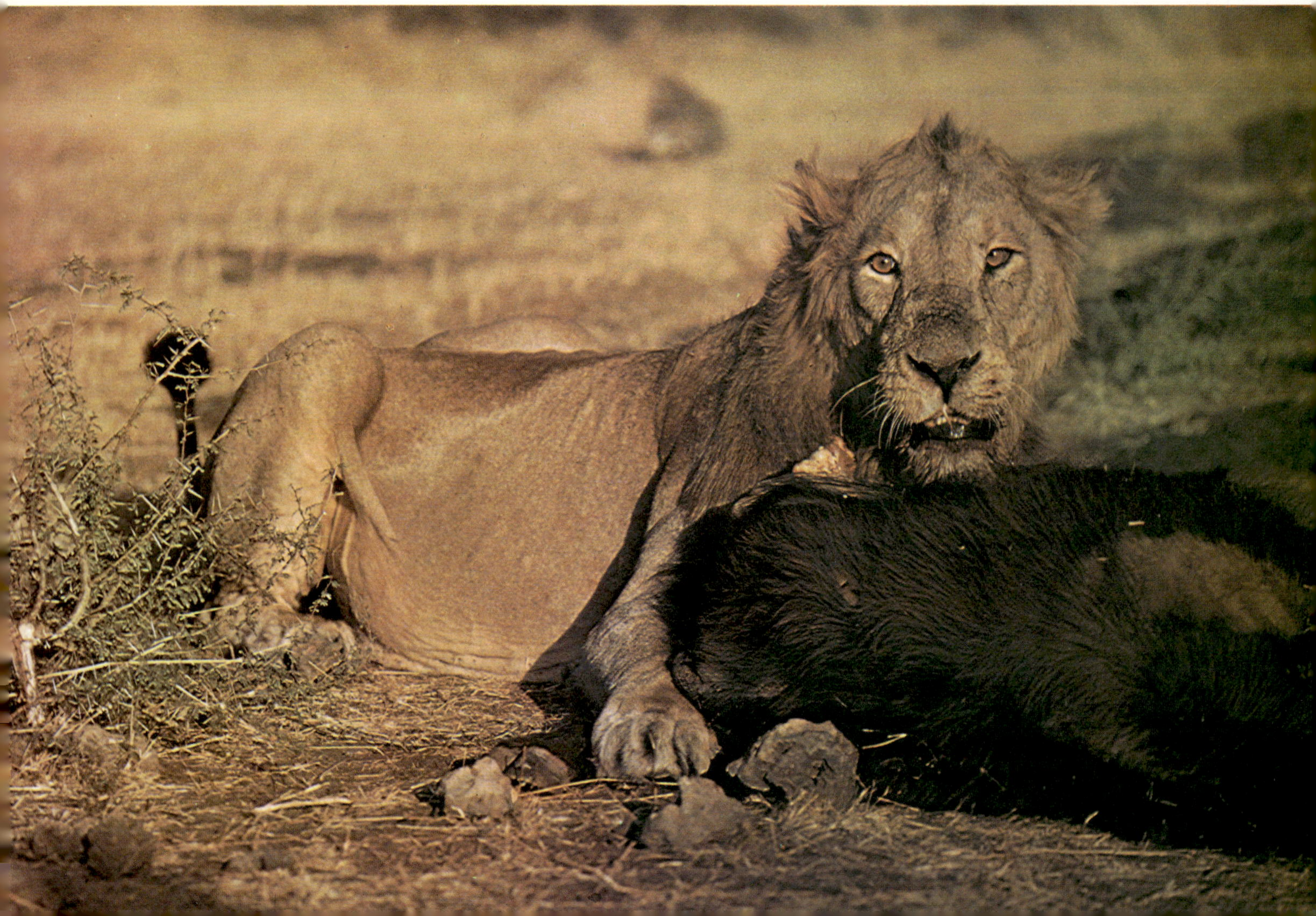

villagers, the latter are not so forgiving and strike back by setting poisoned bait for the carnivores.

In 1963, alarmed by the number of lions that were dying of poison, the Indian authorities decided to adopt urgent measures which involved counting the lions in the reserve and then setting up a fund to compensate the local shepherds for any animals killed by the carnivores. The census indicated that quite a number of lions must have been poisoned in the immediately preceding years.

Today, as a result of joint efforts by the Indian government and the World Wildlife Fund, the Asiatic lion, although its name still appears in the IUCN Red Data Book, seems safe for the time being in the Gir Forest sanctuary. But it would be premature to claim – in spite of the fact that numbers appear fairly stable at present – that the future of the species is assured.

Has the cheetah vanished from Asia ?

The cheetah (*Acinonyx jubatus*) was in years gone by one of the most characteristic animals of the steppes and deserts of southern Asia. Ancient legends recount how Oriental potentates and nobles would tame the splendid cats and train them for hunting. In a previous chapter dealing with the cheetah population of East Africa we described the methods used in such countries as Persia, Arabia and India to train cheetahs for coursing, a practice that lasted until the 19th century. Some royal households kept 'stables' of cheetahs. It is recorded that the great Mongol emperor Akbar, who reigned from 1556 until 1605, owned a thousand of the animals.

Today things are very different. In 1935 the Bombay Natural History Society, in a report on the various animal species of the Indian Empire, indicated that the cheetah was on the verge of extinction. During the decades that preceded this paper, reports of cheetahs sighted in the wild were few and far between. The Society's report for 1919 had been the last to declare that there were any cheetahs left on the Asian continent. In 1928 a female cheetah with her cubs was seen in Iraq. The sad story came to an end in 1947 when, as already mentioned, a poacher bagged three male cheetahs, although no trace was found of his victims. Despite subsequent rumours, the rest was silence. The Asiatic cheetah had completely vanished.

Looking back, it is interesting to note that the cheetah's residence in Asia was a comparatively short one, for the species only found its way into India, by way of Persia and Baluchistan, together with the tiger in the course of the last Ice Age. It extended its range to the Deccan in the south and to what is now Bangla Desh in the east.

In both appearance and behaviour the Asiatic cheetah could not be distinguished from its African counterpart. The only difference was in the type of prey captured. In Africa the cheetah's principal victims are Thomson's gazelles, Grant's gazelles, impalas, gnus and warthogs. In the Oriental region favourite prey used to include Indian gazelles, axis deer and blackbucks. The last named species was a particularly worthy

Above and facing page : The lions of the Gir Forest sanctuary still retain the majestic air of a free-hunting predator; but lack of sufficient natural prey in the reserve has compelled the authorities to provide them with food in an endeavour to obviate attacks on domestic stock.

The cheetah, which in bygone years was one of the most characteristic predators of the Oriental region, is now assumed to have vanished from Asia as a result of intensive hunting, the destruction of the herbivore population and failure to breed in captivity.

adversary. To capture such an antelope the cheetah had to summon up its maximum resources of speed (over short distances it is said to travel at something like 70 miles per hour) as well as all its cunning, so as to catch its victim by surprise. Over a short distance of, say, a hundred yards the great spotted cat undoubtedly had the advantage, but if it failed to make a kill quickly the blackbuck usually got away, since the odds switched in the latter's favour with every passing second. Over a long distance it was stamina that told, and in this respect the leaping, bounding blackbuck could prove more than a match for its heavier pursuer. The antelope was capable of keeping up a steady speed without slowing down, and after a while the cheetah, overcome by fatigue, would be compelled to abandon the chase. Realising the hopelessness of further exertion it would stretch out on the ground to rest and recover its energy, perhaps succeeding on another occasion.

Man's intervention had drastic repercussions on the population of Asiatic cheetahs. Those that he captured for coursing proved to be sterile and would not breed, while others starved to death as a result of the massacre of their traditional deer and antelope prey.

The deplorable consequence of this double-edged pressure on both hunting animals and their victims has been to destroy the Asiatic cheetah and to decimate the numbers of Indian gazelles and antelopes. Unfortunately the damage had already been done many years before conservationists arrived on the scene, so that there is no hope of the cheetah benefiting from the protective measures which have perhaps come to the rescue, in the nick of time, of the Asiatic lion.

The dhole: wild dog of Asia

The Indian wild dog or dhole (*Cuon alpinus*), a once-familiar inhabitant of the Asiatic steppes, has been the subject of many strange legends. In size and general appearance it resembles a wolf but may be distinguished from the latter by more rounded ears, a shorter muzzle and the four, not five, toes on the forefeet. Its tail is bushier, with streaks of black, and the coat is also of a different colour, sparse and yellowish-brown in the warm southern climes, thicker and yellowish- to brownish-grey in northern regions during the winter, browner in summer.

Ancient tales that have filtered down to us through the centuries speak of the dhole as a creature reputed to possess evil powers. It used to be claimed, for example, that the urine of the animal gave out corrosive fumes. Thus when a pack closed in on their prey the dogs would drive the victim towards the spot where they had previously urinated. The animal would be temporarily blinded and could then be finished off by its pursuers before managing to regain its sight. There are numerous other legends about dholes in popular Indian literature and it is surprising how often this theme recurs. Close study of the species shows how the odd belief may have originated. The dogs are in fact territorial animals and the placing of scent posts therefore plays an important role in group defence and recognition. Simple peasants recognised the habit but of course could not discern its true significance.

The Indian wild dog is a social hunter. Like the African hunting dog it lives in packs consisting of leaders and subordinates, the number in each group depending on the amount of prey locally available. It must be emphasised, however, that whereas we know a great deal about the behaviour of the hunting dog in the wild, thanks to detailed first-hand observations and reports, there have been few similarly objective surveys of the dhole, and consequently our information about the species is severely restricted. Nevertheless, by combining the sometimes unreliable facts supplied by hunters and travellers with observations of the species in captivity it has been possible to come to certain conclusions about the animals' predatory habits, these being similar to those of hunting dogs and wolves.

Dholes roam around in groups in search of food, apparently concentrating their efforts on game which is relatively easy to capture, such as muntjacs, axis deer or sambars. Although swift of foot they are more notable for their remarkable stamina. Like the majority of carnivores that hunt in groups they choose their victim in advance, preferably either a heavy animal handicapped by its weight, a youngster which is not yet steady on its feet, or a sick individual. As soon as the prey shows obvious signs of exhaustion the dogs close in, snapping at its flanks, its rump and its hind legs until it sinks to the ground. The entire pack then leap in for the kill, ripping the still-breathing herbivore to pieces. It is a most unpleasant sight, reminiscent of hunting dogs in similar situations.

Although local villagers have claimed that dholes will attack larger animals such as water buffaloes, Himalayan black bears

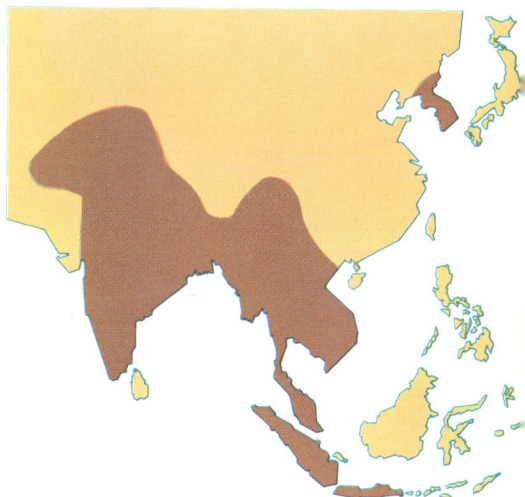

Geographical distribution of the Indian wild dog.

INDIAN WILD DOG OR DHOLE
(*Cuon alpinus*)

Class: Mammalia
Order: Carnivora
Family: Canidae
Length of head and body: 30–39½ inches (76–100 cm)
Length of tail: 9½–19 inches (24–48 cm)
Weight: 31–46 lb (14–21 kg)
Gestation: about 9 weeks
Number of young: 2–6, sometimes more

Coat light chestnut, pale and thicker in winter in northern races. Pointed muzzle, large ears. Forty teeth instead of forty-two as in other Canidae; the last molar on either side of the lower jaw is missing.

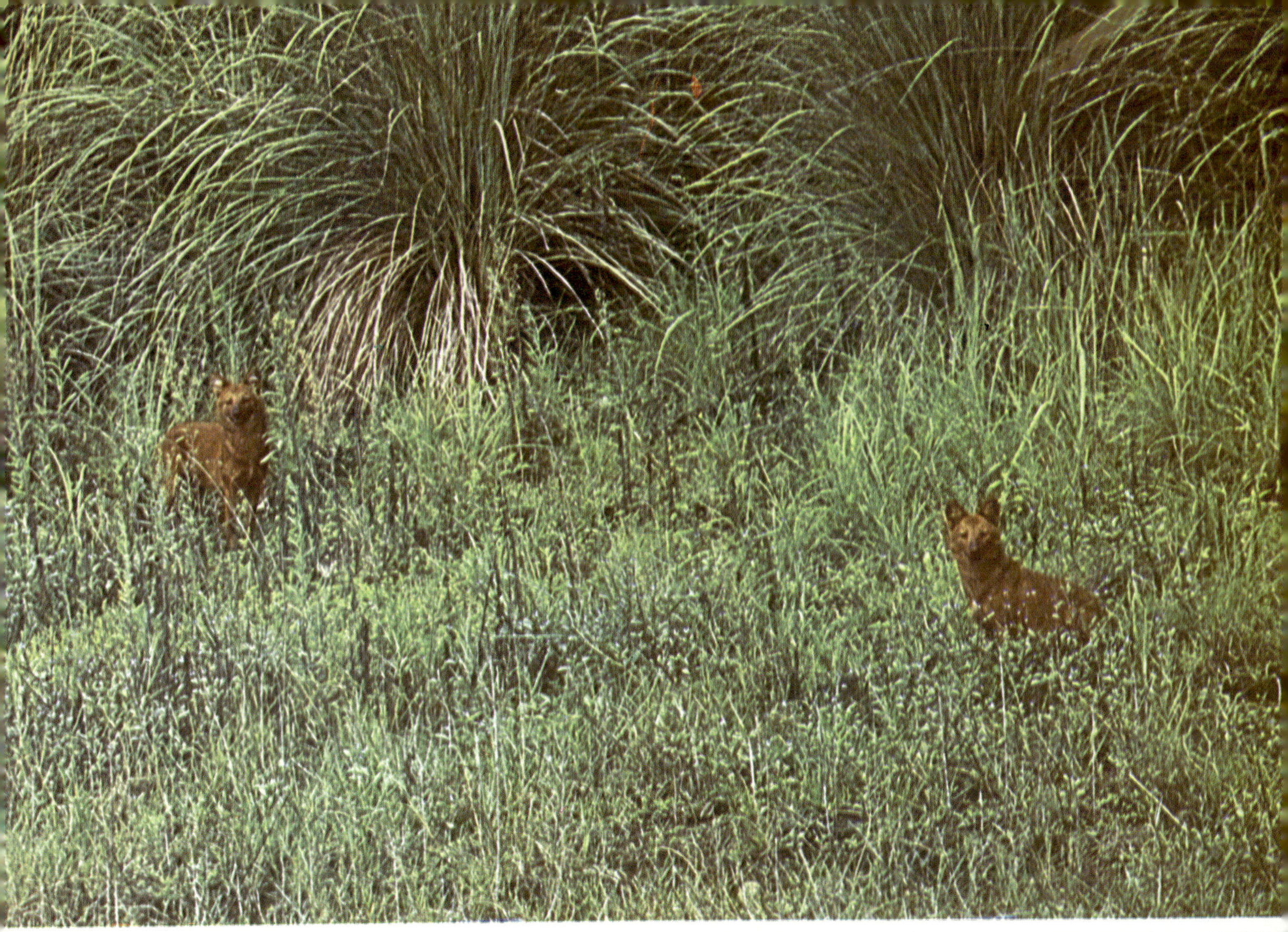
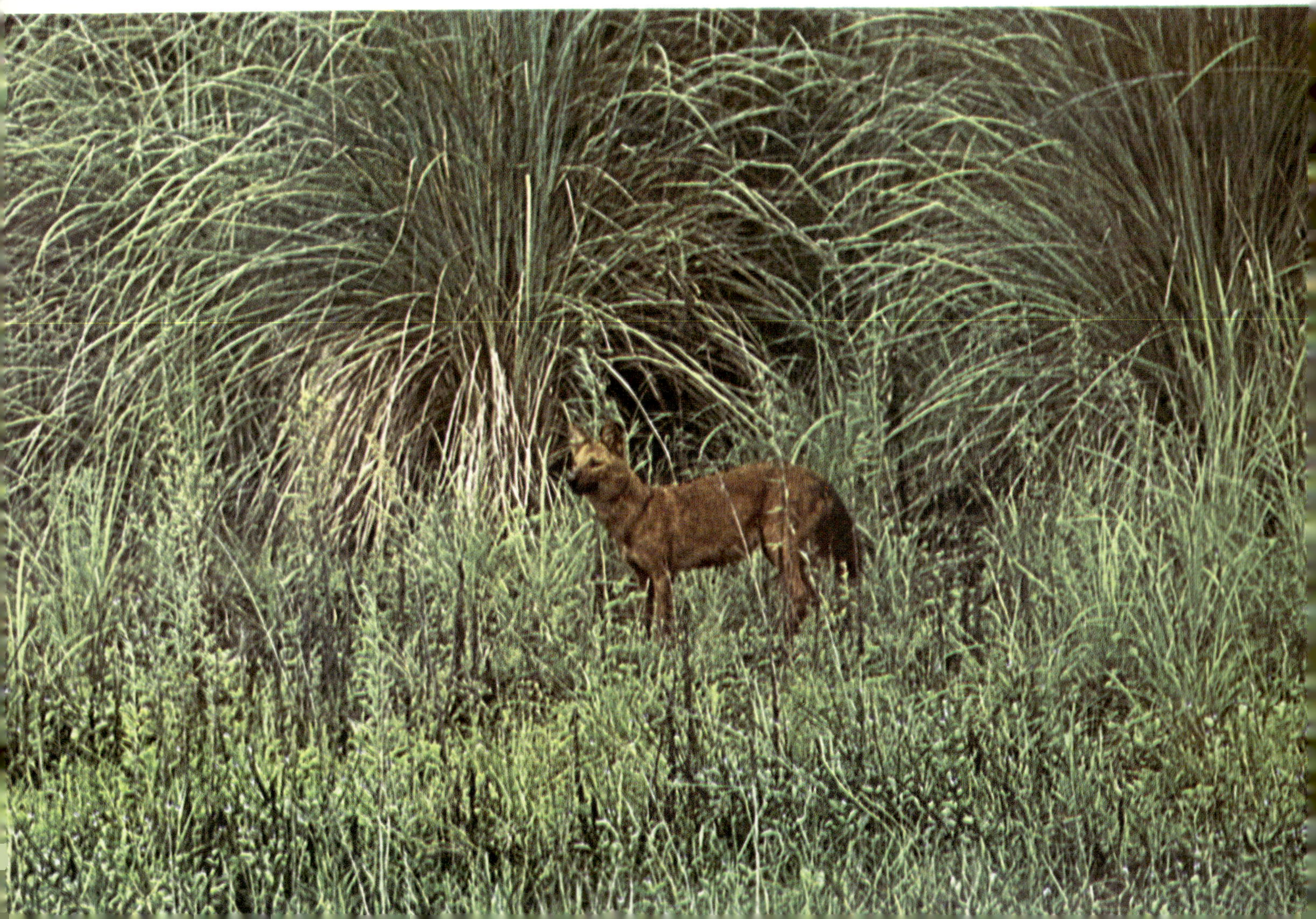

and sloth bears, this probably occurs only when food is very scarce for in the ordinary way this would involve them in considerable waste of energy. Allegations that tigers and leopards stand in dread of these dogs and give them a wide berth may or may not be true.

The enormous packs of wild dogs which once roamed the jungle clearings, hillsides and open plains of Asia are now part of legend. The steady decimation of the numbers of wild cattle and deer has had a direct effect on the dhole population. Another cause of their slow disappearance has been the relentless hostility of local peasants who have made use of traps, poisoned bait and firearms to destroy what they have always looked upon as a harmful animal. Persecution has had its inevitable effect. Whereas in former times a typical pack might consist of twenty or thirty individuals, it is now rare to see more than two or three dogs together – another sad example of the way in which the wildlife of Asia has suffered at the hands of man.

Wolf-child: fact or fable ?

Mowgli, the famous character in Rudyard Kipling's *Jungle Book*, was raised in the heart of a wolf pack led by Akela, and personifies a local tradition whose origin goes back to the Stone Age – that of a human child reared by wolves. Such legends are found in the mythologies of other nations as well and in recent times there have been a number of reputedly factual accounts in newspapers of so-called wolf-children. Most of these stories have been greeted with disbelief or scepticism although, as we shall see later in this section, not all can be dismissed as pure invention. Arguments advanced in refutation of the idea of wolf-children point to the lactation period of a she-wolf – surely too short to satisfy the needs of a human infant – and to the fact that a newborn baby could not possibly survive without even the most elementary hygienic attentions. After two or three months, when wolf cubs would normally be expected to accompany their mother on expeditions outside the lair, human babies would still be helpless and unable to fend for themselves. Once abandoned they would inevitably die of starvation.

Dr Felix Rodriguez de la Fuente, whose experiments with tame wolves have already been described, is not one of those who scoffs at the idea. In his six-year survey of the species he was particularly intrigued by certain features of group behaviour, notably those relating to adoption and upbringing. His reports contained a number of observations about the relationship between she-wolves and human babies. Thus when the doctor held out his four-month-old baby daughter for the inspection of a pair of adult wolves, he was struck by the way in which both animals assumed what he termed an 'adoptive attitude'. Male and female alike, uttering little yelps and generally displaying every sign of acute nervousness, cautiously approached the strange object. Encouraged by the doctor, they soon overcame their initial doubts and began to lick the baby. Then they tried tugging gently at her clothing as if to entice her away to their lair. Both animals, reputed to be so belligerent, appeared to be drawn

Facing page : Indian wild dogs or dholes are social hunters. In former times they were found in large packs but today they are seldom seen in groups of more than two or three.

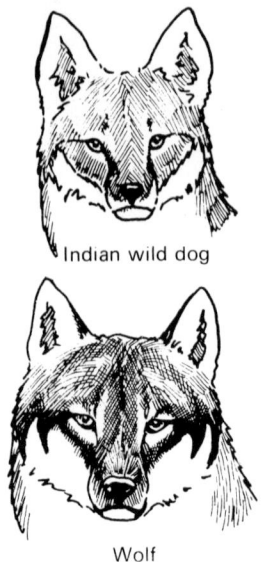

Indian wild dog

Wolf

The essential anatomical difference between the Indian wild dog and the wolf is in the size and shape of the head. In the former species the ears are larger and somewhat rounder, and the muzzle is shorter.

Facing page : The Indian wolf is a variety of the common species, recognisable by its smaller size, more slender shape and paler coat.

instinctively towards this tiny defenceless creature, giving out the same cries and tendering the identical attentions that they would have lavished on a cub of their own.

Dr de la Fuente came to the following conclusions. In an over-populated country such as India it is by no means unlikely that abandoned human babies could have been 'adopted' by she-wolves that had perhaps lost their own cubs, and carried off to their lairs, given the fact that the mere sight of a newborn creature, of whatever species, is enough to stimulate the adoptive instincts of the carnivore and set in train the necessary reactions. The short term of lactation would not seem to be a major problem since the baby, in addition to suckling the mother's milk, could derive nourishment from predigested food regurgitated by the she-wolf. Such substances can certainly be assimilated by the human organism. Those who believe that a she-wolf, once she has finished rearing an adopted baby, would necessarily abandon it, fail to take account of the fact, frequently observed, that suckling and other maternal cares are stimulated by the very presence of the cubs and that such attentions will be prolonged should the babies be physically incapable of looking after themselves. Thus in the eyes of the she-wolf a human baby would simply be another cub but one which developed more slowly than usual. Because it is clearly unable to move about on its own at an early age and because it shows, by crying and other means, its need for protection, the she-wolf, sensing that it is still dependent on her, would therefore continue to care for it as long as might be necessary, for months or even for years.

The most celebrated case history of babies that have apparently been reared by wolves is that of the two little Indian girls who were given the names Amala and Kamala. It happened in 1920 in the village of Midnapore, south-west of Calcutta. A group of villagers were panic-stricken when one night they discovered in the lair of a wolf (an animal which they greatly feared and to which they attributed evil powers) two strange four-legged creatures with human faces. The latter were romping and playing with a family of wolves consisting of two adults and three cubs. It so happened that the discovery coincided with a pastoral visit to the area of the Reverend R. P. Singh who requested one of the villagers to conduct him to the lair so that he could check the authenticity of this unlikely rumour. Initially sceptical, he confessed himself quite astonished to see the five wolves emerging from a hole shortly before dawn, followed by two small, dirty, dishevelled children on all-fours.

The minister immediately decided to explore the lair in which this incredible family evidently lived, in order to identify its various members. But the villagers of Midnapore strenuously refused his invitation to participate in what they considered a profane act and he was forced to seek help from another village. He described in his diary how, when digging began, two wolves managed to flee while another, a female, was killed. At the bottom of the lair he came across two cubs and two little girls. The elder seemed to be about seven or eight years of age and the younger hardly more than two years old. They were both taken to the local orphanage and baptised. The younger girl, Amala, died a

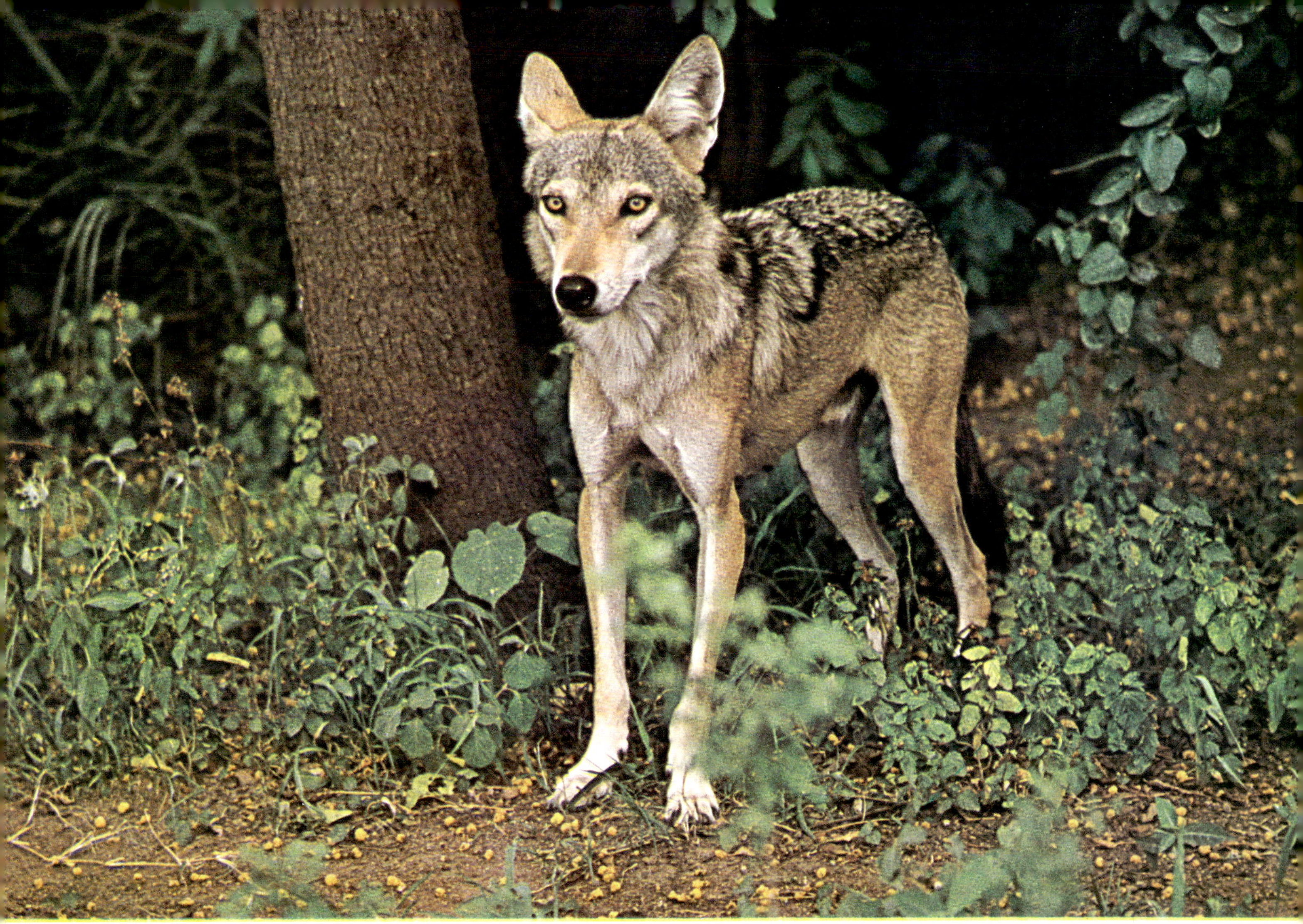

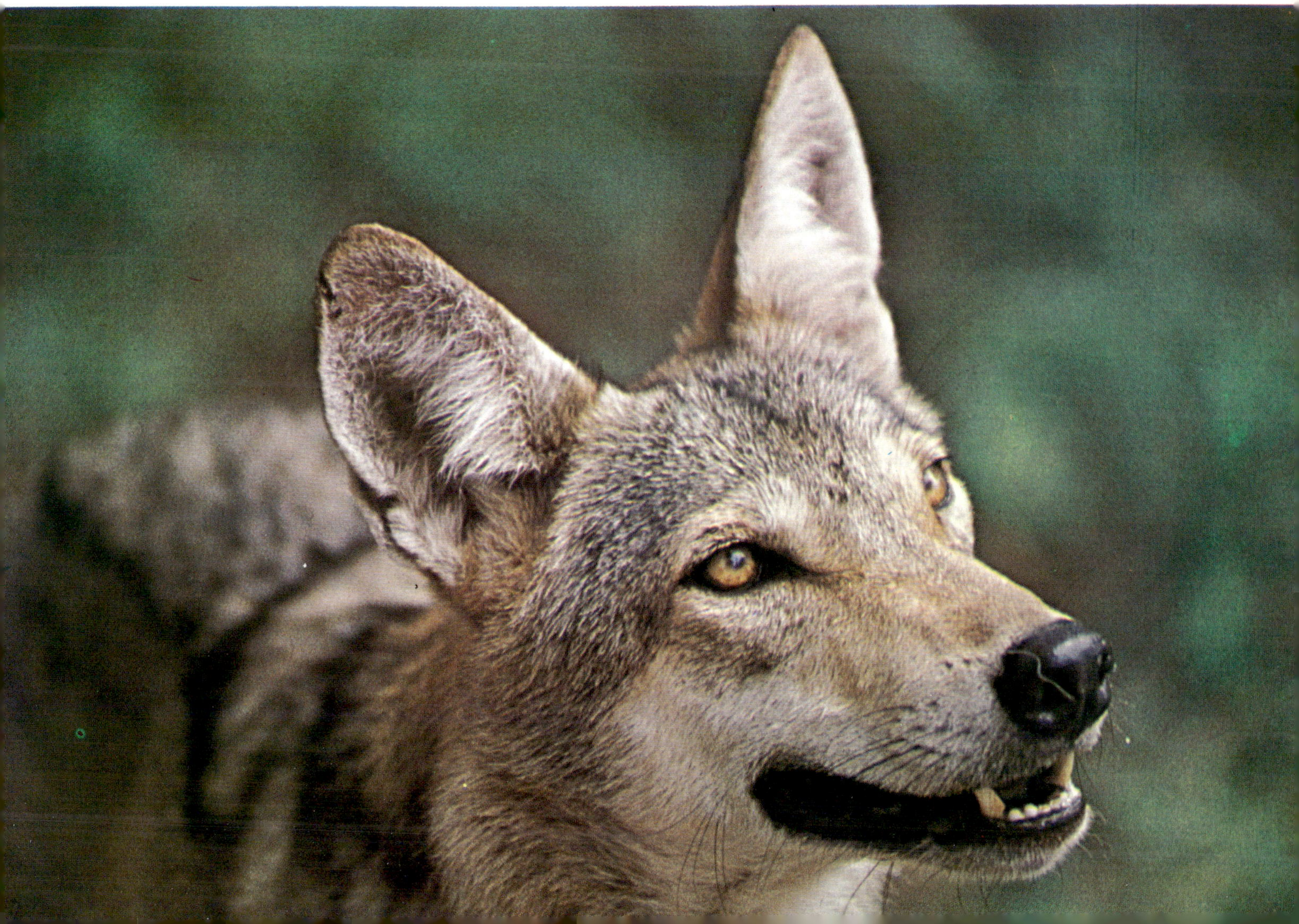

Despite its unusual appearance the black panther is simply a melanistic form of the common species of leopard. The animal is found more frequently in Asia than in Africa.

Facing page : The Indian jackal is both a predator and a scavenger. The species is still present in fairly large numbers in tropical Asia.

few months later but Kamala lived until 1929. During the nine years she lived with the nuns Dr Singh watched her daily activities and kept a record of her progress.

Kamala behaved exactly like an animal in the first few months, crawling about on all-fours, normally on hands and knees, but, when she wanted to get along more rapidly, on hands and tips of toes. When 'running' in this manner she could outpace anyone on two feet. She pushed away all vegetable foods and ate only raw meat, gnawing the bones. At night she frequently howled. She showed the greatest tenderness for little Amala and when the latter died she rejected all food for several days, spending her time sniffing at the bed where her companion had slept. When she was first removed to the orphanage she had a mental age of a six-month-old baby. Just before she died, at the age of seventeen years, she was able to pronounce a few words, to stand upright and to use her hands for feeding. At that stage her mental age was equivalent to that of a four-year-old.

This remarkable story would seem to prove that wild animals with strong maternal feelings are capable of rearing the young of other species. There have been cases of other strange liaisons – bitches suckling piglets, kittens and lion cubs – indicating that as far as adoptive behaviour is concerned a domesticated dog and a she-wolf will act in a similar way. But the case of Kamala shows that human development depends on the cultural environment. It was only as a result of patient effort on the part of those who subsequently cared for her that she was able to make a little slow progress in the rudiments of human culture.

The Indian wolf

The carnivore which features so prominently in this respect in Asiatic legend is the Indian wolf (*Canis lupus pallipes*), a variety of common wolf which differs from the latter in being smaller and slenderer, and in the colour of its coat.

In the opinion of certain authors the Indian wolf is the ancestor of the domestic dog, or at any event of some breeds. The discovery by a group of palaeontologists from the University of Pennsylvania of the fossil remains of a primitive dog in a cave in Iran showed, by radiocarbon testing, that they were more than 11,500 years old. Similar remains found in a cave on Mount Carmel in Israel were pronounced to be 10,000 years old. The scientists who reconstructed the latter skeleton came up with an animal very similar to the Arabian wolf, a close relative of the Indian wolf. So it is reasonable to suppose that in the Middle East, around the beginning of the Neolithic, wolves which had developed alongside man since early times might have been captured when still very young and become accustomed to life in villages. The dog was certainly the first domesticated animal.

The Indian wolf roamed the plains of India in the north from Bengal to Sind, to Baluchistan in the south, to Iraq and northern Arabia in the west. The fact that the species was not subjected to intensive hunting leads to the conclusion that its modern decline has been due to the dwindling numbers of its prey and to a variety of epizootic diseases, including rabies.

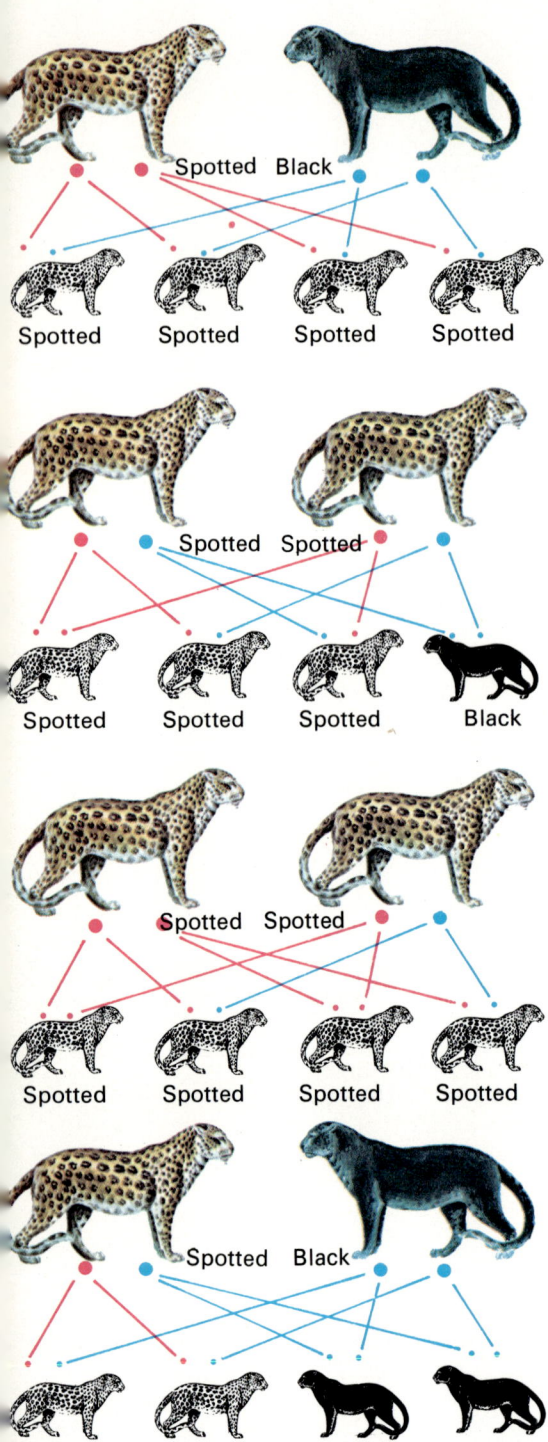

Coat colour in the leopard is hereditary, spots being a dominant genetic factor and black recessive. The former is represented here by a red dot, the latter by a blue dot. Thus the offspring of a pure spotted individual and a black individual will be spotted even though all carry the recessive melanistic factor. But in the later generations offspring inheriting the melanistic factor from both parents will be black, as these diagrams show.

Hunters of small prey

The decreasing numbers of large herbivores and the steady expansion of man's dominion over virgin land have favoured the growth of the populations of medium-sized predators. Although they too have been subjected to the same kinds of pressures as the larger carnivores, they have suffered less, partly because they do not require so much food, partly because they are experts at concealment. Among the smaller feline hunters are the leopard cat (*Felis bengalensis*), with its lovely soft, spotted coat, which is found both in the jungle and on the savannah where it chiefly attacks birds; and the bush cat (*Felis libyca*), a native of Africa which preys on birds and small mammals. Another characteristic species of Asia and North Africa is the jungle cat (*Felis chaus*) which looks rather like a lynx, with tufts of black hairs on the tips of its ears. Unlike other Felidae it is not a particularly nimble climber and spends more time on the ground than up in the trees. As for the leopard (*Panthera pardus*) this handsome carnivore is found in a variety of habitats, from dense jungle to sparsely wooded plains. Those leopards that live on the steppes are markedly paler in colour and more slender than those inhabiting the forests.

The common fox (*Vulpes vulpes*) and the Bengal fox (*Vulpes bengalensis*) are both widely distributed throughout the Oriental region, the latter being the smaller and lighter coloured of the two. The scope of their range is due to their broad choice of diet which includes vegetables as well as animal food. These hunters help to control the large rodent population and also prey on lizards and amphibians. So too does the small Indian civet (*Viverricula indica*) which in addition eats nestlings, eggs, fruit, tender shoots and roots.

Hyenas and jackals

The striped hyena (*Hyaena hyaena*) plays a dual role in the ecological pattern of the arid regions of India because it is both a predator and a scavenger. It has a rather plain coat, grey with black stripes, and a stiff mane of longer hairs along the back. Its forefeet, used for excavating holes, are powerful but the hind legs are weak and ineffectual by comparison and the rump is slightly lower than the forequarters. The hyena has a massive head and the jaws are much stronger than those of other carnivores, ideal either for killing prey (usually a newborn herbivore) or for gnawing the bones of its victims and of carrion.

The ambling gait of the striped hyena only serves to reinforce its already rather ungainly appearance. As with the spotted hyena of Africa ugliness has tended to be equated in the popular mind with evil, low cunning and cowardice. So much has been said and written about the alleged misdemeanours of the species that it is difficult to find anyone to speak up on its behalf. Those who are prepared to consider the matter objectively will admit that the striped hyena population performs a valuable service in helping to keep the countryside clean and relatively disease-free.

Unlike its African relative the striped hyena is seldom sighted

abroad during the day, taking advantage of the cover of darkness to hunt for carrion abandoned by the larger carnivores and the vultures.

Another predator-cum-scavenger with an even wider range of food is the Indian jackal (*Canis aureus*). It eats fresh and putrefying meat as well as vegetable matter. Very common throughout the Indo-Malayan region, this nocturnal animal frequently follows in the tracks of shepherds and strays into villages to pick up waste products of all descriptions.

Although normally solitary, jackals band together to hunt young herbivores. While one of the group distracts the mother a second will attack and carry off the baby.

Jackals are not hunted for their skin or flesh, which are worthless, but simply because, like hyenas, they inspire the instinctive contempt of local villagers. Yet they have remained fairly numerous, thanks to the diversity of their eating habits; and they too perform a valuable hygienic function in many overpopulated areas. Thus the howl of the jackal continues to echo across the plains at night and it is to be hoped that common sense will outweigh prejudice to guarantee its survival.

The striped hyena, like the spotted hyena in Africa, is mistakenly accused of being a cruel cowardly animal. In fact it hunts in addition to feeding on carrion. Furthermore its scavenging activities help to keep the countryside clean and to check the spread of disease.

CHAPTER 94

The tiger: lord of the Asiatic jungle

Many people consider the wildlife of Asia unrivalled. Although it has to yield pride of place to Africa as far as immense concentrations of plant-eating animals are concerned, it can boast two of the largest predators in the world, the tiger and the lion. Both species—particularly the latter—are unhappily in decline but at one time they were abundant, undisputed lords of jungle and plain respectively.

Today the tiger rules supreme in the wild, for the lion, as already noted, has been rescued from extinction only by being afforded strict protection in a reserve. But there is no good reason to assume that the one species survived only at the expense of the other. It is unlikely that they ever entered into direct rivalry for the tiger's domain—woodland zones and clearings as well as thick forest and jungle—has seldom been invaded by the lion, an inhabitant of plains and savannahs covered at best by stunted trees, bushes and thorns. It is possible that the two species may at times have confronted each other among tall grasses on the savannah but the lion would hardly have lingered long in this biome which represents a handicap rather than an aid to its specialised form of hunting.

The diversity of forms among plant-eating animals—ranging from mice to buffaloes—has doubtless affected the evolutionary development of the predators hunting them. Among the Felidae, for example, which are found in all continents and include the tiger, the lion, the leopard and the jaguar, there is an astonishing variety in individual appearance and habit. But of all the Big Cats the tiger can claim to have reached the acme of all-round perfection. The lion, it is true, is capable of killing animals as large as buffaloes and giraffes, but to do so it has to hunt in groups. The tiger, on the other hand, is essentially a lone hunter

Facing page : The tiger is for many people the most handsome and imposing of all the Big Cats. Now that the lion has all but disappeared from Asia the tiger, a much later arrival, is undisputed lord of the jungle.

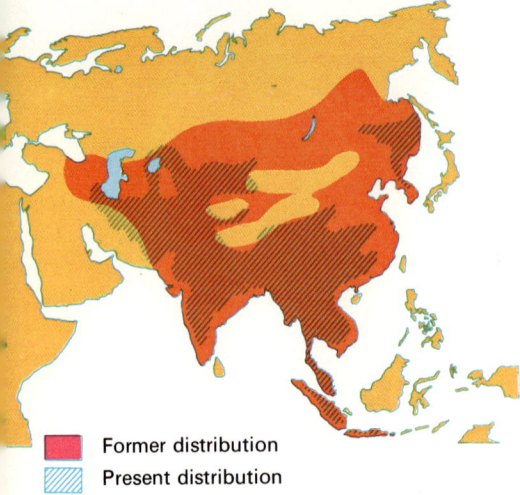

Former and present distribution of the tiger.

TIGER
(Panthera tigris)

Class: Mammalia
Order: Carnivora
Family: Felidae
Length of head and body: 90–120 inches (230–300 cm)
Length of tail: 29½–36 inches (75–91 cm)
Height to shoulder: 35½ inches (90 cm)
Weight: 495–600 lb (225–275 kg)
Diet: flesh
Gestation: 105–112 days
Number of young: 1–6, usually 3 or 4
Longevity: 20 years

Adults
Largest of living Felidae. Coat yellow or golden above, sometimes creamy-white underparts, and striped brown, black or grey. Female smaller than male.

Young
Weigh 2–3½ lb at birth; striped coat. Open eyes at fourteen days.

Facing page : The tiger frequently conceals itself in tall grass where its striped coat blends perfectly with the surroundings. Note too how the face, with its brownish-yellow, black and white markings, blends with the background as it crouches in the undergrowth, waiting for prey.

and there is no limit to the size of its potential victim. Perhaps for this reason it is somewhat larger and a good deal stronger than the lion, although in other physical respects, including skeleton structure, they are very alike. All the minor anatomical variations, such as colour, presence or absence of mane and so forth, as well as the major contrasts of behaviour, are certainly due to the basic fact that the animals occupy distinct ecological niches.

Solitary action is perfectly understandable for a predator which operates in an enclosed environment where it is patently impossible to organise group hunting; and in any event cooperation of this kind is unnecessary where concealed approach and ambush present no problems. On the open plains where the technique is to pursue, encircle and wear down the victim, results can best be achieved by coordinated action; in the jungle, where the attack and kill must be swift and sudden, individual action is the rule. This is the traditional hunting method of the cat tribe and jungle life has not modified their behaviour.

The tiger is generally rather heavier and stockier than the lion, for it is not by nature a runner. Certainly the animal moves very swiftly into action but it is incapable of keeping up a burst of speed for long. Lions, on the other hand, are capable of maintaining a steady, moderate pace for a considerable length of time. Dr de la Fuente points out that in his experience a lioness is a hundred times as fast as a tiger over a distance, say of fifty yards, but that the swift-as-lightning attack of a tiger pouncing on prey five or ten yards away is unmatched by the most agile lion.

In spite of these fundamental differences in behaviour which have come about principally as a result of adaptation to contrasted habitats, the tiger will not necessarily conform to the expected pattern in all circumstances. As many zoologists have already pointed out, and as Dr George Schaller has confirmed after a long series of studies in India, tigers do sometimes join together for hunting excursions. The reason that a tiger generally hunts on its own, by night, is that this technique is best suited to the jungle environment and for the capture of available prey. But there are exceptions. When a tigress is instructing her cubs in the rudiments of hunting she can get the message across much more quickly by venturing out with them in a group and by taking the initiative as the youngsters watch and imitate. Furthermore, in areas where there is plenty of game, several tigers may cooperate in attacking a large ungulate, just as a lioness will depart from her usual habits and tackle a herbivore on her own if the opportunity occurs unexpectedly. This does not contradict the accepted view of the tiger as a solitary hunter–in the majority of cases it is just that–but simply indicates that this animal, like all highly evolved predators, is capable of modifying its normal behaviour when special situations arise.

It is worth pointing out too that the behaviour gap between the two species becomes appreciably narrower when the animals are removed from the wild. In captivity, for example, lions and tigers have been known to mate with each other and produce cubs, although such hybrids are themselves infertile.

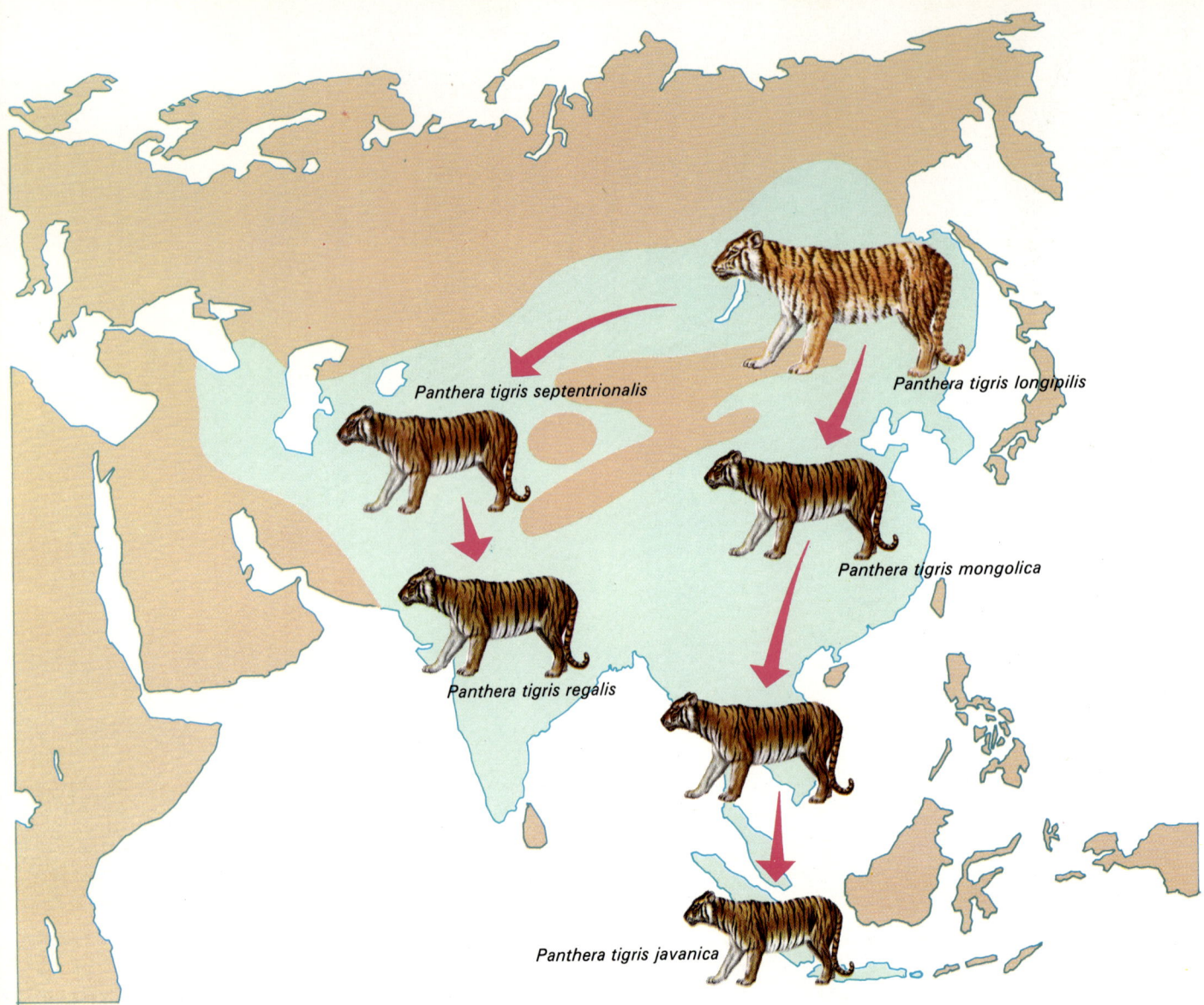

Panthera tigris septentrionalis

Panthera tigris longipilis

Panthera tigris mongolica

Panthera tigris regalis

Panthera tigris javanica

Originally an inhabitant of temperate northern climes, the tiger migrated southward after the last Ice Age. The species fanned out in two directions, to either side of the Himalayas, adapting to warmer climates and new environments, and in time becoming differentiated in size and colour to form a number of distinct races.

Facing page : The tiger usually hunts by night and retires to a sheltered spot during the day in order to rest. The fact that it is also an expert swimmer has been one factor in determining its present wide range of distribution.

Invaders from Siberia

The tiger is usually regarded as a typical inhabitant of tropical climes and it is difficult to imagine it ever having lived far from the steamy jungles of South-east Asia. Yet in prehistoric times the small hunting population of the Indian peninsula had no experience of the celebrated man-eating tiger; and later, in the Stone Age, when sheep-grazing and farming spread throughout the region, the great striped predator posed no danger to domestic livestock for the simple reason that it was a stranger to these parts, a later arrival from the north.

Palaeontologists are now satisfied that the tiger originally came from northern Siberia at the commencement of the Pleistocene epoch. At that time the climate of the Siberian region was temperate and its flourishing forests were the homes of large numbers of ungulates. But when in the course of the last glaciation Siberia was covered by an immense ice sheet, there were dramatic upheavals affecting the region's flora and fauna. As a result of the rapid drop in temperature many species that survived the cold migrated southwards. A few tigers managed to adapt to the new conditions and remained where they were, and in fact the descendants of these hardy individuals are still to be found, though in small numbers, in Siberia. They are very large,

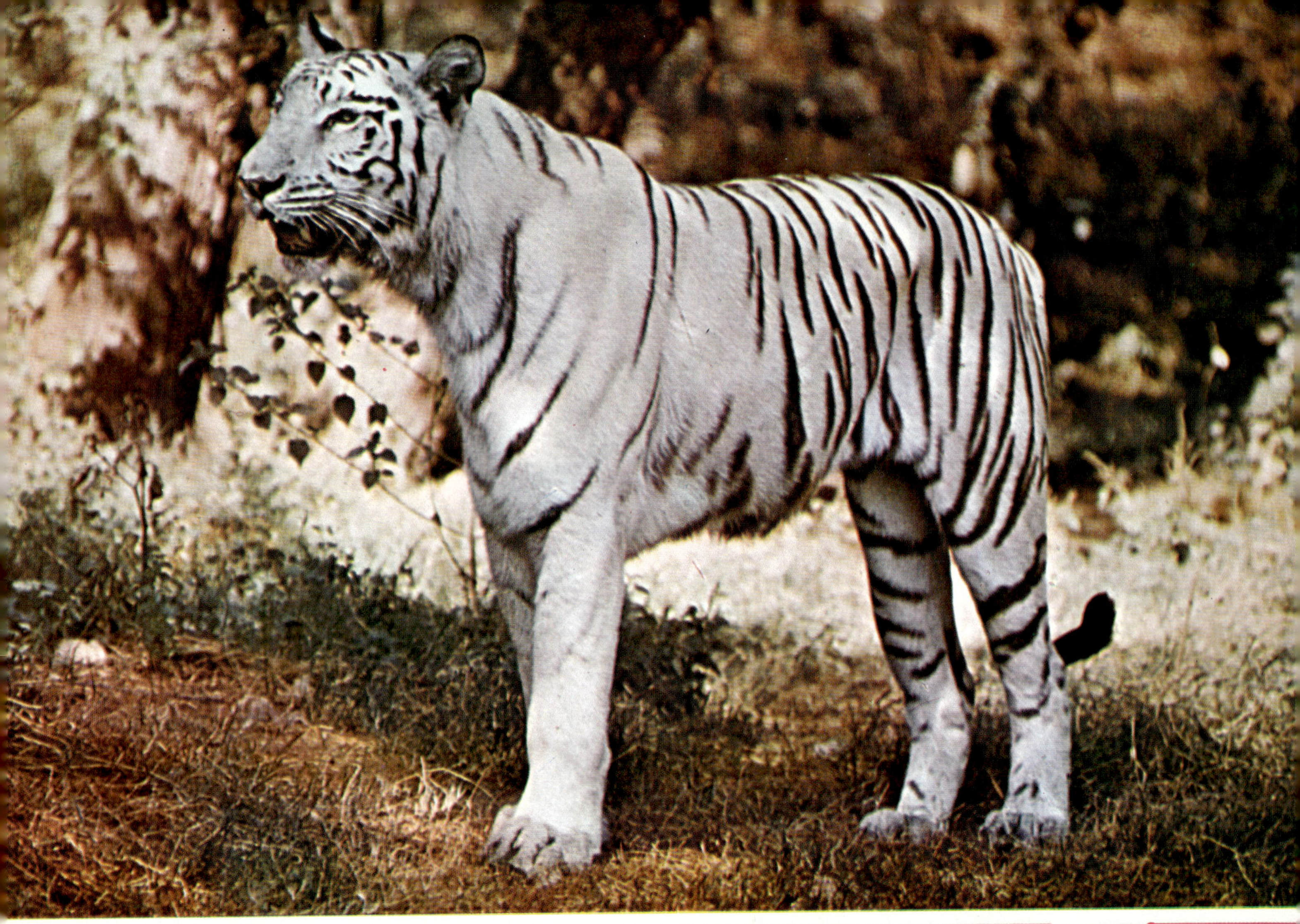

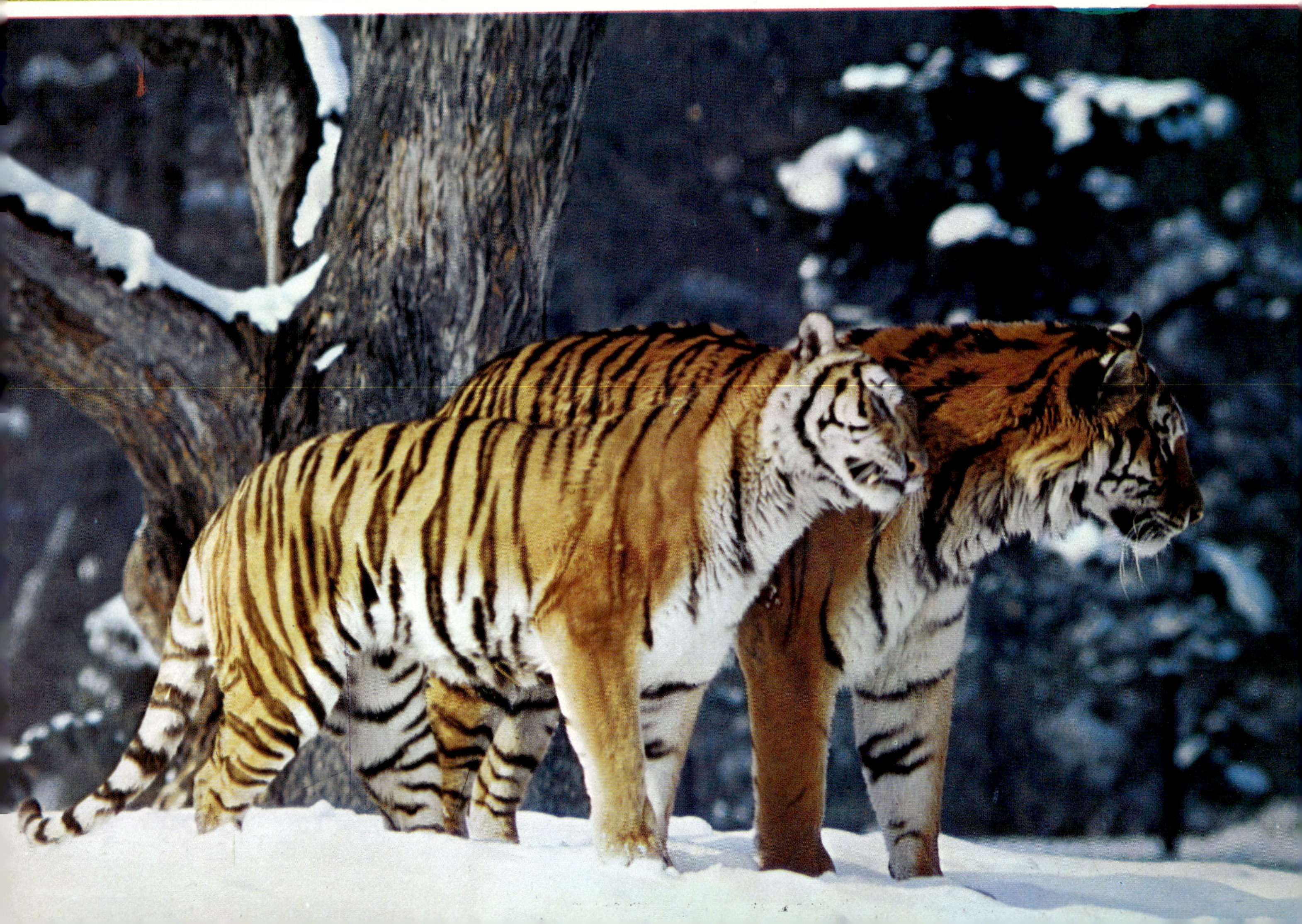

Deer and wild boars

Rodents, frogs, lizards, fishes

Man

Female and young antelopes

Domestic animals

Food preferences of the tiger.

weighing over 650 lb and measuring up to 11 feet long. Their fur is thick, pale in colour and with few stripes. Other felines retreated before the cold front and moved south in two broad waves which bypassed the Himalayas and the steppes and deserts of central Asia. The branch that headed west reached the eastern outposts of the Caucasus, journeying on to Persia and Afghanistan; the other, moving eastwards, crossed Manchuria and then made its way into Korea, China and Indo-China. The latter route was the better of the two for the migrating tigers, and a number of them succeeded in penetrating the Indian peninsula by way of Burma. Here they prospered and continued to maintain their numbers until comparatively recent times.

Naturally the tigers from the north only acclimatised very gradually to their new surroundings for the migration process was relatively slow. In due course the adaptation of the species to diverse habitats occasioned new forms, sizes and colours, enabling modern zoologists to distinguish eight varieties, ranging from the large pale-coloured Siberian subspecies to the small race from Bali. In between are the tigers of Mongolia, Iran, India, China, Sumatra and Java. The animals living on the islands are in general less heavy than their mainland counterparts; those inhabiting hot, moist regions with dense tree cover are darker than those living farther north in the mountains and on the high plateaux. Broadly speaking, the coat colour ranges from pale brown to reddish-yellow. The stripes, which are never symmetrical, also differ according to the race and the individual, the fewest markings being found on the tigers living at altitude in the north, and the most prominent patterns on those of the southern islands. Notable exceptions to this rule, however, are the celebrated and very beautiful white tigers of India, which are

Facing page : Although the tiger is generally associated with tropical countries it is still often found at altitude where it has adapted well to snow and intense cold. These individuals (*below*) are normally somewhat paler and less prominently striped than their jungle counterparts. An interesting exception to the colour rule is the Indian white tiger (*above*), found in fairly large numbers in the former principality of Rewa. The stripes of this race are grey or black on a creamy-white ground.

The tiger, with only rare exceptions, is a solitary hunter and a typical foray will follow more or less the same general pattern. Roaming its territory, the carnivore keeps its ears pricked for unusual sounds that may betray a victim (1). Having detected its prey the tiger quietly moves off in its direction (2). When a few yards away it crouches low in the grass and slinks slowly forward (3). Once within striking range the tiger leaps at its victim, bowling the latter to the ground (4). Holding the stunned animal firmly, the predator sinks its fangs into the victim's throat, snapping the vertebrae of the neck and severing the blood vessels (5).

especially common in the former principality of Rewa. These magnificent beasts are not albinos but possess a creamy-white coat with a variable number of grey and black stripes.

The standard form of Indian tiger may be considered typical of the species for its measurements are between the extremes of the Siberian and island races. The adult male stands approximately 3 feet high and its average length is about 9 feet, seldom exceeding 10 feet. The weight ranges from 500-600 lb. The female is much lighter, only in exceptional cases measuring 9 feet and weighing usually 100 lb. less.

The solitary yet sociable tiger

The individual tiger usually lives alone. Males and females hunt on their own account and devour their prey separately. The best known subspecies, the Bengal tiger, inhabiting regions that have been profoundly modified by human activities, is a shy, easily frightened animal. Even the so-called man-eaters which live in the most dense and impenetrable parts of the jungle, never make a frontal attack on a person rash enough to stray unprotected into their territory. There can be no doubt that this most spectacular of all Indian carnivores has been compelled to modify its fierce temperament as a result of its parallel development alongside its arch-rival and super-predator—man.

As a rule the tiger conceals itself in inaccessible areas during the day and emerges to hunt at nightfall, silently roaming its territory in quest of prey. After a kill it does not immediately begin to devour its victim but drags it off to a well sheltered hiding place where it can tear the animal up at leisure. Even then it will only make a gash in the rump and eat a small part of the prey, after which it lopes away, not returning to its lair until the following evening. It makes a stealthy approach to the scene of the feast and then proceeds to eat another meal, this time on the flanks and forelegs. On the third occasion it devours the neck and the head, stripping the bones clean. After satisfying its appetite the carnivore wanders off to drink and bathe nearby.

Secrecy, solitude and silence are the ideal conditions for the carnivore. So long as its hiding places remain undisturbed—no trees hacked down or bushes uprooted—it is content to live in heavily populated areas. But the behaviour pattern is very varied, especially in wildlife reserves in India and Siberia, for example, where it is not obliged to hunt for a living. George Schaller, who has studied the tiger in such situations in India, points out that the animals that grow up in such peaceful surroundings tend to venture out much more by day. He also noted that two families, consisting of tigresses and cubs, shared the same prey together without any hint of aggressiveness. One day a male joined one of the groups and waited some distance apart until the females and young had finished their meal before himself coming forward for his share.

Judging by such examples it seems clear that tigers which have become accustomed to human presence are much more sociably inclined than the solitary hunters of outlying regions. It is noticeable too that social behaviour changes at different stages of